Instrumentation and Measurements in Compressible Flows

Instrumentation and Measurements in Compressible Flows presents detailed information on experiments in compressible fluid flows including technical information pertaining to a wide variety of applications and the experimental basis for compressible flows. A step-by-step procedure is given to estimate the measurement errors as well as the uncertainty. Computational fluid dynamics data can be validated with the experimental results presented in the book. Further, it answers most pertinent queries related to conducting experiments and measuring the data at very high speeds. This volume also includes MATLAB® programs for selected topics.

Features:

- Presents detailed coverage of instrumentation, measurements, and experiments in compressible flows
- Covers both experimental and applied aspects of gas dynamics
- Provides a real-time exposure to the modern supersonic and hypersonic wind tunnel applications
- Explains supersonic and hypersonic shock/boundary-layer interactions and their control
- Includes real-time experimental problems and their analysis

This book is aimed at researchers and graduate students in aerospace, and mechanical engineering.

Control Theory and Applications

Series Editor: Dipankar Deb, *Institute of Infrastructure, Technology, Research and Management (IITRAM)*

This book series is envisaged to add to the scholarly discourse on high-quality books in all areas related to control theory and applications. The book series provides a forum for the control scientists and engineers to exchange related knowledge and experience on contemporary research and development in control and automation. This includes aircraft control, adaptive control, sliding mode control, evolutionary control, fuzzy theory and control, robotic manipulators, and even control applications in areas such as the Internet of Things and Big Data. The scope includes all aspects of control engineering needed to implement practical control systems, from analysis and design, through simulation and hardware, with a special emphasis on bridging the gap between theory and practice. It aims to explore the latest research findings and provide attention to emerging topics in control theory and its applications to diverse domains of engineering and technology, to expand the knowledge base and applications of this rapidly evolving and interdisciplinary field. The series will include textbooks, references, handbooks, and short-form books.

Control Strategies of Permanent Magnet Synchronous Motor Drive for Electric Vehicles
Chiranjit Sain, Atanu Banerjee, and Pabitra Kumar Biswas

Distributed Energy Systems
Design, Modeling, and Control
Edited by Ashutosh K. Giri, Sabha Raj Arya, and Dmitri Vinnikov

Instrumentation and Measurements in Compressible Flows
Mrinal Kaushik

For more information about this series, please visit: www.routledge.com/Control-Theory-and-Applications/book-series/CRCCTA

Instrumentation and Measurements in Compressible Flows

Mrinal Kaushik

CRC Press
Taylor & Francis Group
Boca Raton London New York

CRC Press is an imprint of the
Taylor & Francis Group, an **informa** business

First edition published 2024
by CRC Press
6000 Broken Sound Parkway NW, Suite 300, Boca Raton, FL 33487-2742

and by CRC Press
4 Park Square, Milton Park, Abingdon, Oxon, OX14 4RN

CRC Press is an imprint of Taylor & Francis Group, LLC

ISBN: 978-0-367-68875-2 (hbk)
ISBN: 978-0-367-68876-9 (pbk)
ISBN: 978-1-003-13944-7 (ebk)

DOI: 10.1201/9781003139447

Typeset in Nimbus font
by KnowledgeWorks Global Ltd.

Publisher's note: This book has been prepared from camera-ready copy provided by the authors.

Contents

SECTION II Measurements in Compressible Flows

About the Author

Dr. Mrinal Kaushik is an Associate Professor in the Department of Aerospace Engineering at IIT Kharagpur. Prior to joining this institute in 2013, he worked at the Defence Institute of Advanced Technology in Pune, the General Motors Technical Center in Bangalore, the Indian Space Research Organization in Trivandrum, and Tata Consultancy Services (Mumbai). Dr. Kaushik earned his Ph.D., M.Tech., and B.Tech. in Aerospace Engineering from IIT Kanpur in 2012, 2003, and 1999, respectively. His research interests include Shock-Wave/Boundary-Layer Interactions and Flow Control & Jet Acoustics, for which he has published several research articles in peer-reviewed international journals and conference proceedings. Dr. Kaushik has established an indigenously designed fully automated Supersonic Jet Research Laboratory at IIT Kharagpur to investigate jet characteristics. He is a member of numerous national and international professional societies. He has written three other textbooks for engineering students, namely *Fundamentals of Gas Dynamics* (Springer Nature, Singapore, 2022); *Theoretical and Experimental Aerodynamics* (Springer Nature, Singapore, 2019); and *Essentials of Aircraft Armaments* (Springer Nature, Singapore, 2016). He has also published a reference book, *Innovative Passive Control Techniques for Supersonic Jet Mixing* (Lambert Academic Publishing, Saarbrücken, Germany, 2012).

Preface

This book results from the author's extensive experience teaching aerodynamics courses at the undergraduate and graduate levels for over a decade. It is primarily intended for undergraduate aerospace and mechanical engineering undergraduate students as an introductory course in learning experimental techniques in compressible flows. However, it can also benefit professional engineers who require a fundamental understanding of the subject for their current work. This book is also meant for independent and unassisted readers who need a reasonable foundation in experiments.

The present text covers various topics in a logical and easy-to-understand manner. Many illustrations are provided in each chapter to clarify the physics behind these phenomena while covering many aspects of flow phenomena. The author also feels that to have a strong foundation on the basic concepts, the theoretical treatment must be reinforced with worked-out examples highlighting the application of underlying principles. As a result, each chapter contains many solved problems based on the theory presented in that chapter. In addition, to evaluate the reader's understanding, exercise problems are also provided at the end of each chapter. To provide a solid foundation for the experimental methodology, the purpose and goal of the experiments, as well as several important fluid properties, are examined first. The fundamental devices in experimental compressible flows, especially wind tunnels, are then introduced. The design, operation, and application methods of subsonic and supersonic tunnels are comprehensively covered, beginning with the principles. In addition, a description of hypersonic and hypervelocity facilities is provided to give a general idea of what such equipment is capable of. Following that is a detailed chapter on the author's self-developed supersonic open jet facility at the Department of Aerospace Engineering at the Indian Institute of Technology Kharagpur, highlighting each component's design, operation, and use. Subsequently, the techniques of pressure, Mach number, temperature, and density measurements in compressible flows are discussed. The hot wire anemometer is covered in depth, including the working and application procedures and the benefits and drawbacks. A brief overview of Particle Image Velocimetry and the Laser Doppler Anemometer is given, emphasizing their operating principles and special capabilities. Flow visualization techniques for high-speed flows are demonstrated using schematic and photographic views. Finally, certain experimental studies conducted in the author's laboratory are presented, emphasizing the results and the physics behind such phenomena.

In an elementary text like this one, the author understandably does not claim any innovation aside from the presentation and structure of existing approaches. Indeed, the author relied on works considered to be masterpieces in the subject while writing the manuscript. Some of the classics mentioned are listed at the end of the manuscript.

I thank God for his kindness, without which this work would not have been possible. I thank all my undergraduate and graduate students for their direct or indirect support in bringing this book to its current form.

Mrinal Kaushik

Nomenclature

α	divergence angle
ℓ	characteristic length
η_i	isentropic efficiency or diffuser effectiveness
η_p	polytropic efficiency
γ_{sw}	specific weight
κ	compressibility
λ	mean free path
β_k	kinetic energy flux or correction factor
β_s	screen porosity
β	shock angle
χ	screen's open area ratio
δ	boundary layer thickness
Δ	material roughness
η_F	fan efficiency
η	diffuser efficiency
$\forall_T$	volume of high-pressure tank
κ_s	isentropic compressibility
κ_T	isothermal compressibility
λ_p	power factor
μ_M	Mach angle
Ψ	ratio of honeycomb cell length to its diameter
ρ_0	stagnation density
A^*	choked flow throat area
$A^*_{2,min}$	minimum value of second throat area at choked flow
A^*_2	second throat area at choked flow
a_c	contraction ratio
a_d	diffuser's area ratio
AR	aspect ratio
A	cross-sectional area
C_p	pressure coefficient
c_p	specific heat at constant pressure
c	characteristic length
D_0	diameter of the cylindrical test section
D_1	inlet diameter of the round diffuser
D_2	exit diameter of the round diffuser
D_h	honeycomb cell hydraulic diameter
Eu	Euler number

E	bulk modulus of elasticity
f_{avg}	average friction factor
Fr	Froude number
g	acceleration due to gravity
H	enthalpy
h	convective heat transfer coefficient
I_t	intensity of turbulence
K_d	pressure loss coefficient for a subsonic diffuser
K_i	pressure drop coefficient for a wind tunnel component
K_s	pressure drop coefficient for the screens
k	thermal conductivity
L_h	honeycomb cell length
L_{TS}	test section length
M	Mach number
Nu	Nusselt number
n	polytropic exponent
p_0	stagnation pressure
P_{input}	power supplied to the fan
p_b	backpressure
Pr	Prandtl number
p	static pressure
Q	heat loss due to convection
R_w	probe resistance
Re	Reynolds number
r	recovery factor
R	gas constant
S.G.	specific gravity
Stk	Stokes number
S	Standard deviation
s	entropy
T_0	stagnation temperature
T_∞	freestream temperature
t_p	pumping time
T_r	recovery temperature
T	Static temperature
U_∞	freestream velocity
v	flow velocity
σ_s	screen solidity
K_{exp}	flow expansion coefficient
K_c	pressure drop coefficient for a contraction cone
K_f	wall skin friction coefficient
K_g	pressure drop coefficient for the corners with guide vanes

K_h	pressure drop coefficient for honeycomb structures
K_m	mesh factor
K_{TS}	pressure drop coefficient for a test section
μ	dynamic viscosity
ν	kinematic viscosity
ρ	density
ρ_{Hg}	density of mercury
τ	shear stress
υ	specific volume
F_c	compressibility force
F_g	gravity force
F_i	inertia force
F_p	pressure force
F_v	viscous force
M_{TS}	test section Mach number
L_c	contraction cone length

Section I

Instrumentation in Compressible Flows

1 Experiments in Fluids

1.1 NEED AND AIM OF EXPERIMENTS

Because of advances in computer tools, it is now possible to numerically simulate practically all fluid flow processes at rates ranging from extremely slow to very fast. These techniques are cost-effective due to the ease of studying various flow configurations, however complex they may be. Nevertheless, despite the advantages, computational results are not yet free from errors because of several approximations involved in the simulation. Therefore, the outcomes of a computational study need to be supplemented by the experiments. In many areas, information of fundamental nature can be derived from the experiments, which is otherwise not possible or cumbersome from the computations.

Experiments to investigate aerodynamic problems can be carried out in various approaches, including flight testing, missile launches, drop test water tunnels, ballistic ranges, and wind tunnel investigations. Wind tunnel tests are among the most important measurement techniques in experimental aerodynamics.

1.2 WIND TUNNELS

A wind tunnel is an experimental environment that creates airflows under controlled conditions to test the models of interest, simulating the actual flight conditions. Although most wind tunnels are designed to test scaled models of aerospace vehicles, some are large enough to accommodate complete (prototype) vehicles. The movement of air past an object causes it to appear to be flying in the atmosphere. From an operational perspective, wind tunnels are categorized into low-speed, high-speed, or specific-purpose tunnels.

A wind tunnel is essential for the construction of aerodynamics research tools. It investigates the effects of airstream passing over a solid body. Despite significant developments in the field of computational fluid dynamics, wind tunnel testing results remain critical for guiding specific design judgments for various engineering problems.

1.2.1 CLASSIFICATION OF WIND TUNNELS

Wind tunnels can be divided into several categories on the basis of their specific use. However, they are usually classified based on the following aspects.

DOI: 10.1201/9781003139447-1

1.2.1.1 Test Section Mach Number

The wind tunnels are divided into the following categories based on the designed Mach number in the test section:

- Subsonic wind tunnels ($M < 0.8$)
- Transonic wind tunnels ($0.8 < M < 1.2$)
- Supersonic wind tunnels ($1.2 < M < 5$)
- Hypersonic wind tunnels ($M > 5$)

The Mach number is an important parameter for characterizing wind tunnels due to compressibility effects. It is well known that compressibility effects are often ignored for Mach numbers less than 0.3. However, the flow must be presumed compressible for Mach numbers greater than 0.3, achieved in transonic and supersonic wind tunnels. The compressibility of the working fluid significantly impacts tunnel design. For instance, the test section is the minimum area region in a subsonic tunnel. On the other hand, the minimum area region in a supersonic tunnel is the nozzle throat.

1.2.1.2 Tunnel Geometry

A wind tunnel is also specified by its geometry. An open-circuit wind tunnel is one that is open at both ends and gets clean air from the surrounding environment. A closed-circuit wind tunnel, on the other hand, is a tunnel where the same fluid recirculates through the test section in a specified path. This arrangement is commonly employed in subsonic tunnels. It should be noted that both open- and closed-circuit tunnels can be used with either open or closed test section walls.

Wind tunnels can also be intermittent (with a small test time) or continuously operated (long test time). A blowdown tunnel is an intermittent tunnel that can be used to create supersonic flows in the test section. It directs air from a high-pressure reservoir placed ahead of the test section into a low-pressure tank or the surrounding environment. A shock tube is a blowdown tunnel variation used to generate supersonic flow. Supersonic and hypersonic tunnels have significantly less testing time than subsonic tunnels.

1.2.1.3 Working Fluid

Tunnels are also designated on the basis of the working fluid utilized in tunnel operations. In low-speed wind tunnels, air is the common working fluid. Water is utilized in high-speed aircraft to see compression waves and to examine flow patterns around submarines and undersea vehicles. Because of the extreme temperatures encountered in hypersonic (or hypervelocity) facilities, more stable gases such as nitrogen or helium are employed. Liquid nitrogen is used to test the models in transonic flows at high Reynolds numbers.

1.2.1.4 Specific Purpose

Wind tunnels are also known as specific purpose tunnels since they are created and built for a certain function. They are not the same as low- or high-speed wind tunnels. Some common instances of this category include:

Spinning Tunnels

Because stalling airplanes tend to go into spin mode, these facilities are used to investigate such events.

Stability Tunnels

Because it is occasionally important to evaluate an aircraft's stability, these tunnels enable the model to move freely in the test section.

Low Density Tunnels

In order to accurately re-create high-temperature effects at hypersonic speeds or rarefied states of gases in the atmosphere's outer layer, specific low-density tunnels are developed.

1.2.2 CALIBRATION OF A WIND TUNNEL

A wind tunnel calibration entails calculating the mean values and homogeneity of different flow parameters in the test section. Stagnation pressure and temperature, Mach number, and flow angularity are the basic characteristics of any wind tunnel calibration. Static pressure and temperature, turbulence, and the quantity of condensation or liquefaction are all important flow conditions to consider. However, in the absence of shock waves, condensation of water vapor, and air liquefaction, the airflow from the nozzle to the test section in all the wind tunnels can be assumed isentropic. This assumption simplifies wind tunnel calibration to a great extent. Isentropic flow is supposed to exist when collisions, condensation, and liquefaction are absent. The total pressure in the test section is equivalent to a value in the stagnant air, which is directly measured in the settling chamber. The temperature remains constant except in heated tunnels, wherein convection losses in the settling chamber become significant. Because the ratios of total pressure and temperature to flow rates are distinct functions of Mach number, the calibrator can measure each test section parameter to define all others once the settling chamber conditions are known.

1.3 EXPERIMENTAL MODELS AND SIMILITUDE

The physics behind the model and prototype flow systems working under different conditions must be the same for accurate quantitative results from the model research. The ratio of the given physical parameters calculated in each system determines the similarity of flow systems. The similarity is called geometric similarity if the specified physical quantity is a geometric dimension, kinematic similarity if the magnitude represents motion, and dynamic similarity if the magnitude represents force. Two fluid flows are considered similar when all three similarities are present.

1.3.1 GEOMETRIC SIMILARITY

This resemblance relates to the similarity of geometric shapes. Geometrically similar systems are those in which the length in one system to the corresponding length in the other is constant throughout. The scale factor is the ratio of lengths in these systems. These systems have similar shapes but differ in size. Prototypes are full-scale systems, while models are scale-size systems. It is worth noting that a model does not necessarily have to be smaller than the prototype. Some applications, such as microfluidics, may be larger than the prototype. Using the same liquid for the model and the prototype is also unnecessary. Geometric similarity necessitates surface roughness similarity; for example, if the model is one-tenth the prototype's size in each linear dimension, the roughness projections will be in the same ratio.

1.3.2 KINEMATIC SIMILARITY

Kinematic similarity is the term used to describe the similarity of motion. Because distance and time define motion, kinematic similarity relates to the similarity of lengths (geometrical similarity) and time intervals. Kinematic similarity guarantees geometric similarity; however, the contrary is not always true. When the fluid flows are kinematically similar, the streamline patterns at corresponding instants are geometrically similar.

1.3.3 DYNAMIC SIMILARITY

It is fundamentally a force similarity, which means that the dynamic pressure ratio at related locations must be constant. A body moving in a fluid is subject to numerous forces and moments caused by the viscosity, inertia, medium elasticity, and gravity of the fluid. At the corresponding points in the model and prototype, the ratios of different forces must be the same dynamic pressure ratio. These forces are as follows:

- Inertia force (F_i)
- Viscous force (F_v)
- Gravity force (F_g)
- Pressure force (F_p)
- Compressibility force (F_c)

In most cases, the inertia force is used to characterize these ratios. The Reynolds number is the ratio of inertia force to viscous force. Similarly, the Euler number is the ratio of inertia force to pressure force. Furthermore, the Mach number and the Froude number are the ratios of inertia force to compressibility force and inertia force to gravitational force, respectively.

$$\text{Reynolds number} \quad \text{Re} = \frac{F_i}{F_v} \tag{1.3.1}$$

$$\text{Euler number} \quad \text{Eu} = \frac{F_i}{F_p} \tag{1.3.2}$$

$$\text{Mach number} \quad \text{M} = \frac{F_i}{F_c} \tag{1.3.3}$$

$$\text{Froude number} \quad \text{Fr} = \frac{F_i}{F_g} \tag{1.3.4}$$

It should be noted that not all dimensionless parameters are altered at the same time in each experiment. It is determined by the nature of the forces created and the specific study. In a low-speed flow domain, for example, where the effect of gravitational force is insignificant, Reynolds number simulation is critical to ensuring dynamic similarity; the similarity of the Mach number is not essential. On the other hand, Mach number similarity is necessary for high-speed flows.

1.4 PERFORMANCE PARAMETERS IN MEASUREMENTS

This section discusses some important performance parameters associated with the measurements.

1.4.1 PRECISION

The precision of experimental measurements represents how close repeated measurements are to one another. It is estimated by standard deviation, which gives an idea of the relative proximity of the results (Figure 1.4.1). The results, however, may be very far from the actual value.

1.4.2 ACCURACY

The accuracy is how close a measured value is to the true value. It is measured in terms of the percentage error. Nevertheless, the results need not be close, as illustrated in Figure 1.4.1(c).

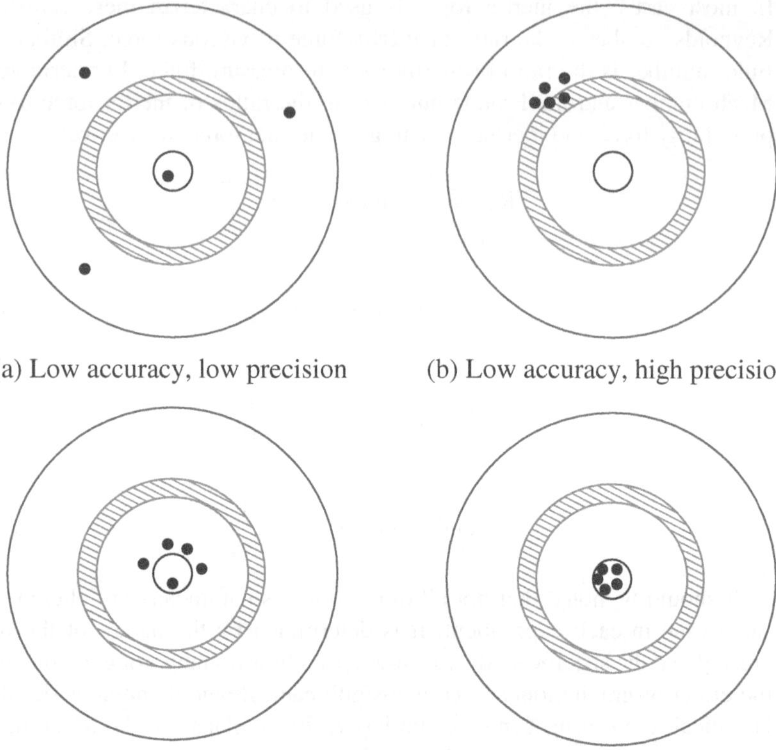

(a) Low accuracy, low precision (b) Low accuracy, high precision

(c) High accuracy, low precision (d) High accuracy, high precision

Figure 1.4.1 Accuracy and precision.

1.4.3 ABSOLUTE ERROR

The absolute error represents the difference between a measured value and the true or accepted value. Thus,

$$\text{Absolute error} = \text{Measured value} - \text{True value} \tag{1.4.1}$$

1.4.4 RELATIVE ERROR

The relative error, commonly known as the percentage error, indicates how far a measured value deviates from the true value. It is the ratio of absolute error to the true value of the measured quantity. That is,

$$\text{Relative error }(\%) = \frac{\text{Absolute error}}{\text{True value}} \times 100 \tag{1.4.2}$$

For example, we intended to measure 10 mℓ of a liquid using a metered test tube; however, the volume measured by the tube was 10.04 mℓ. In this case, the absolute error in the liquid volume measurement is 0.04 mℓ, and the relative error is 0.4%.

1.4.5 RANDOM ERROR OR INDETERMINATE ERROR

Random error is one that cannot be easily controlled or even identified. It causes measured values to fluctuate somewhat consistently around a mean value.

1.4.6 SYSTEMATIC ERROR OR DETERMINATE ERROR

Systematic error, also known as determinate error, has a known source. It might occur as a result of faulty or inaccurately calibrated devices, human error, chemical behavior, or side reactions. For example, we wanted to weigh 5 g of iron dust with a weighing balance, but the observed value was less than 5 g since some of the iron was oxidized. Because the reason of such a measurement error is known, it is referred to as a systematic error.

1.5 CONCEPT OF SIGNIFICANT DIGITS IN MEASUREMENTS

1. The nonzero integers always count as significant digits. For example, the measured values of 1465 and 21 in an experiment have 4 and 2 significant digits, respectively.
2. Zeros that precede nonzero digits don't count as significant digits. For instance, 0.0034 and 0.89 have two significant digits, but 0.00255 has three significant digits.
3. If the zeros are placed between two nonzero digits, they are called captive zeros and always count as significant digits. For example, the number 1.005 has four, and the number 409 has three significant digits.
4. If the zeros are at the end of a number, they are significant only if a decimal is present. Thus, the number 100.0 has four significant digits, the number 20 has one significant digit, and 1.00×10^2 has three significant digits.

1.5.1 RULES FOR ROUNDING OFF THE MEASURED VALUES

1. In a series of calculations, carry forward the extra digits till the final result and then round off the value.
2. In rounding off the values, if the digit to the right of the final significant digit is less than 5, the preceding digit remains the same. For example, the number 6.22 may be rounded off as 6.2.

3. However, if the digit to the right of the final significant digit is equal to or greater than 5, the final significant digit should be increased by 1. Thus, the number 4.57 can become 4.6 when rounded off.

1.6 ERROR AND UNCERTAINTY IN EXPERIMENTS

The deviation of the measured value from the true value is called the error. The error can arise from faulty or poorly calibrated instruments, human intervention, chemical behavior, or side reactions. It is an unavoidable mistake that can occur despite the utmost prevention and cannot be eliminated. However, we can minimize the errors by using best measurement practices. Measuring any quantity involves two things: the first is the true value of the quantity, which cannot be measured accurately, and the other is the uncertainty in measurement associated with it. The key difference between error and uncertainty is that an error is a difference between the actual value and the measured value while uncertainty is an estimate of the range between those values that represents the reliability of the measurement. That is, the uncertainty describes the reliability of the statement that the specified measured value really represents the value of the measured variable. In other words, uncertainty is the quantification of the doubt calculated from the result of a measurement.

1.6.1 UNCERTAINTY IN EXPERIMENTAL DATA

We can never measure a physical quantity with absolute accuracy because experimental measurements are prone to errors. However, a sophisticated measurement technique with greater care can reduce the amount of error to a large extent. In an experiment, the deviation of the measured value from the true value is termed the experimental error. The error value confirms the accuracy level of the experimental data; a lower error implies a more accurate result. In practical situations, the possible range of errors is often described instead of estimating an exact error value. This error range is usually called the uncertainty in the measurement. That is, uncertainty is typically defined as the range within which one is 95% convinced that the true value lies.

Experimental errors are inherent in a measurement process that repeated measurements can never eliminate. Both accuracy and precision involved in an experiment can effectively determine the associated error. The accuracy of a result indicates the deviation of the measured value from the true value, whereas precision measures the closeness between the results of the repeated measurements. Usually, two sources are found to be responsible for the experimental errors: systematic error and random error. Systematic errors are termed one-sided errors caused due to faulty calibration of the instruments involved in the measurements. In contrast, random errors show fluctuating, unpredictable up-and-down variations from the true value. We minimize the systematic error by careful calibration and then estimate the random error statistically. The general procedure for estimating the uncertainties in the calculated quantities using measurement data is described below.

It is seen that the values of a variable obtained during the repeated measurements are slightly different from each other. Generally, the average of the measured data set is considered the best estimate of the true value because they are essentially grouped around a mean value. Suppose the measurement of a quantity is repeated N number of times and the values obtained are A_1, A_2, A_3, ..., A_N. The average or mean value of the quantity is, therefore,

$$A = \frac{A_1 + A_2 + A_2 + ... + A_N}{N} \quad (1.6.1)$$

However, this average value is inaccurate since there will still be some systematic errors due to the measuring equipment. Since it is quite tedious to calibrate the equipment perfectly, the uncertainty associated with the average value is generally expressed in terms of average deviation, d, defined as

$$d = \frac{|A_1 - A| + |A_2 - A| + |A_3 - A| + ... + |A_N - A|}{N} \quad (1.6.2)$$

However, the most convenient way to characterize the measurement uncertainty is utilizing the standard deviation, S, which is identical to the following:

$$S = \sqrt{\frac{(A_1 - A)^2 + (A_2 - A)^2 + (A_3 - A)^2 + ... + (A_N - A)^2}{N - 1}} \quad (1.6.3)$$

The uncertainty associated with a variable is usually expressed as

$$\sigma = \frac{S}{\sqrt{N}} \quad (1.6.4)$$

Example 1. A barometer at one location indicates that the height of the mercury column is 750 mm. Let the barometer's vernier scale has a least count of 0.1 mm. Determine the level of uncertainty in atmospheric pressure measurement.

Solution. Given,

The apparent height of the mercury column is:

$$h = 750 \text{ mm}$$

The barometer's vernier scale has a least count of LC = 0.1 mm. Because it is well known in the literature that if the scale divisions are close together, we may only be able to estimate the least count to the nearest some fraction

$(1/2, 1/5, 1/10)$. Given that the fraction is $1/2$, the estimated measurement error in the vernier reading is,

$$\begin{aligned} e_h &= \frac{1}{2} LC \\ &= 0.05 \text{ mm} \end{aligned}$$

The relative uncertainty in determining the barometric height is

$$\begin{aligned} u_h &= \pm\frac{e_h}{h} = \pm\frac{0.05 \text{ mm}}{750 \text{ mm}} \\ &= \pm 6.67 \times 10^{-5} \\ &= \pm 0.00667\% \end{aligned}$$

1.6.1.1 Propagation of Uncertainty

Any experimental measurement determines the final result by measuring a combination of variables. In such cases, the uncertainty associated with the final result is determined using the root-sum-square (RSS) combination of the uncertainties of all variables measured. This method, developed by Kline and McClintock (1953) [16], states that if the function F depends upon a single variable x such that $F = F(x)$ and σ_x is the uncertainty in the independent variable x, the uncertainty in the result σ_F can be determined from the partial derivative:

$$\sigma_F = \frac{\partial F}{\partial x}\sigma_x \tag{1.6.5}$$

However, if F is a function of several independent variables, $x_1, x_2, x_3, \ldots, x_N$ such that $F = F(x_1, x_2, x_3, \ldots, x_N)$, and σ_{x_1}, σ_{x_2}, σ_{x_3}, ..., σ_{x_N} are the uncertainties in the independent variables, the overall uncertainty in the result σ_F can be determined using a root-sum-square estimate:

$$\sigma_F = \left[\left(\frac{\partial F}{\partial x_1}\sigma_{x_1}\right)^2 + \left(\frac{\partial F}{\partial x_2}\sigma_{x_2}\right)^2 + \left(\frac{\partial F}{\partial x_3}\sigma_{x_3}\right)^2 + \ldots + \left(\frac{\partial F}{\partial x_N}\sigma_{x_N}\right)^2\right]^{\frac{1}{2}} \tag{1.6.6}$$

or

$$\sigma_F = \left[\sum_{i=1}^{N}\left(\frac{\partial F}{\partial x_i}\sigma_{x_i}\right)^2\right]^{\frac{1}{2}} \tag{1.6.7}$$

where the term σ_{x_i} is the uncertainty associated with the variable x_i. Equation (1.6.7) is the general equation for calculating the uncertainty in the final result. Note that this calculation is statistically much more probable than adding the various uncertainties linearly σ_{x_i}, making the unlikely assumption that all variables simultaneously attain maximum error.

After the uncertainty with equation (1.6.7) has been calculated, the value of F is expressed in the standard form:

$$F = F(\text{measured}) \pm \sigma_F \tag{1.6.8}$$

Example 2. The Mach number in compressible flows is often calculated using the stagnation pressure p_0 measured from a pitot probe and the ambient pressure p_a from a barometer. Obtain an expression for the uncertainty in the Mach number calculation.

Solution. The expression for Mach number can be found by slightly modifying the isentropic relation (equation (2.7.15)) for the measured values of p_0 and p_a:

$$M = \sqrt{\frac{2}{\gamma-1}\left[\left(\frac{p_0}{p_a}\right)^{\frac{\gamma-1}{\gamma}} - 1\right]}$$

which on differentiation yields

$$dM = \frac{\partial M}{\partial p_0} dp_0 + \frac{\partial M}{\partial p_a} dp_a$$

where

$$\frac{\partial M}{\partial p_0} = \frac{1}{2}\frac{1}{\sqrt{\frac{2}{\gamma-1}\left[\left(\frac{p_0}{p_a}\right)^{\frac{\gamma-1}{\gamma}} - 1\right]}}\left[\frac{2}{\gamma-1}\frac{\gamma-1}{\gamma}\left(\frac{p_0}{p_a}\right)^{-\frac{1}{\gamma}}\frac{1}{p_a}\right]$$

$$= \frac{1}{\sqrt{\frac{2}{\gamma-1}\left[\left(\frac{p_0}{p_a}\right)^{\frac{\gamma-1}{\gamma}} - 1\right]}}\left[\frac{1}{\gamma p_a}\left(\frac{p_0}{p_a}\right)^{-\frac{1}{\gamma}}\right]$$

$$\frac{\partial M}{\partial p_a} = -\frac{1}{2}\frac{1}{\sqrt{\frac{2}{\gamma-1}\left[\left(\frac{p_0}{p_a}\right)^{\frac{\gamma-1}{\gamma}} - 1\right]}}\left[\frac{2}{\gamma-1}\frac{\gamma-1}{\gamma}\left(\frac{p_0}{p_a}\right)^{\frac{\gamma-1}{\gamma}}\frac{p_0}{p_a^2}\right]$$

$$= -\frac{1}{\sqrt{\frac{2}{\gamma-1}\left[\left(\frac{p_0}{p_a}\right)^{\frac{\gamma-1}{\gamma}} - 1\right]}}\left[\frac{1}{\gamma p_a}\left(\frac{p_0}{p_a}\right)^{\frac{\gamma-1}{\gamma}}\right]$$

Therefore, we get

$$dM = \frac{1}{\sqrt{\frac{2}{\gamma-1}\left[\left(\frac{p_0}{p_a}\right)^{\frac{\gamma-1}{\gamma}} - 1\right]}}$$

$$\left\{\left[\frac{1}{\gamma p_a}\left(\frac{p_0}{p_a}\right)^{-\frac{1}{\gamma}}\right]dp_0 - \left[\frac{1}{\gamma p_a}\left(\frac{p_0}{p_a}\right)^{\frac{\gamma-1}{\gamma}}\right]dp_a\right\}$$

The uncertainty in M can be expressed as

$$u_M = \pm\left[\left(\frac{p_0}{M}\frac{\partial M}{\partial p_0}u_{p_0}\right)^2 + \left(\frac{p_a}{M}\frac{\partial M}{\partial p_a}u_{p_a}\right)^2\right]$$

where u_{p_0} and u_{p_a} are the uncertainty in calculating p_0 and p_a, respectively.

1.7 SUMMARY

This chapter emphasizes the need for and aim of an experimental investigation from both basic and practical research perspectives. Because the experimental data is realistic, it validates the numerical simulation results.

A wind tunnel is an experimental environment that creates airflows under controlled conditions to test the models of interest, simulating the actual flight conditions. Although most wind tunnels are designed to test scaled aircraft and spacecraft models, some are large enough to accommodate a complete (prototype) vehicles. The movement of air past an object causes it to appear to be flying in the atmosphere. Wind tunnels are classified as low-speed, high-speed, or special-purpose tunnels from an operational viewpoint. They are also classified according to the required flow speed in the test section, such as subsonic wind tunnel, transonic wind tunnel, supersonic wind tunnel, and hypersonic wind tunnel.

Compressibility effects influence the categorization of wind tunnels depending on the Mach number. According to the literature, compressibility effects are often neglected for Mach numbers less than 0.3. However, the flow must be assumed to be compressible for Mach values greater than 0.3, as obtained in transonic and supersonic tunnels. The fluid compressibility significantly impacts tunnel design; the test section has the smallest area in a subsonic tunnel, whereas the throat has the smallest area in a supersonic wind tunnel.

A wind tunnel calibration entails calculating the mean values and homogeneity of different flow parameters in the test section. Stagnation pressure and temperature, Mach number, and flow angularity are the basic characteristics of any wind tunnel calibration. Static pressure and temperature, turbulence, and the quantity of condensation or liquefaction are all important flow conditions to consider.

The following are some useful performance parameters for the experiments:

- The precision of experimental measurements shows the similarity of repeated measurements. It is estimated using standard deviation, which gives a notion of how close the results are.
- The accuracy of a measurement is how near it is to the true value. It is calculated as a percentage error.
- The absolute error represents the difference between a measured value and the true or accepted value.
- The relative error, commonly known as the percentage error, indicates how far a measured value deviates from the true value. It is the ratio of absolute error to the true value of the measured quantity.
- Random error is one that cannot be easily controlled or even identified. It causes measured values to fluctuate somewhat consistently around a mean value.
- Systematic error, also known as determinate error, has a known source. It might occur as a result of faulty or inaccurately calibrated devices, human error, chemical behavior, or side reactions.

The error is defined as the difference between the true and measured values, whereas the uncertainty is defined as the reported value that falls within the range of values that the true value is meant to fall inside. Any experimental measurement calculates the final result by measuring a set of variables. In this case, the uncertainty associated with the final result is calculated using the RSS combination of the uncertainties of all the variables measured.

EXERCISE PROBLEMS

Exercise 3. In each of the following numbers, how many significant digits are there?

(a) 0.00041 (b) 0.24900 (c) 3.8×10^7 (d) 389.060×10^{-19} (e) -187.04×10^{-23}

Exercise 4. The Grashof number, an important dimensionless parameter in natural convection, is given by

$$\mathrm{Gr} = \frac{g\beta\rho^2 L^3 \Delta T}{\mu^2}$$

where g is the acceleration due to gravity, β is the coefficient of thermal expansion, ρ is the density, L is the characteristic length, ΔT is the temperature difference, and μ is the viscosity. If the uncertainty in the measurement of each variable is $\pm 3\%$, what is the uncertainty in computing Gr?

2 Review of Compressible Flows

2.1 INTRODUCTION

The domain of compressible flow deals with the motion of fluids where the density change caused by a pressure change is significant. Such fluids are commonly referred to as barotropic fluids. Further, the domain of fluid flows with noticeable temperature change is referred to as fluid mechanics, a domain of science that studies constant enthalpy (isenthalpic) flows. Continuity and momentum equations and the second law of thermodynamics govern a fluid dynamic stream. The energy equation is redundant in this case. A fluid dynamic stream is described as a flow with a Mach number less than 0.5 at standard sea-level conditions, where a temperature change of less than 5% is deemed negligible. A fluid dynamic stream might be compressible or incompressible. Temperature and density differences are minor in an incompressible flow, whereas temperature and density variations are always finite in a compressible flow.

In many practical applications, including flying aircraft, missiles, and satellite launch vehicles, the Mach number is typically larger than 0.5. Temperature and density changes caused by the flow become considerable in these instances. The study of flow in which pressure changes produce density and temperature variations is known as gas dynamics. If the Mach number is more than one, the flow is supersonic, and Mach waves, expansion waves, and shock waves comprise the flow field. Only these waves cause the flow properties to change from one state to another. The following three mechanisms in gas dynamics produce a change in state or flow properties:

- Utilizing area change, assuming the flow to be inviscid and the duct to be frictionless. Such a flow is called the isentropic flow.
- Utilizing friction, in the absence of heat transfer between the flow system and surroundings. This type of flow is known as the Fanno flow.
- Utilizing heat transfer, assuming the fluid to be inviscid, the duct may or may not be frictionless. This type of flow is called the Rayleigh flow.

The following articles discuss some of the fundamental aspects of compressible flows described above. However, bearing in mind the experimental methodology presented in this book, we restrict our discussion to inviscid flows moving through frictionless ducts, with property changes driven solely

DOI: 10.1201/9781003139447-2

by area change. We also examine the flows dominated by the Mach waves, shock waves, and expansion waves; and summarize the laws governing their application. But, before we go into compressible flows, let's go over some basic fluid properties.

2.2 CONCEPT OF CONTINUUM

A fluid (a liquid or a gas) consists of a large number of molecules in constant random motion and frequently collide with each other, and the wall if stored in a container. The forces within the fluid or experienced by the wall arise due to these complex movements and collisions. One way of estimating these pressures is by modeling the behavior of each molecule in a fluid. However, such an approach is cumbersome and not feasible even with the latest computers due to the extremely large quantity of molecules even in a small mass of the fluid. Apart from the fact that such an approach is unmanageable, it is not at all necessary in analyzing the liquid because the strong intermolecular forces compel the fluid to behave as a continuous mass of the substance. Even though the intermolecular cohesive forces are weak in gases, the number of molecules is large. Thus, it is convenient to avoid the cost-ineffective modeling of molecular motions by considering the average effects of all the molecules within the gas. However, we can do this only if the gas has sufficient molecular density (number of molecules in a unit volume of the gas) so that the gas can be treated as a continuous media. Consider, for example, the estimation of pressure in a fluid. Any surface in a fluid or the container wall enclosing the fluid is constantly bombarded by several molecules experiencing a change of momentum. The rate of net momentum exchange of these molecules with the surface (fluid surface or the wall) results in a force per area called pressure. Even if we consider the area of the surface as low as 10^{-12} m^2, the number of colliding molecules is about 10^7 per second in a gas at the normal pressure and temperature. This is such a large number that the individuality of molecular interactions does not matter and we observe an overall effect.

The validity of the continuum hypothesis is determined by two crucial elements. The first is the distance a molecule travels between two successive collisions with other molecules, which is not the same for all gas molecules at any given time. As a result, an average distance, known as the mean free path (λ), is defined to represent the molecular density. It is the average distance traveled by a molecule between two successive collisions. If the mean free path is less than a characteristic length (ℓ) of the fluid, i.e., $\lambda \ll \ell$, the gas can be considered a continuous medium. However, if $\lambda > \ell$, the gas can no longer be viewed as a continuous media and must be studied at the molecular (microscopic) level. The second factor is the duration between collisions, which must be short enough to retain molecular motion's random, statistical aspect.

In this book, we will assume the fluid to be a continuous media. It will also be assumed that the elastic properties are the same at all points in the fluid and that they are identical in all directions; the fluids are assumed to be homogeneous and isotropic, respectively.

2.3 FUNDAMENTAL PROPERTIES OF FLUIDS

This section describes some basic fluid properties that are useful in flow analysis.

2.3.1 DENSITY, SPECIFIC VOLUME, AND SPECIFIC WEIGHT

The density, denoted by the Greek letter rho (ρ), of a fluid is defined as the mass of the fluid contained in a unit volume. It has the units of kg/m^3 and dimensions ML^{-3}. Mathematically, the density of fluid at a point is expressed as

$$\rho = \lim_{\delta\forall \to 0} \frac{\delta m}{\delta\forall} \ (kg/m^3) \tag{2.3.1}$$

where δm is the infinitesimal mass of the fluid surrounding the point, and $\delta\forall$ is the infinitesimal volume enclosing the point. Note that $\delta\forall$ is infinitesimal but sufficiently large to hold the continuum hypothesis.

The reciprocal of density is known as specific volume (υ), which is defined as volume per unit mass and is equivalent to:

$$\upsilon = \frac{1}{\rho} \tag{2.3.2}$$

Thus, the specific volume has units of m^3/kg and dimensions $M^{-1}L^3$.

In general, the density of a system depends on pressure and temperature. For most gases, the density is proportional to the temperature, but inversely proportional to the pressure. For liquids and solids, on the other hand, the change in density with pressure is usually very small and thus neglected for all practical purposes. For instance, at 20°C, the density of water is 998 kg/m^3 at 1 atm, which changes to 1000 kg/m^3 at 100 atm, i.e., only 0.5% change in density even by increasing the pressure 100 times. Indeed, the density of liquids and solids depends strongly on the temperature and is rather weak on the pressure. At 1 atm, for example, the density of water changes from 998 kg/m^3 at 20°C to 975 kg/m^3 at 75°C; a change of 2.3%, which, however, can still be neglected in most of the practical situations. Thus, the density of the water remains nearly constant even with a low to moderate increase in pressure or temperature. This result, however, holds for most of the liquids and all the solids. Consequently, both liquids and solids are considered to be incompressible substances in almost all engineering analyses.

Closely related to density is the specific weight (γ_{sw}), defined as the weight of fluid per unit volume. Mathematically, it is expressed as

$$\gamma_{sw} = \rho g \tag{2.3.3}$$

where g is the acceleration due to gravity in m/s^2. Thus, γ_{sw} has the units of $kg/(m^2.s^2)$ or N/m^3, and dimensions $ML^{-2}T^{-2}$. The specific weight essentially represents the gravitational force acting on the unit volume of the fluid.

2.3.2 SPECIFIC GRAVITY

It is sometimes convenient to express the density of a fluid relative to the density of a standard fluid at a specified temperature. This ratio is called the specific gravity of the given fluid. Water is usually considered as the standard fluid which has a density of 1000 kg/m^3 at 4°C. Thus, the specific gravity is defined as

$$\text{S.G.} = \frac{\rho_f}{\rho_{H_2O}} \tag{2.3.4}$$

where ρ is the density of the given fluid. Equation (2.3.4) shows that the specific gravity is a dimensionless quantity. Consider, for example, the density of mercury (ρ_{Hg}) at 20°C is 13.6×10^3 kg/m^3. The specific gravity of mercury therefore becomes

$$\begin{aligned} \text{S.G.} &= \frac{\rho_{Hg}}{\rho_{H_2O}} \\ &= \frac{13.6 \times 10^3}{1 \times 10^3} = 13.6 \end{aligned}$$

It is worth noting that substances with a specific gravity greater than one are heavier than water and hence sink in it. Substances with a specific gravity of less than one, on the other hand, will float on the water.

Example 5. If the specific gravity of gasoline is 0.72, find the density, weight, and specific weight of 2.0 liters gasoline

Solution. Given,

$$\begin{aligned} \forall_{\text{gasoline}} &= 2.0\ \ell \\ &= 2 \times 10^{-3}\ m^3 \\ \text{S.G.} &= 0.72 \end{aligned}$$

Since

$$\text{S.G.} = \frac{\rho_{\text{gasoline}}}{\rho_{H_2O}}$$

The density of gasoline is

$$\rho_{\text{gasoline}} = \text{S.G.} \times \rho_{\text{H}_2\text{O}}$$
$$= 0.72 \times 10^3 \ {}^{kg}/_{m^3}$$

and the specific weight of the gasoline is

$$(\gamma_{sw})_{\text{gasoline}} = \rho_{\text{gasoline}} \times g$$
$$= 0.72 \times 10^3 \times 9.81$$
$$= 7063.1 \ {}^{N}/_{m^3}$$

Therefore, the weight of the gasoline is

$$W = (\gamma_{sw})_{\text{gasoline}} \times \forall_{\text{gasoline}}$$
$$= 7063.1 \times 2 \times 10^{-3}$$
$$= 14.112 \text{ N}$$

Example 6. If 12 liters of a liquid A of specific gravity 1.2 is mixed with 8 liters of another liquid B of specific gravity 0.9. Find the density, weight, and specific weight of the mixture.

Solution. Given,

$$\forall_A = 12\,\ell = 12 \times 10^{-3} \text{ m}^3$$
$$(\text{S.G.})_A = 1.2$$
$$\forall_B = 8\,\ell = 8 \times 10^{-3} \text{ m}^3$$
$$(\text{S.G.})_B = 0.9$$

The total volume of the mixture is

$$\forall_T = \forall_A + \forall_B = 12 \times 10^{-3} + 8 \times 10^{-3} = 20 \times 10^{-3} \text{ m}^3$$

From equation (2.3.4), we get

$$\rho_A = 1.2 \times 10^3 \ {}^{kg}/_{m^3}$$
$$\rho_B = 0.9 \times 10^3 \ {}^{kg}/_{m^3}$$

The masses of the liquids A and B are

$$m_A = \rho_A \times \forall_A = 1.2 \times 10^3 \times 12 \times 10^{-3} = 14.4 \text{ kg}$$
$$m_B = \rho_B \times \forall_B = 0.9 \times 10^3 \times 8 \times 10^{-3} = 7.2 \text{ kg}$$

Thus, the total mass of the liquids is

$$m_T = m_A + m_B = 14.4 + 7.2 = 21.6\ \text{kg}$$

The density of the mixture is

$$\rho_m = \frac{m_T}{\forall_T} = \frac{21.6}{20 \times 10^{-3}} = 1080\ {}^{kg}/_{m^3}$$

which gives the specific weight of the mixture as

$$(\gamma_{sw})_m = \rho_m g = 1080 \times 9.81 = 10594.8\ {}^{N}/_{m^3}$$

and the weight

$$W_m = (\gamma_{sw})_m \times \forall_T = 10594.8 \times 20 \times 10^{-3} = 211.89\ \text{N}$$

2.3.3 VISCOSITY

Fluid parameters like density and specific weight efficiently quantify a fluid's heaviness. These parameters, however, are insufficient to characterize fluid behavior because two fluids (such as water and oil) can have comparable density values but flow differently. This means that more fluid parameters can be used to characterize the mobility of the fluid. The property that characterizes the resistance a fluid offers to the applied shear stress is called. Viscosity represents a fluid's internal resistance to motion or fluidity. Molasses and tar are highly viscous fluids; water and air have minimal viscosity. The resistance offered by a fluid to the motion does not depend upon the deformation (unlike the case of solids) but rather on the rate of deformation. Consider a fluid layer between two very large plates separated by a small distance with the lower plate stationary and the upper plate moving with velocity U_o (Figure 2.3.1). The upper plate sets the fluid in motion with a velocity u which is a function of y, the vertical distance measured from the lower plate. Further, it has been demonstrated through several experiments that for all real fluids possessing viscosity, however small, the fluid properties in immediate contact with any solid surface move with the velocity of the surface itself. That is, there is no relative motion between the fluid near the surface and the solid surface. Fluid sticks to the surface, and there is no slip. This is known as the no-slip condition and is valid for all fluids except supercooled helium.

In Figure 2.3.1, the no-slip condition requires that $u = 0$ at $y = 0$ and $u = U_0$ at the moving plate. In steady laminar flows, the fluid velocities between the plates varies linearly from 0 to U_0. In a small time interval, δt, an imaginary

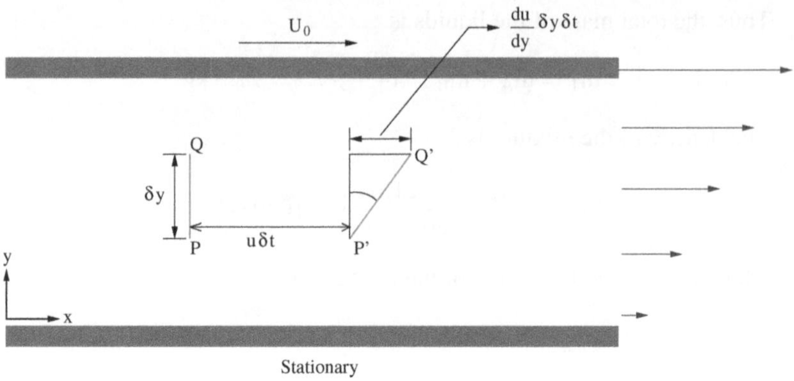

Figure 2.3.1 Fluid layer between two large parallel plates.

vertical line PQ in the fluid moves to $P'Q'$ resulting an angular rotation (or shear strain) β. In this case, the rate of deformation of the fluid is

$$\begin{aligned}\text{Rate of deformation} &= \text{Rate of shear strain}\\ &= \dot{\gamma}\\ &= \frac{\partial \gamma}{\partial t}\\ &= \frac{\partial}{\partial t}\left[\frac{1}{\partial y}\left\{\left(u+\frac{du}{dy}\delta y\right)\delta t - u\delta t\right\}\right]\\ &= \frac{du}{dy}\end{aligned} \tag{2.3.5}$$

Thus, we conclude that the rate of deformation of a fluid element, $\frac{\partial \gamma}{\partial t}$, is equivalent to the velocity gradient, $\frac{du}{dy}$. Extensive experiments have shown that for most fluids the rate of deformation or rate of shear strain (and thus the velocity gradient) is proportional to the applied shear stress, τ. That is,

$$\tau \propto \frac{\partial \gamma}{\partial t} \implies \tau \propto \frac{du}{dy} \tag{2.3.6}$$

Newtonian fluids are those in which the rate of deformation is directly proportional to the shear stress, as first expressed by Sir Isaac Newton in 1687. The linear relationship is used to express shear stress and the rate of shearing strain (velocity gradient) for common Newtonian fluids such as water, oil, gasoline, and air and is similar to:

$$\tau = \mu \frac{du}{dy} \tag{2.3.7}$$

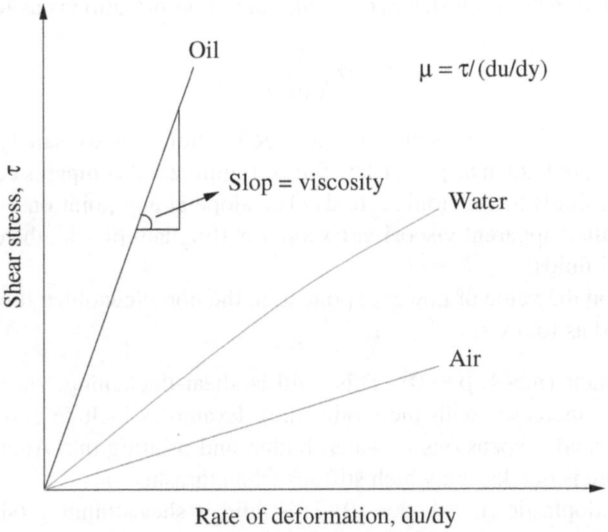

Figure 2.3.2 Shear stress variation with rate of deformation for Newtonian fluids.

where the constant of proportionality μ is called the absolute viscosity, dynamic viscosity or simply the viscosity of a fluid. Equation (2.3.7) is Newton's law of viscosity, also known as Newton's law of fluid friction, asserts that the stresses opposing fluid shearing are proportional to the shear strain rate.

The SI unit of viscosity is $kg/(\mathrm{m.s})$ or equivalently $(N.s)/\mathrm{m}^2$ and its dimension is $\mathrm{ML}^{-1}\mathrm{T}^{-1}$. Viscosity is also expressed in poise or centipoise (one-hundredth of a poise). The unit of viscosity from poise to SI units and vice versa may be converted using the following conversion factors.

$$1 \text{ poise} = 0.1\ (N.s)/\mathrm{m}^2 = 0.1\ kg/(\mathrm{m.s}) \tag{2.3.8}$$

According to equation (2.3.7), the change of shear stress with the rate of deformation (velocity gradient) for a Newtonian fluid is a straight line with a slope equal to the viscosity of the fluid, as illustrated in Figure 2.3.2. It should be noted that the actual value of viscosity of Newtonian fluids depends on the particular fluid and is independent of the rate of deformation.

2.3.3.1 Non-Newtonian Fluids

Certain fluids (including blood and liquid polymers) have a non-linear relation between shear stress and deformation rate. This type of fluid is known as a non-Newtonian fluid. Lubricating oil, clay dispersion in water, printer ink, butter, and sewage sludge are all non-Newtonian fluids. For non-Newtonian fluids, the general relation between shear stress and rate of deformation is as follows:

$$\tau = \alpha \left(\frac{du}{dy} \right)^n + \beta \tag{2.3.9}$$

where α and β are constants for a specific fluid; fluids satisfying equation (2.3.9) are known as power-law fluids. Figure 2.3.2 compares certain non-Newtonian fluids to Newtonian fluids. The slope at any point on a non-linear curve is called apparent viscosity. Except for Bingham plastic, the constant is zero for all fluids.

Based on the value of power exponent, n, the non-Newtonian fluids may be categorized as follows:

- Dilatant ($n > 1, \beta = 0$): This fluid is shear-thickening (shear resistance increases with the strain rate). Examples include corn starch and sand suspensions in water, butter, and printing ink. Another example is quicksand, which stiffens when thrashed in.
- Pseudoplastic ($n < 1, \beta = 0$): This fluid is shear-thinning (shear resistance decreases with the strain rate). Plastic is a highly strong thinning. Polymer solutions, colloidal dispersion, paper pull-in water, latex paint, blood plasma, syrup, and molasses are some common examples – the perfect example is paints, which are thick when poured but thins when stroked quickly.
- Bingham Plastic ($n = 1, \beta \neq 0$): This fluid is the ultimate instance of a plastic that requires a finite yield stress before flowing. Figure 2.3.3

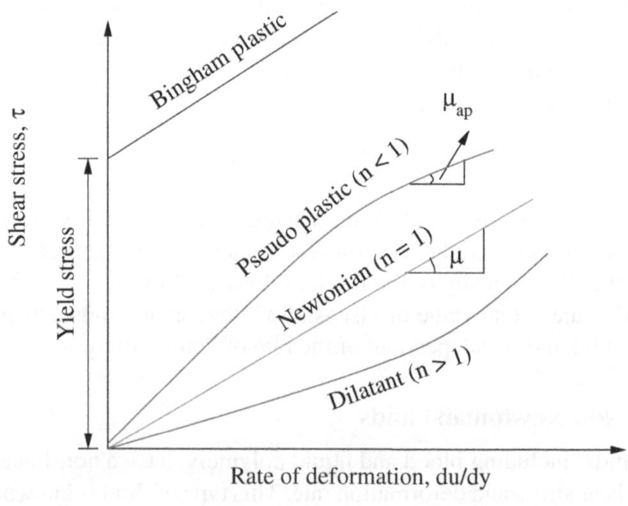

Figure 2.3.3 Shear stress variation with rate of deformation for non-Newtonian fluids.

depicts yielding followed by linear behavior; however, non-linear flow is possible. Clay dispersion, drilling mud, toothpaste, mayonnaise, chocolate, and mustard are a few examples. A good example is cat-sup, which will not come out of the bottle unless you shake it.

Note that, a Newtonian fluid is a special case of power law fluid having n = 1 and $\beta = 0$, and $\alpha = \mu$ varies only with the fluid type.

2.3.3.2 Time-Independent Fluids

The strain rate (or velocity gradient) in these fluids is determined only by shear stress and is a mono-valued function of it. The viscosity of a Newtonian fluid is independent of shear stress, whereas it is dependent on it in non-Newtonian fluids.

2.3.3.3 Time-Dependent Fluids

For these fluids, the rate of deformation (rate of shear strain) and the viscosity depends on both shear stress and duration of its application to the fluid. Thus, the transient effect further complicates the non-Newtonian behavior, as shown in Figure 2.3.4.

Some fluids, known as rheopectic, require a steadily increasing shear stress over time to maintain a constant strain rate. Thixotropic fluids, on the other hand, thin out with time and require less stress. Rheopectic fluids include gypsum suspensions in water and bentonite solutions; thixotropic fluids include paints and enamels.

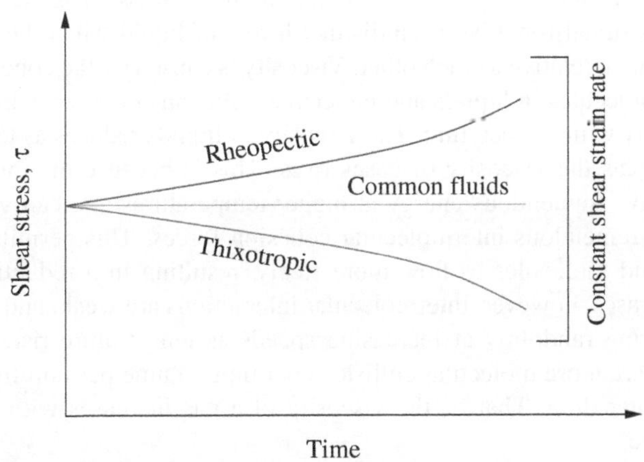

Figure 2.3.4 Effect of time on applied shear stress.

2.3.3.4 Viscoelastic Fluids

There are also some substances which cannot be classified as either fluids or solids but show intermediate behavior. These are called viscoelastic fluids. Both non-Newtonian and viscoelastic fluids are beyond the scope of this book. We will continue our attention to Newtonian fluids only.

2.3.3.5 Kinematic Viscosity

In fluid flow applications, a dynamic viscosity-to-density coupling is frequently observed. This ratio is known as the kinematic viscosity and is denoted as:

$$\nu = \frac{\mu}{\rho} \tag{2.3.10}$$

The common units of ν are m^2/s and stoke (1 stoke = 1 cm^2/s = 0.0001 m^2/s). The dimensions of kinematic viscosity are L^2T^{-1}.

2.3.3.6 Effects of Temperature and Pressure on Viscosity

The viscosity of a fluid is often affected by pressure and temperature, albeit the influence of pressure is feeble. Both dynamic and kinematic viscosities of liquids are pressure independent, and any tiny fluctuation in pressure is generally ignored except at extremely high-pressure values. This is also true for dynamic viscosity (at low to moderate pressures) but not for kinematic viscosity in gases because the density of a gas is proportional to its pressure.

The viscosity of a fluid is a measure of its internal resistance to the deformation rate generated by applying external shear stress. It is caused by the internal frictional force between distinct layers of liquid when they are compelled to move relative to each other. Viscosity is caused by the cohesive forces between molecules in liquids and molecular collisions in gases, which change dramatically with temperature. The viscosity of liquids reduces as temperature rises, whereas the viscosity of gases rises. This is because the molecules in a liquid have tremendous energy at higher temperatures and can vehemently reject the tremendous intermolecular cohesion forces. This permits the electrified liquid molecules to flow more freely, resulting in a reduction in viscosity. In gases, however, intermolecular interactions are weak, and molecules begin moving randomly at increasing speeds as temperature rises. As a result, there are more molecular collisions per unit volume per unit time, resulting in greater drag. That is, the viscosity of a gas increases with increasing temperature.

The variation of viscosity of gases with temperature can be approximated by either of two laws, called respectively, the power law and Sutherland's law, and expressed as follows:

$$\frac{\mu}{\mu_0} = \begin{cases} \left(\frac{T}{T_0}\right)^n & \text{Power law} \\ \left(\frac{T}{T_0}\right)^{3/2}\left(\frac{T_0+S}{T+S}\right) & \text{Sutherland's formula} \end{cases} \tag{2.3.11}$$

where μ_0 is some known viscosity at absolute temperature T_0, and the constants S and n are determined by curve fitting. For air, $n \approx 0.7$ and $S \approx 110$ K.

For air, Sutherland's formula of viscosity can also be written as [16]:

$$\mu = 1.46 \times 10^{-6}\left(\frac{T^{3/2}}{T+111}\right) \, kg/(m.s) \tag{2.3.12}$$

where T is the temperature in Kelvin. Equation (2.3.12) is valid for static pressures ranging from 0.01 atm to 100 atm (typical in atmospheric flight) and temperatures up to 3000 K. The dynamic viscosity in this equation is solely determined by temperature since air behaves as a perfect gas in this pressure and temperature range, with negligible intermolecular forces. Furthermore, viscosity is a momentum transport phenomenon caused by random molecule motion in the presence of thermal energy or temperature.

As with liquids, viscosity decreases with temperature and is roughly exponential of the form:

$$\mu = Ae^{-BT} \tag{2.3.13}$$

where A and B are constants found again by curve fitting data for a particular liquid. However, a better fit is the following empirical relation:

$$\ln\frac{\mu}{\mu_0} = a + b\left(\frac{T_0}{T}\right) + c\left(\frac{T_0}{T}\right)^2 \tag{2.3.14}$$

For water, with $\mu = 0.001792 \, kg/(m.s)$ at $T_0 = 273.16$ K, the constants are $a = -1.94$, $b = -4.8$, and $c = 6.74$.

Example 7. The viscosity of carbon dioxide at 50°C and 200°C are 1.61×10^{-5} Pa.s and 2.27×10^{-5} Pa.s, respectively. If the viscosity of carbon dioxide is expressed by the relation: $\mu = \frac{aT^{1/2}}{1+b/T}$, find the constants a and b. Compute the viscosity of carbon dioxide at 100°C.

Solution. Given,

$$T_1 = 50^\circ C + 273 = 323 \text{ K}$$

$$\mu_1 = 1.61 \times 10^{-5} \text{ Pa.s}$$

$$T_2 = 200^\circ C + 273 = 473 \text{ K}$$

$$\mu_2 = 2.27 \times 10^{-5} \text{ Pa.s}$$

At 50°C, we get

$$\mu_1 = \frac{aT_1^{1/2}}{1+{}^{b}/_{T_1}} = \frac{a \times (323)^{1/2}}{1+{}^{b}/_{323}}$$

$$1.61 \times 10^{-5}\,(1+{}^{b}/_{323}) = a \times (323)^{1/2} \tag{1}$$

At 200°C, we have

$$\mu_1 = \frac{aT_2^{1/2}}{1+{}^{b}/_{T_2}} = \frac{a \times (473)^{1/2}}{1+{}^{b}/_{473}}$$

$$2.27 \times 10^{-5}\,(1+{}^{b}/_{473}) = a \times (473)^{1/2} \tag{2}$$

Dividing equation (2) by equation (3) yields

$$\frac{1.61 \times 10^{-5}\,(1+{}^{b}/_{323})}{2.27 \times 10^{-5}\,(1+{}^{b}/_{473})} = \frac{(323)^{1/2}}{(473)^{1/2}}$$

which on solving gives

$$b \approx 262$$

By substituting this value in equation (2) and solving, we get

$$a = 1.622 \times 10^{-6}$$

Therefore, the given law of viscosity becomes

$$\mu = \frac{1.622 \times 10^{-6} T^{3/2}}{T+262} \tag{3}$$

Now, at 100°C, the viscosity of carbon dioxide is

$$\mu = \frac{1.622 \times 10^{-6}\,(373)^{3/2}}{373+262}$$
$$= 1.84 \times 10^{-5} \text{ Pa.s}$$

It is interesting to note that equation (3) is indeed Sutherland's law of viscosity of carbon dioxide.

2.3.4 THERMAL CONDUCTIVITY

The thermal conductivity of the air determines the heat transmission between the high-temperature air and the vehicle. For temperatures ranging from 1500

K to 2000 K, C. F. Hansen's [17] approximate relation for calculating the thermal conductivity of air, which is similar to Sutherland's formula for the viscosity of air, is as follows:

$$k = 1.993 \times 10^{-3}\left(\frac{T^{3/2}}{T+112}\right) W/(m.K) \tag{2.3.15}$$

where k is the thermal conductivity of air, and T is the temperature in Kelvin. Equation (2.3.15) is valid for static pressures ranging from 0.01 atm to 100 atm. It is worth noting that the thermal conductivity of air depends only on temperature for the same reasons those stipulated for its viscosity in Section 2.3.3.6.

Example 8. What is the thermal conductivity of air at 500 K?

Solution. Given,

$$T = 500\ K$$

The thermal conductivity of air is

$$\begin{aligned} k &= 1.993 \times 10^{-3}\left(\frac{T^{3/2}}{T+112}\right) \qquad (2.3.15)\\ &= 1.993 \times 10^{-3}\left(\frac{500^{3/2}}{500+112}\right)\\ &= 0.0364\ W/(m.K) \end{aligned}$$

2.3.5 COMPRESSIBILITY

The compressibility of a substance is the measure of its change in volume un der the action of external forces, particularly the normal compressive forces. These forces compress the fluid element causing a decrease in its volume. The normal compressive stress of any fluid at rest is known as the hydrostatic pressure, which arises due to abundant molecular collisions occurring in the entire fluid. The compressibility of a fluid is characterized in terms of the bulk modulus of elasticity, E, defined as

$$E = \lim_{\delta\forall \to \delta\forall_0} \frac{\delta p}{(-\delta\forall/\forall)} \tag{2.3.16}$$

where $-\delta\forall/\forall$ and δp are the fractional change in volume and applied pressure (normal compressive stress), respectively. Note that $\delta\forall_0$ is the minimum possible fluid volume over which the continuum assumption remains valid. The negative sign in equation (2.3.16) implies a decrease in volume with an increase in pressure. For a given mass of the substance, the density and volume are related as

$$\frac{\delta\forall}{\forall} = -\frac{\delta\rho}{\rho} \tag{2.3.17}$$

Combining equation (2.3.16) and equation (2.3.17) yields

$$E = \lim_{\delta\forall \to \delta\forall_0} \frac{\delta p}{\delta\rho/\rho} \tag{2.3.18}$$

which in differential form can be written as

$$E = \rho\frac{dp}{d\rho} \tag{2.3.19}$$

The value of E for liquids are very high compared to gases (except at exceedingly high pressures). Thus, the liquids at normal conditions are considered as incompressible. The reciprocal of bulk modulus of elasticity is called the compressibility, κ. Thus,

$$\kappa = \frac{1}{E} = \frac{1}{\rho}\frac{d\rho}{dp} = -\frac{1}{\forall}\frac{d\forall}{dp} \tag{2.3.20}$$

It follows from equation (2.3.20) that the compressibility of a purely incompressible substance is zero. Although the liquids at moderate pressures are incompressible, a small change in the density of liquids can cause an interesting phenomenon in a pipe network such as a water hammer. It is characterized by a sound that resembles the sound produced when a pipe is hammered. It occurs when a liquid flowing through a pipe system is suddenly restricted, such as by a closing valve, and is locally compressed. Eventually, the sound waves are produced, which hit the pipe surfaces, bends, valves, etc., and reflect, causing the pipe to vibrate and generate an irritating sound. Sometimes the water hammer can be so destructive that it can lead to structural damage. This problem can be eliminated by using a water-hammer-arrestor; it is a volumetric chamber with bellows or pistons to absorb the shocks.

When compressed by increasing pressure, a gas experiences a rise in its temperature. This results in heat exchange across the system (gas) boundary. If the temperature is kept constant by an appropriate heat transmission system, the isothermal compressibility (κ_T) of the gas can be expressed as follows:

$$\kappa_T = \frac{1}{\rho}\left(\frac{d\rho}{dp}\right)_T = -\frac{1}{\forall}\left(\frac{d\forall}{dp}\right)_T \tag{2.3.21}$$

where the subscript T denotes that the partial derivative is taken at constant temperature. On the other hand, if the system is isolated and no heat exchange with the surrounding environment is feasible, the compression proceeds isentropically. In this situation, κ is called the isentropic compressibility (κ_S), expressed as:

$$\kappa_S = \frac{1}{\rho}\left(\frac{d\rho}{dp}\right)_s = -\frac{1}{\forall}\left(\frac{d\forall}{dp}\right)_s \tag{2.3.22}$$

The subscript s indicates that the partial derivative is evaluated at constant entropy.

Gases have a compressibility that is several orders of magnitude greater than liquids. According to equation (2.3.20), for a given pressure change dp, the accompanying change in density $d\rho$ will be infinitesimally small for liquids, so liquids are classified as incompressible. However, for a given pressure change, the equivalent change in density for gases will be high (due to large κ); consequently, gases are generally compressible. However, in the low-speed flow of a gas, the change in pressure from one location to another is insignificant compared to the pressure itself, despite considerable κ, and the value of $d\rho$ is dominated by small dp.

2.4 THE ZONE OF ACTION AND THE ZONE OF SILENCE

There are three types of compressible flows: incompressible, subsonic, and supersonic. Because of the medium's compressibility, subsonic and supersonic flow patterns and general behavior change considerably. This section qualitatively investigates the physical differences between subsonic and supersonic flow patterns by depicting the pressure field created by the moving point source in the resting fluid.

Each solid surface element tends to divert air from the route it would otherwise travel when a body moves through a stationary fluid or flows around a stationary object. These small disturbances operate as a point source by sending the spherically spreading pressure waves into the ambient air. These waves can be represented at any time as a superposition of all prior waves released by the same source. The resulting wave pattern, however, can be symmetric or asymmetric depending on whether the source is stationary or moving. Consider the following examples:

1. When the point source generating the sound waves in a compressible fluid is stationary ($M = 0$), the pressure disturbances appear to spread in concentric circles at different intervals. At integer multiples of Δt, the intersection of these wavefronts with a plain containing the source is represented in Figure 2.4.1(i).
2. When the point source travels, the wavefronts differ dramatically because the source releases each pressure disturbance at a different

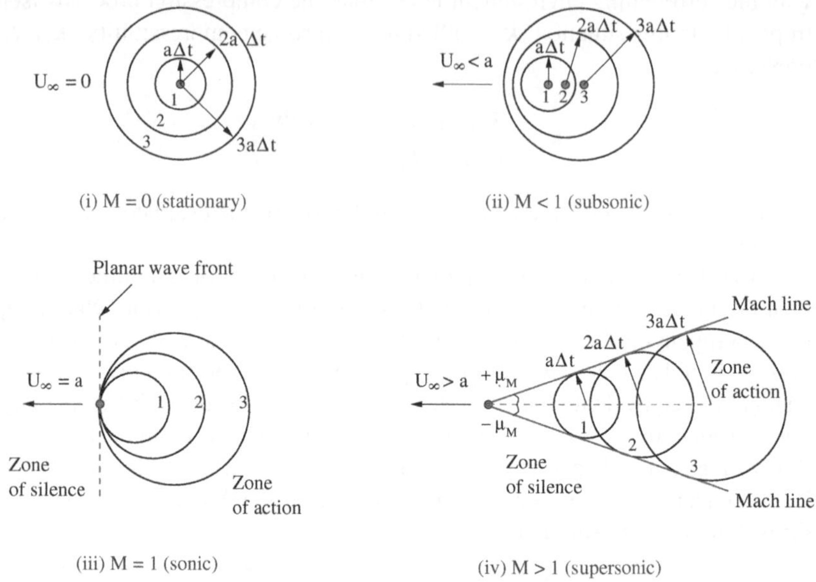

Figure 2.4.1 Acoustic waves transmission from a single point source.

location in the fluid. If the source is moving slower than the speed of sound ($M < 1$), the wavefronts will look like in Figure 2.4.1(ii). Because the pressure disturbance moves faster than the source, a series of circles, one inside the other but with separate centers, is generated. These wavefronts do not cross because, $U_\infty < a$. Wavefronts propagate horizontally upstream and downstream, vertically uphill and below, just like a stationary source. Thus, an object moving at subsonic speeds has an impact on the entire flow field.

3. When a point source proceeds at the speed of sound ($M = 1$), the resulting wavefronts are the same as those seen in Figure 2.4.1(iii). Each pressure disturbance strengthens the previous one in this case, resulting in a planar wavefront. Because this is a sound wavefront, it is isentropic by definition.
4. When a point source moves faster than the speed of sound ($M > 1$), the pressure disturbances it emits are enclosed within a cone with the vertex at the body. Disturbances are only relayed downstream inside the cone and are not detected upstream of the cone vertex. The Mach cone is the cone that contains the disturbances, and the Mach angle is the half angle of the cone (Figure 2.4.1(iv)). The zone of action is the area limited by the Mach cone, whereas the zone of silence is the

area outside the cone. The Mach angle can be computed as follows:

$$\sin \mu_M = \frac{\text{Distance traveled by the point source in a particular time interval } \Delta t}{\text{Distance traveled by the sound wave in the same time interval } \Delta t} \tag{2.4.1}$$

$$= \frac{U_\infty \Delta t}{a \Delta t} = \frac{1}{M} \tag{2.4.2}$$

$$\Longrightarrow \quad \mu_M = \sin^{-1}\left(\frac{1}{M}\right) \tag{2.4.3}$$

It should be noted that when the Mach number reduces, the Mach cone spreads and forms a planar wavefront ($\mu_M = 90^o$) at $M = 1$, signifying that the object is moving at sonic speed. Hence, point 3 is effectively a subset of point 4.

Example 9. At 350 K, air is moving at Mach 2.5. Determine the flow velocity and the associated Mach angle.

Solution. Given,

$$T_\infty = 350 \text{ K}$$
$$M_\infty = 2.5$$

The speed of sound is

$$a_\infty = \sqrt{\gamma R T_\infty} 2.7.6$$
$$= \sqrt{1.4 \times 287 \times 300} = 375 \, {}^{m}/_{s}$$

Therefore, the flow velocity is

$$V_\infty = M_\infty a_\infty 2.7.7$$
$$= 2 \times 375 = 750 \, {}^{m}/_{s}$$

The corresponding Mach angle at Mach 2 is

$$\mu_M = \sin^{-1}\left(\frac{1}{M}\right) \tag{2.4.3}$$
$$= \sin^{-1}\left(\frac{1}{2.5}\right) = 23.58^o$$

2.5 STREAMLINE FLOW

When air flows at subsonic speeds through a duct, the increase in Mach number is accompanied by an increase in velocity and a decrease in density. However, velocity rises more quickly than density drops; a 10% increase in velocity results in an 8% decrease in density. In such cases, the mass flow rate at every channel cross-section increases as the Mach number increases. If the channel continues to be filled with the flow, its area must decrease.

The behavior is reversed when the speed exceeds Mach 1.0. A 10% increase in speed could result in a 12% drop in density as the Mach number rises. As a result, the mass flow rate at every cross-section decreases, necessitating channel expansion to accommodate the accelerating flow. To decelerate a supersonic flow, the channel area must be reduced.

In the case of freestreams, we refer to streamtubes rather than channels. Streamtubes are fictitious channels through which a continuous fluid mass is imagined to flow. Streamlines are the imaginary curves that define the two-dimensional stream tubes. Streamtubes, and hence streamlines, have the shape of meticulously planned channels that drop in cross-section while accelerating a subsonic flow and rise in cross-section when accelerating a supersonic flow. When the flow speed exceeds the speed of sound, the subsonic laws of streamlined flow are reversed.

2.6 SHOCK AND EXPANSION WAVES

When a particle flows through the air at Mach 1.0 or higher speeds, concentrated wavefronts occur, as seen in Figure 2.4.1. If these waves are so faint, they create infinitely little changes in the properties of the air while passing through it, they are called Mach waves. A Mach wave is inclined to the flow direction at an angle of $\sin^{-1}\left(\frac{1}{M}\right)$, termed the Mach angle. A concentration of Mach waves known as shock waves produces finite-strength waves (substantially modifying air characteristics).

The Mach wave generation induced by supersonic flow compression is compared with that caused by supersonic flow expansion in Figure 2.6.1. Changing the flow direction takes a short but finite period of time. As a result, air approaching a corner prefers to avoid it. When a corner compresses the flow, disturbance waves occur, as illustrated in Figure 2.6.1(a). A minor disturbance (Mach wave) forms, slowing the flow and turning it slightly. Many more minor disturbances will accomplish the same. Each successive perturbation wave has a greater inclination to the original direction due to the reduced flow velocity and changing flow direction. A shock wave is formed when many weak compression disturbances merge. When the air expands at the corner (Figure 2.6.1(b)), it produces a series of minor disturbance waves, each occurring at a higher Mach number. Due to increased Mach number and changed flow

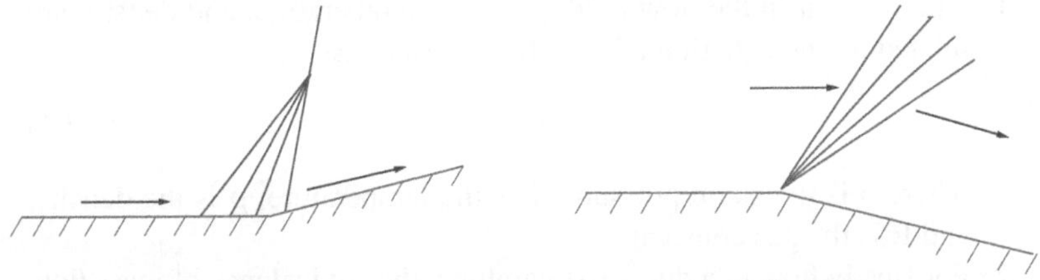

Figure 2.6.1 Compression and expansion waves in supersonic flows.

direction, these waves diverge, resulting in a series of expansion waves known as Prandtl-Meyer expansion fans.

The shock wave in Figure 2.6.1(a) is known as an oblique shock because of its angle to the flow direction. If the flow turning angle is low, the downstream velocity is less than the upstream velocity while being supersonic. When the turning angle is large enough, the shock becomes normal to the flow and separates from the wedge, resulting in a speed downstream of the normal shock that is subsonic. Between these extreme turning angles, the oblique shock steepens, and downstream velocity decreases as the turning angle increases.

Now that we've examined normal and oblique shock waves, we'll delve deeper into shock properties. When traveling through a shock wave, the properties of air change almost instantly. Pressure, temperature, and density increase when speed decreases. Entropy rises because the total pressure before the shock is not recovered. The air can be returned to its pre-impact stagnation temperature, but only at a lower-stagnation pressure than earlier.

When the flow over an object gets supersonic, a shock occurs. If the flow over the object is somewhat faster than Mach 1.0, the shock will be normal, and the losses will be minor. If the flow velocity is larger, the shock can be oblique or normal, depending on the angle at which the body bends the air. At any given Mach number, the losses from a normal shock are always greater than the losses from an oblique shock. When conditions compel a fall in speed to a subsonic level, normal shocks occur in the supersonic flow flowing through a duct, such as a wind tunnel test section.

2.7 GOVERNING EQUATIONS OF COMPRESSIBLE FLOWS

We will develop equations governing compressible flow using the above discussion of compressible flow theory. The following five laws govern airflow in general.

1. At any point in the flow field, pressure, temperature, and density are related by the equation of state for a perfect gas.

$$p = \rho RT \tag{2.7.1}$$

where p is the static pressure, T is the temperature, ρ is the density, and R is the gas constant.

2. For steady flow in a duct or streamtube, the equivalence of mass flow at any two points 1 and 2 in the duct is given by the following one-dimensional continuity equation:

$$\dot{m} = \rho_1 A_1 v_1 = \rho_2 A_2 v_2 \tag{2.7.2}$$

where $\dot{m}$ is the mass flow rate, ρ is the flow density, v is the flow velocity, and A is the nozzle cross-sectional area.

3. For an adiabatic flow of a fluid between two locations 1 and 2 in a duct or streamtube, the following energy equation holds:

$$c_p T_1 + \frac{v_1^2}{2} = c_p T_2 + \frac{v_2^2}{2} = c_p T_0 \tag{2.7.3}$$

where c_p is the specific heat at constant pressure, and T_0 is the stagnation temperature.

4. If a fluid's state changes during its movement from one point to another are isentropic, the following relationship holds:

$$\frac{T_1}{p_1^{\frac{\gamma-1}{\gamma}}} = \frac{T_2}{p_2^{\frac{\gamma-1}{\gamma}}} \tag{2.7.4}$$

where γ denotes the ratio of specific heats.

5. For a fluid flow through a constant area streamtube or duct without friction, the momentum equation between two stations 1 and 2 gives

$$p_1 + \rho_1 v_1^2 = p_2 + \rho_2 v_2^2 \tag{2.7.5}$$

6. Besides the above equations, the following relationships are required to develop the governing equations for compressible flow:

$$a = \sqrt{\gamma RT} \tag{2.7.6}$$

$$v = Ma = M\sqrt{\gamma RT} \tag{2.7.7}$$

$$c_p = \frac{\gamma R}{\gamma - 1} \tag{2.7.8}$$

where a is the speed of sound, M is the Mach number, and c_p is the specific heat at constant pressure.

7. From the energy equation (equation (2.7.3)) and the relations (equation (2.7.6)–equation (2.7.8)), the ratio of temperatures between two locations is expressed as

$$\frac{T_1}{T_2} = \frac{1+\frac{\gamma-1}{2}M_2^2}{1+\frac{\gamma-1}{2}M_1^2} \tag{2.7.9}$$

8. Combining equation (2.7.9) with the isentropic flow relation (equation (2.7.4)) yields the pressure ratio:

$$\frac{p_1}{p_2} = \left\{\frac{1+\frac{\gamma-1}{2}M_2^2}{1+\frac{\gamma-1}{2}M_1^2}\right\}^{\frac{\gamma}{\gamma-1}} \tag{2.7.10}$$

9. Combining equation (2.7.9) and equation (2.7.10) with the equation of state for a perfect gas (equation (2.7.1)) gives the density ratio:

$$\frac{\rho_1}{\rho_2} = \left\{\frac{1+\frac{\gamma-1}{2}M_2^2}{1+\frac{\gamma-1}{2}M_1^2}\right\}^{\frac{1}{\gamma-1}} \tag{2.7.11}$$

10. Combining equation (2.7.11) with equation (2.7.2), we get the area ratio:

$$\frac{A_1}{A_2} = \frac{M_2}{M_1}\left\{\frac{1+\frac{\gamma-1}{2}M_2^2}{1+\frac{\gamma-1}{2}M_1^2}\right\}^{-\frac{\gamma+1}{2(\gamma-1)}} \tag{2.7.12}$$

11. The dynamic pressure $\frac{1}{2}\rho v^2$ can be represented in terms of Mach number M and static pressure p as follows:

$$\begin{aligned} q &= \frac{1}{2}\rho v^2 \\ &= \frac{1}{2}\frac{p}{RT}v^2 \\ &= \frac{1}{2}\gamma p M^2 \end{aligned} \tag{2.7.13}$$

12. By substituting $M_2 = 0$ for stagnation conditions in the preceding equations, we get the following isentropic relations

$$\frac{T_0}{T} = 1+\frac{\gamma-1}{2}M^2 \tag{2.7.14}$$

$$\frac{p_0}{p} = \left[1+\frac{\gamma-1}{2}M^2\right]^{\frac{\gamma}{\gamma-1}} \tag{2.7.15}$$

$$\frac{\rho_0}{\rho} = \left[1+\frac{\gamma-1}{2}M^2\right]^{\frac{1}{\gamma-1}} \tag{2.7.16}$$

where T_0 denotes the stagnation temperature, p_0 denotes the stagnation pressure, and ρ_0 denotes the stagnation density.

13. Using an area A^* at $M = 1$ (corresponding to the choked throat of a supersonic nozzle), the following relation for area ratio is obtained:

$$\frac{A}{A^*} = \frac{1}{M}\left[\frac{2}{\gamma+1}\left(1+\frac{\gamma-1}{2}M^2\right)\right]^{\frac{\gamma+1}{2(\gamma-1)}} \tag{2.7.17}$$

14. There is an entropy change across a normal shock because of its irreversibility. As a result, the earlier isentropic flow equations become invalid. Normal shock equations are derived using the equations of state (equation (2.7.1)), continuity (equation (2.7.2)), energy (equation (2.7.3)), and momentum (equation (2.7.5)) equations. Let subscripts 1 and 2 represent conditions upstream and downstream of a normal shock, respectively. Now, combining equation (2.7.1) and equation (2.7.5) yields

$$\frac{p_2}{p_1} = \frac{1+\gamma M_1^2}{1+\gamma M_2^2} \tag{2.7.18}$$

The combination of equation (2.7.2) with equation (2.7.9) and equation (2.7.18) results in

$$M_2^2 = \frac{\frac{2}{\gamma-1}+M_1^2}{\frac{2\gamma}{\gamma-1}M_1^2-1} \tag{2.7.19}$$

Eliminating M_2 in equation (2.7.9), equation (2.7.18) and equation (2.7.11) using equation (2.7.19) gives

$$\frac{p_2}{p_1} = \frac{2\gamma M_1^2-(\gamma-1)}{\gamma+1} \tag{2.7.20}$$

$$\frac{T_2}{T_1} = \frac{\left[1+\frac{\gamma-1}{2}M_1^2\right]\left[\frac{2\gamma}{\gamma-1}M_1^2-1\right]}{\frac{(\gamma+1)^2}{2(\gamma-1)}M_1^2} \tag{2.7.21}$$

$$\frac{\rho_2}{\rho_1} = \frac{(\gamma+1)M_1^2}{2+(\gamma-1)M_1^2} \tag{2.7.22}$$

The stagnation pressure downstream of a normal shock is less than that upstream. The ratio of downstream and upstream stagnation pressures across a normal shock is obtained as follows:

$$\frac{p_{02}}{p_{01}} = \frac{p_1}{p_{01}}\frac{p_2}{p_1}\frac{p_{02}}{p_2} = \left[\frac{\gamma+1}{2\gamma M_1^2-(\gamma-1)}\right]^{\frac{1}{\gamma-1}}\left[\frac{(\gamma+1)M_1^2}{(\gamma-1)M_1^2+2}\right]^{\frac{\gamma}{\gamma-1}} \tag{2.7.23}$$

For air ($\gamma = 1.4$), this reduces to

$$\frac{p_{02}}{p_{01}} = \left(\frac{6M_1^2}{5+M_1^2}\right)^{3.5}\left(\frac{6}{7M_1^2-1}\right)^{2.5} \tag{2.7.24}$$

Example 10. Consider the isentropic airflow past an airfoil. The freestream conditions are 45 kPa and 250 K. At a point on the airfoil, where the pressure is 36 kPa, find the flow density.

Solution. Given,

$$p_\infty = 45 \text{ kPa}$$
$$T_\infty = 250 \text{ K}$$
$$p_1 = 36 \text{ kPa}$$

From isentropic relation,

$$\frac{T_\infty}{p_\infty^{\frac{\gamma-1}{\gamma}}} = \frac{T_1}{p_1^{\frac{\gamma-1}{\gamma}}} 2.7.4$$

$$\frac{250}{(45)^{\frac{1.4-1}{1.4}}} = \frac{T_1}{(36)^{\frac{1.4-1}{1.4}}}$$

$$\implies T_1 = 234.55 \text{ K}$$

Now, from the equation of state for the perfect gas,

$$\rho_1 = \frac{p_1}{RT_1} = \frac{36 \times 10^3}{287 \times 234.55}$$

$$\implies \rho_1 = 0.534 \ \text{kg/m}^3$$

Example 11. An aircraft is cruising at Mach 0.85 at an altitude of 10,000 m. The temperature and pressure at flying altitude are 26.5 kPa and 225 K, respectively. Determine the temperature and pressure at the leading edge of the wing where the relative air velocity is negligible.

Solution. Given,

$$M_\infty = 0.85$$
$$p_\infty = 26.5 \text{ kPa}$$
$$T_\infty = 225 \text{ K}$$

The stagnation point is the leading edge of the wing, where relative air velocity is negligible. From the isentropic relation,

$$\frac{T_0}{T_\infty} = 1 + \frac{\gamma - 1}{2} M_\infty^2 \tag{2.7.14}$$
$$= 1 + \frac{1.4 - 1}{2} (0.85)^2$$
$$= 1.1445$$

The stagnation temperature is

$$T_0 = 1.1445 \times T_\infty$$
$$= 1.1445 \times 225 = 257.51 \text{ K}$$

Similarly, from the relation,

$$\frac{p_0}{p_\infty} = \left[1 + \frac{\gamma - 1}{2} M_\infty^2\right]^{\frac{\gamma}{\gamma - 1}} \tag{2.7.15}$$
$$= \left[1 + \frac{1.4 - 1}{2} (0.85)^2\right]^{3.5}$$
$$= 1.6038$$

The stagnation pressure is

$$p_0 = 1.6038 \times p_\infty$$
$$= 1.6038 \times 26.5 = 42.5 \text{ kPa}$$

Example 12. A variable area duct allows air to move isentropically. The area at a particular location is 600 cm^2, and the Mach number is 0.5. Calculate the Mach number at a downstream location with an area of 500 cm^2. What is the area where the Mach number is one?

Solution. Given,

$$A_1 = 600 \text{ cm}^2$$
$$M_1 = 0.5$$
$$A_2 = 500 \text{ cm}^2$$

(a) From isentropic table, for $M_1 = 0.5$,

$$\frac{A_1}{A^*} = 1.3398$$

At the downstream location,

$$\frac{A_2}{A^*} = \frac{A_2}{A_1} \times \frac{A_1}{A^*}$$
$$= \frac{500}{600} \times 1.3398 = 1.1165$$

From isentropic table, for $\frac{A_2}{A^*} = 1.1165$,

$$M_2 = 0.67 \text{ (Subsonic) and } M_2 = 1.4 \text{ (Supersonic)}$$

Because no additional information is provided in the problem, both of the above values of M_2 are possible.

(b) It is worth noting that the throat is the position in a duct when $M = 1$. If $M_2 = 0.67$, there is no throat and the duct area steadily reduces from 600 cm^2 to 500 cm^2. However, if $M_2 = 1.4$, a throat with an area equal to A^* (throat choked) exists, as given by

$$A^* = \frac{A^*}{A_1} \times A_1$$

$$= \frac{1}{1.3398} \times 600 = 447.82 \text{ cm}^2$$

2.8 SUMMARY

This chapter briefly reviews the basic concepts of compressible flows. The domain of compressible flow deals with the motion of fluids in which the density change caused by a pressure change is noticeable. Such fluids are commonly referred to as barotropic fluids.

The temperature and density variations are insignificant for an incompressible flow, whereas for a compressible flow, the temperature change may be small, but density change is always finite.

A gas is treated as a continuous medium, if the mean free path (λ) is very small compared to a characteristic length (ℓ) of the fluid, i.e., $\lambda \ll \ell$.

The density, denoted by the Greek letter rho (ρ), of a fluid is defined as the mass of the fluid contained in a unit volume. It has the units of kg/m^3 and dimensions ML^{-3}. However, it is sometimes convenient to express the density of a fluid relative to the density of a standard fluid at a specified temperature. This ratio is called the specific gravity of the given fluid. Water is usually considered as the standard fluid which has a density of 1000 kg/m^3 at 4°C.

The properties which characterizes the resistance offered by a fluid to the applied shear stress is called the viscosity. That is, viscosity represents a fluid's internal resistance to motion or fluidity. Molasses and tar are examples of highly viscous fluids; water and air have very small viscosity. The resistance offered by a fluid to motion does not depend upon the deformation (unlike the case of solids) rather on the rate of deformation. For air, Sutherland's law of viscosity is

$$\mu = 1.46 \times 10^{-6} \left(\frac{T^{3/2}}{T + 111} \right) kg/(m.s)$$

where T is the temperature in Kelvin. The viscosity of a gas increases with temperature, whereas the viscosity of a liquid decreases with temperature.

An approximate relation for calculating the thermal conductivity of air is

$$k = 1.993 \times 10^{-3}\left(\frac{T^{3/2}}{T+112}\right) W/(m.K)$$

where k is the thermal conductivity of air, and T is the temperature in Kelvin.

The compressibility of a substance is the measure of its change in volume under the action of external forces, particularly the normal compressive forces. It is characterized in terms of the bulk modulus of elasticity, E, defined as

$$E = \lim_{\delta\forall \to \delta\forall_0} \frac{\delta p}{(-\delta\forall/\forall)}$$

The value of E for liquids are very high compared to gases (except at exceedingly high pressures). Thus, the liquids at normal conditions are considered as incompressible. The reciprocal of bulk modulus of elasticity is called the compressibility, κ. Thus,

$$\kappa = \frac{1}{E} = \frac{1}{\rho}\frac{d\rho}{dp} = -\frac{1}{\forall}\frac{d\forall}{dp}$$

There are three types of compressible flows: incompressible, subsonic, and supersonic. Because of the medium's compressibility, subsonic and supersonic flow patterns and general behavior change considerably. When a body travels through a stationary fluid or flows past a stationary body, each element of the solid surface attempts to divert the air from its normal route. These microscopic perturbations operate as a point source, generating the spherically propagating pressure waves in the surrounding air. These waves can be represented every time as a superposition of all previous waves generated by the same source.

1. The pressure disturbance wavefronts appear to radiate in concentric circles at different time intervals when a point source releases sound waves in a stationary compressible fluid.
2. When the point source travels, the wavefronts differ dramatically because the source emits each pressure disturbance at a different location in the fluid. If the source moves slower than the speed of sound, a succession of circles is generated, one inside the other but with distinct centers.
3. When a point source moves at the speed of sound, each pressure disturbance reinforces the previous one, forming a planar wavefront. Because this is a sound wavefront, it is by definition isentropic.

4. When a point source moves faster than the speed of sound, the pressure disturbances it emits are contained within a cone, the vertex of which is at the body. Disturbances are only relayed downstream inside the cone and are not noticed upstream of the cone vertex. The Mach cone is the cone that contains the disturbances, and the Mach angle is the half angle of the cone. The zone of action is defined as the area bounded by the Mach cone, and the zone of silence is the area beyond the cone. The Mach angle of the Mach cone is:

$$\mu_M = \sin^{-1}\left(\frac{1}{M}\right)$$

When air flows at subsonic speeds through a duct, the increase in Mach number is accompanied by an increase in velocity and a decrease in density. However, velocity rises more quickly than density drops; a 10% increase in velocity results in an 8% decrease in density. In such cases, the mass flow rate at every channel cross-section increases as the Mach number increases. If the channel continues to be filled with the flow, its area must decrease.

The behavior is reversed when the speed exceeds Mach 1.0. A 10% increase in speed could result in a 12% drop in density as the Mach number rises. As a result, the mass flow rate at every cross-section decreases, necessitating channel expansion to accommodate the accelerating flow. To decelerate a supersonic flow, the channel area must be reduced.

When a particle flows through the air at Mach 1.0 or higher speeds, concentrated wavefronts occur. If these waves are so faint, they create infinitely little changes in the properties of the air while passing through it, they are called Mach waves. A Mach wave is inclined to the flow direction at an angle of $\sin^{-1}\left(\frac{1}{M}\right)$, termed the Mach angle. A concentration of Mach waves known as shock waves produces finite-strength waves (substantially modifying air characteristics). Because of its angle to the flow direction, the shock wave in is called an oblique shock. If the flow turning angle is small, the downstream velocity is smaller than the upstream velocity but still supersonic. When the turning angle is considerable, the shock becomes normal to the flow and separates from the wedge, resulting in a subsonic speed downstream of the normal shock.

EXERCISE PROBLEMS

Exercise 13. Calculate the dynamic viscosity of air at 0°C, 25°C, and 500°C.

Example 14. Determine the thermal conductivity of air at 350 K.

Exercise 15. What is the Mach angle for air traveling at a velocity of 610 m/s, assuming the local speed of sound for this air is 290 m/s?

Exercise 16. Determine the static pressure at a location in an airstream where the Mach number is 1.8 while the static pressure and Mach number at another location in the stream are 138 kPa and 0.6, respectively.

Exercise 17. Air with a reservoir pressure of 35 kPa and a density of 2.08 kg/m^3 is accelerated to a Mach number of 1.7. Find the static pressure, density, and temperature corresponding to this flow.

Example 18. The airflow conditions just upstream of a normal shock are as follows: $p_1 = 86$ kPa, $T_1 = 267$ K, and $M_1 = 3$. Determine the velocity and Mach number immediately ahead of the shock.

3 Subsonic Wind Tunnels

3.1 INTRODUCTION

A wind tunnel is designed to produce a controlled environment with fast-moving air in order to investigate the flow around various objects. At the same time, it is used to compute lift and drag, as well as the forces acting on objects resulting from yaw, pitch, and roll motion. In 1871, Frank H. Wenham constructed the first wind tunnel. The device routed uniform air into a test section to study lift and drag on a body. Wind tunnels exist in various sizes and can be used in various situations. We can use it to calculate the gravitational load on a scaled model of an aircraft rotating at 800 $^{km}/_{h}$ or to examine a golf ball flying through the air at 80 $^{km}/_{h}$. There are two types of wind tunnels: open-circuit, which takes in ambient air, accelerates it, and exhausts it, and closed-circuit, which recirculate the same air. Figures 3.1.1(a) and 3.1.1(b) shows the schematic representations of an open-circuit and a closed-circuit subsonic wind tunnel, respectively.

3.2 COMPONENTS OF A SUBSONIC WIND TUNNEL

A general utility low-speed wind tunnel's four key components are the contraction cone, the test section, the divergent diffuser, and the power unit.

3.2.1 EFFUSER OR CONTRACTION CONE

It is a converging conduit located ahead of the test section in which the fluid is accelerated from rest (or at extremely low speed) and brought to the required conditions at the upstream end of the test section. As a result, the effuser is often referred to as the contraction cone (Figure 3.2.1). The contraction ratio of a contraction cone is given by:

$$a_c = \frac{A_1}{A_2} \tag{3.2.1}$$

where A_1 and A_2 are the effuser's inlet and exit cross-sectional areas, respectively.

In a conventional subsonic wind tunnel, the contraction ratio typically runs between 4 and 20. By utilizing the honeycomb structures and the screens, the effuser minimizes vortical structures to increase flow speed, resulting in a smooth flow at the outlet (sieves). In order to create a uniform and unidirectional flow in the test section, we must eliminate flow abnormalities, which are as follows:

DOI: 10.1201/9781003139447-3

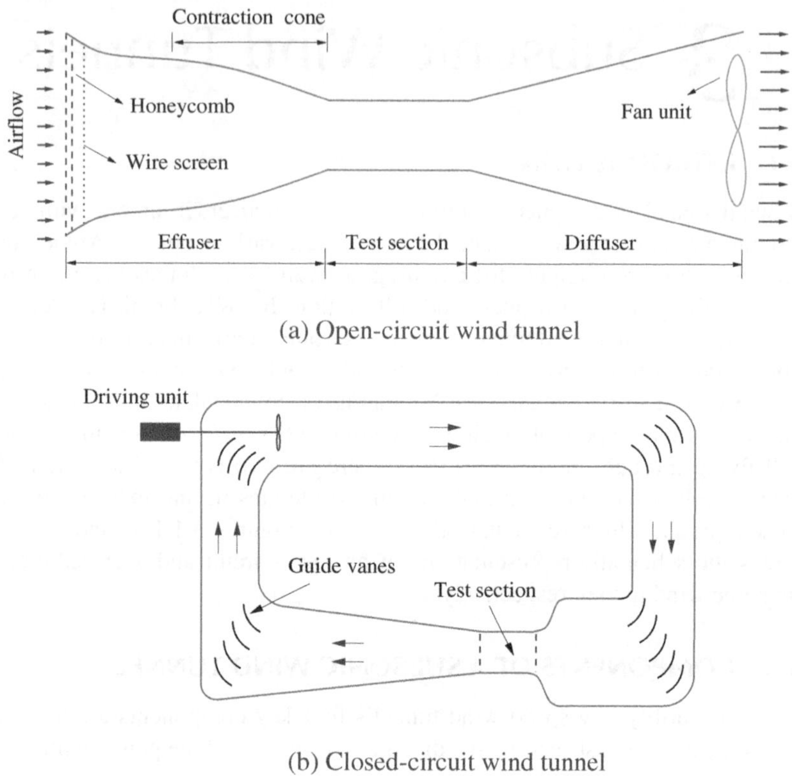

Figure 3.1.1 Schematic layouts of subsonic wind tunnels.

3.2.1.1 Spatial Velocity Variations

The excess total head is moved from high-velocity to low-velocity regions to lessen average velocity variations across a cross-section. Let us determine how effectively an effuser is reducing the spatial variations in flow velocity. Referring to Figure 3.2.1, v_1 and v_2 are the flow velocities at the inlet and exit of the contraction cone, and Δv_1 and Δv_2 are the corresponding changes in velocities at these two locations. Thus, $\frac{\Delta v_1}{v_1}$ is the fractional change in velocity at the inlet to the contraction cone, and likewise, $\frac{\Delta v_2}{v_2}$ is the fractional change in velocity at the exit of the contraction cone. The incompressible Bernoulli's equation between points 1 and 2, with mean velocities v_1 and v_2, yields

$$p_1 + \frac{1}{2}\rho v_1^2 = p_2 + \frac{1}{2}\rho v_2^2 \tag{3.2.2}$$

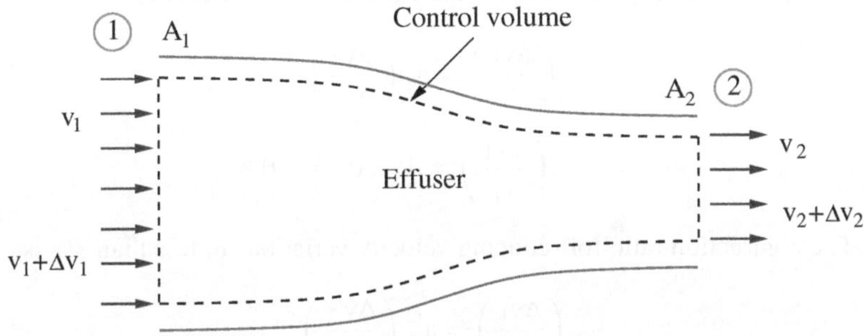

Figure 3.2.1 Schematic layout of an effuser (contraction cone).

and when combined with the total velocities, it gives

$$p_1+\frac{1}{2}\rho\left(v_1+\Delta v_1\right)^2=p_2+\frac{1}{2}\rho\left(v_2+\Delta v_2\right)^2 \tag{3.2.3}$$

$$p_1+\frac{1}{2}\rho v_1^2+\rho v_1\Delta v_1+\frac{1}{2}\rho\left(\Delta v_1\right)^2=p_2+\frac{1}{2}\rho v_2^2+\rho v_2\Delta v_2+\frac{1}{2}\rho\left(\Delta v_2\right)^2 \tag{3.2.4}$$

Substituting equation (3.2.2) and ignoring higher order terms results in

$$v_1\Delta v_1=v_2\Delta v_2 \tag{3.2.5}$$

$$v_1^2\left(\frac{\Delta v_1}{v_1}\right)=v_2^2\left(\frac{\Delta v_2}{v_2}\right) \tag{3.2.6}$$

$$\longrightarrow\quad\left(\frac{\Delta v_1}{v_1}\right)=a_c^2\left(\frac{\Delta v_2}{v_2}\right) \tag{3.2.7}$$

Example 19. A subsonic wind tunnel effuser has a contraction ratio of 4. A 5% spatial velocity variations are observed at the exit of the effuser. What should the contraction ratio be to reduce the velocity variations to less than 3%?

Solution. Given,

$$a_c=8$$

$$\frac{\Delta v_2}{v_2}=5\%$$

$$\frac{\Delta v_2'}{v_2'}=3\%$$

The spatial velocity variations at the entrance to the effuser is:

$$\left(\frac{\Delta v_1}{v_1}\right) = a_c^2 \left(\frac{\Delta v_2}{v_2}\right) \tag{3.2.7}$$

$$\Longrightarrow \quad \left(\frac{\Delta v_1}{v_1}\right) = 4^2 \times 0.05 = 0.8$$

The contraction ratio for reducing velocity variations to less than 3% is:

$$\left(\frac{\Delta v_1}{v_1}\right) = a'^2_c \left(\frac{\Delta v'_2}{v'_2}\right)$$

$$0.8 = a'^2_c \times 0.03$$

$$\Longrightarrow \quad a'_c = 5.16$$

3.2.1.2 Flow Swirl

The flow can revolve around an axis, changing the direction of the flow. The honeycomb structures aid in the prevention of swirling.

3.2.1.3 Low-Frequency Pressure Oscillations

A surge is an unstable flow situation caused by a quickly changing wavefront. Surges happen when the velocity of a fluid changes rapidly and evolves unstable or transient.

3.2.1.4 Turbulence

Almost all turbulent flows exhibit deviation from the mean velocity. These deviations are caused by substantial changes in the perturbation components of velocities along the three coordinate axes. In wind tunnel measurements, the relative magnitude of longitudinal and transverse velocity fluctuations is an important factor. It governs how measurements taken on a model can be applied to the full-scale structure and how measurements taken in different tunnels can be compared. The mesh of a tunnel's screen, grids, or honeycomb structures determines the magnitude of its fluctuations. It is measured in terms of the amount, degree, or intensity of turbulence, which is described as the ratio of the square root of the arithmetic average of the mean square values of fluctuation components of velocities to the mean velocity, which is thus the same as:

$$I_t = \frac{1}{U_\infty}\sqrt{\frac{1}{3}\left(\overline{u'^2} + \overline{v'^2} + \overline{w'^2}\right)} \tag{3.2.8}$$

where u′, v′, and w′ are the RMS (root mean square) values of the time-averaged fluctuation velocity components along the x−, y−, and z−directions, defined as,

$$u' = \sqrt{\frac{1}{t}\int_0^t u'^2 dt} = \sqrt{\overline{u'^2}} \quad (3.2.9)$$

$$v' = \sqrt{\frac{1}{t}\int_0^t v'^2 dt} = \sqrt{\overline{v'^2}} \quad (3.2.10)$$

$$w' = \sqrt{\frac{1}{t}\int_0^t w'^2 dt} = \sqrt{\overline{w'^2}} \quad (3.2.11)$$

In wind tunnels, turbulence becomes isotropic[1] at a given distance from the screens, which means that the time-averaged fluctuation components of velocity are identical in all coordinate directions:

$$\sqrt{\overline{u'^2}} = \sqrt{\overline{v'^2}} = \sqrt{\overline{w'^2}} \quad (3.2.12)$$

In this situation, the x−component of fluctuation velocity solely is sufficient to determine turbulence intensity, and so equation (3.2.8) is reduced to:

$$I_t = \frac{\sqrt{\overline{u'^2}}}{U_\infty} \quad (3.2.13)$$

Because of its simpler form, equation (3.2.13) is widely used in practice, even when the turbulence is not isotropic. Outside the boundary layer, the turbulence in the center zone of a wind tunnel is virtually isotropic. However, it is thought to be homogeneous[2] because turbulence dampens sluggishly downstream of the honeycomb structure.

---

[1]The isotropic turbulence implies that the perturbation velocity components are independent of the reference axes. That is,

$$\sqrt{\overline{u'^2}} = \sqrt{\overline{v'^2}} = \sqrt{\overline{w'^2}}$$

or

$$u' = v' = w'$$

[2]Homogeneous turbulence has the same quantitative structure across the flow field. That is, in homogeneous turbulence, the root mean square values of $u'\left(=\sqrt{\overline{u'^2}}\right)$, $v'\left(=\sqrt{\overline{v'^2}}\right)$, and $w'\left(=\sqrt{\overline{w'^2}}\right)$ might all be different, but each value must be constant over the whole turbulent field. It is worth emphasizing that isotropic turbulence, by definition, is always homogeneous.

Example 20. The perturbation velocity components measured by a hot-wire anemometer in the test section of a subsonic wind tunnel are isotropic. Let the perturbation velocity component along the x−axis to mean velocity ratio to be 0.0005. Calculate the intensity of turbulence in the test section.

Solution. Given,

$$\frac{u'}{U_\infty} = \frac{\sqrt{\overline{u'^2}}}{U_\infty} = 0.0005$$
$$U_\infty = 10\ m/s$$

Since the turbulence is isotropic in the test section,

$$\sqrt{\overline{u'^2}} = \sqrt{\overline{v'^2}} = \sqrt{\overline{w'^2}} \tag{3.2.12}$$

The intensity of turbulence would be

$$I_t = \frac{1}{U_\infty}\sqrt{\frac{1}{3}\left(\overline{u'^2} + \overline{v'^2} + \overline{w'^2}\right)} \tag{3.2.8}$$
$$= \frac{1}{U_\infty}\sqrt{\frac{1}{3}\left(3\overline{u'^2}\right)}$$
$$= \frac{\sqrt{\overline{u'^2}}}{U_\infty} = 0.0005 = 0.05\%$$

3.2.1.5 Influences of Honeycomb Structures on Turbulence

Upon entry into the contraction cone, the honeycomb structures, known as flow straighteners, reduce lateral turbulence in the airstream caused by rotational motion. They can be natural or artificial, with a geometry that uses less material and weights less. The cells of honeycombs can have circular, square, or hexagonal cross-sections, as seen in Figure 3.2.2.

Table 3.1 shows that a hexagonal cell has the lowest pressure loss, and thus, it is particularly effective in removing large-scale eddies. Aside from the cross-sectional shape, the cell length (L_h) to cell diameter (D_h) ratio is an important parameter in determining the optimal cell size, which is equivalent to:

$$\Psi = \frac{L_h}{D_h} \tag{3.2.14}$$

Take note of the following instances:

- When using $\Psi < 1$, honeycomb cells with a small length ratio are used as screens to maintain a uniform profile and lower turbulence intensity.

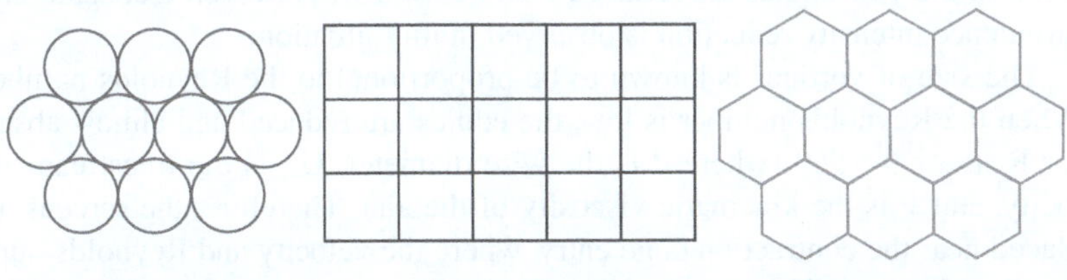

Figure 3.2.2 Different types of cells in honeycomb structures.

Table 3.1
Pressure loss coefficients for various cells in honeycomb structures.

Cell shape	Loss coefficient
Circular cell	0.30
Square cell	0.22
Hexagonal cell	0.20

- When using $\Psi > 1$, to prevent lateral turbulence and flow vortices, honeycomb cells with a high length-to-width ratio are used. The main disadvantage of this technique is that it introduces high-turbulence intensity into the test section.

Aluminum is commonly used in honeycomb structures. Metallic structures can also be used, but burrs must be removed or the structures would introduce more turbulence. Polycarbonate materials provide low-cost structures.

3.2.1.6 Influences of Screen on Turbulence

Bell and Mehta (1989) [17] observed that the honeycomb is particularly efficient at removing flow vortices and lateral velocity components, resulting in a uniform flow as long as the airflow yaw angles are less than 10°. Larger yaw angles than this can cause the honeycomb cell to stall, increasing pressure drop across it and introducing non-uniformity into the flow. To eliminate larger swirl angles, screens with very small mesh sizes are installed in front of the honeycomb structures. The screen enhances flow homogeneity by imposing a static pressure drop and reducing turbulence. The smaller screen mesh breaks up the large vortices into smaller ones, dissipating in the wake region. By using many screens, the yaw angles are reduced well below 10°. However, a considerable turbulence intensity reduction is observed in this situation.

The size of vortices is known to be proportional to the Reynolds number. When the Reynolds number is low, the eddies are reduced and almost absent for $Re_d = \frac{U_\infty d}{\nu} < 60$ (where d is the wire diameter, U_∞ is the freestream velocity, and ν is the kinematic viscosity of the air). Therefore, the screens are placed near the contraction cone entry, where the velocity and Reynolds number are minimum. The pressure drop coefficient is expressed as follows:

$$K_s = \frac{p_1 - p_2}{\frac{1}{2}\rho v_1^2} \tag{3.2.15}$$

where K_s denotes the pressure drop coefficient for screens, p_1 and p_2 are the static pressures ahead and behind the screen, ρ is the flow density, and v_1 is the upstream flow velocity. Equation (3.2.15) implies that the pressure decrease is directly proportional to the square of the local flow velocity, and so the deceleration is larger in places of higher flow velocities. This contrasts with lower-speed regions, where deceleration is observed to be less. The coefficient K_s is essentially a drag coefficient that gauges the normal force applied on the screen's solid surface. Collar (1939) proposed a relationship between the pressure loss coefficient and upstream and downstream turbulence, which is identical to:

$$\frac{u_2}{u_1} = \frac{2 - K_s}{2 + K_s} \tag{3.2.16}$$

where u_1 and u_2 are the fluctuation velocities sufficiently upstream and downstream of the rake. Equation (3.2.16) indicates that a screen can minimize flow non-uniformity when $K_s = 2$ and increase it when $K_s > 2$.

$$K_s = f(Re, \alpha, \chi) \tag{3.2.17}$$

where α represents the flow incidence angle and χ signifies the screen's open area ratio. The porosity (or open area ratio) of a square mesh screen is defined as:

$$\chi = \left(1 - \frac{d}{s}\right)^2 \tag{3.2.18}$$

where d denotes the wire diameter, and s represents the wire spacing (Figure 3.2.3).

The pressure drop coefficient is higher for screens with low porosity. The screens are more effective in reducing axial turbulence than lateral turbulence. This is because the higher downstream pressure drop reduces the higher velocity more than, the lower velocity, resulting in a smoother axial velocity. On

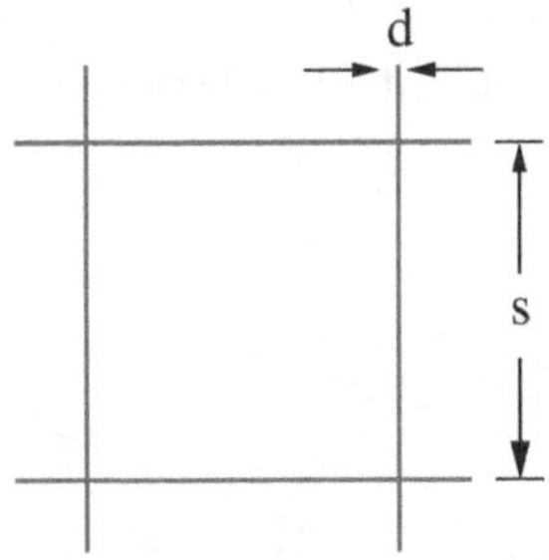

Figure 3.2.3 A square mesh screen.

the other hand, the honeycombs have low-pressure drops in the flow direction and are, therefore, less effective at axial velocity. Due to their length, however, they have lower lateral speeds. By allowing energy to flow between axes, honeycombs and screens reduce lateral and axial turbulence. As a result, the downstream turbulence tends to become isotropic.

3.2.2 TEST SECTION

The test section, also referred to as the working section, is a wind tunnel component with uniform flow properties. It has the same cross-sectional area along its whole length. However, forming a boundary layer over the test section walls reduces net area while raising flow speed. As a result, a horizontal force acts on the model in a downstream direction, known as horizontal buoyancy. Thus, the test section walls must diverge adequately so that the net cross-sectional area remains constant along the length of the test section. The model is placed within the test section, and the relevant measurements and observations are taken. The tunnel is called closed-throat when rigid walls outline the test section.

3.2.3 DIVERGENT DIFFUSER

The diffuser is positioned ahead of the test section and effectively transforms the kinetic energy of the airstream (leaving the test section) into the pressure energy. It is a channel through which a subsonic flow decelerates by increasing the downstream area. Figure 3.2.4 depicts a schematic representation of a subsonic diffuser. The area ratio (a_d) or the divergence angle (α), often known as the equivalent cone angle, defines the diffuser. A diffuser's area ratio is given by:

$$a_d = \frac{A_2}{A_1} \tag{3.2.19}$$

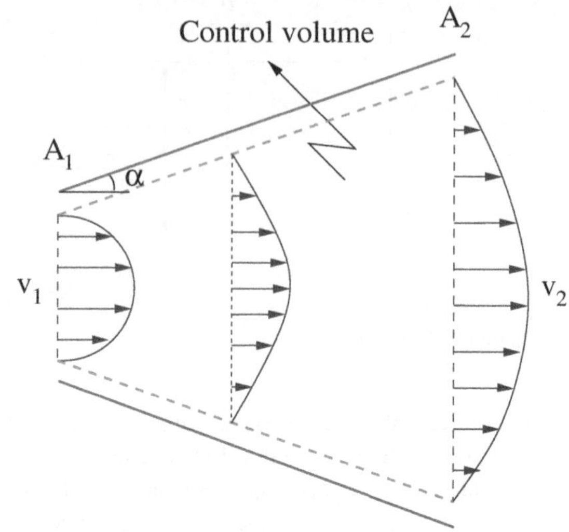

Figure 3.2.4 Schematic diagram of a divergent diffuser.

where A_1 and A_2 are the diffuser's inlet and exit cross-sectional areas, respectively.

The pressure rise in the diffuser can be explained by applying mass conservation to the control volume represented in Figure 3.2.4:

$$A_1 v_1 = A_2 v_2 \tag{3.2.20}$$

Because $A_2 > A_1$, it follows that $v_2 < v_1$, i.e., flow velocity reduces as cross-sectional area increases. Bernoulli's equation in differential form for an ideal diffuser is:

$$\frac{dp}{\rho} + d\left(\frac{v^2}{2}\right) \tag{3.2.21}$$

This equation illustrates that when flow velocity decreases, static pressure rises. Thus, as the flow travels through the diffuser, it meets an adverse pressure gradient, causing boundary layer retardation and rapid thickening on the channel walls, thereby increasing the probability of separation. The flow separation from the diffuser wall is known as the diffuser stall, and it inhibits static pressure recovery. Flow separation causes vibrations, oscillating for loading, oscillations in test section velocities (often referred to as surging), and higher losses in the tunnel downstream. Because energy losses in a subsonic wind tunnel vary as a function of the velocity cube, the velocity reduction should occur at the shortest practical distance to minimize losses. The semi-divergence angle is commonly adjusted between 7 and 8 degrees to reduce boundary layer separation at the diffuser walls. Even with the most efficient design, boundary layer thickening or separation can only be avoided partially. Consequently, kinetic energy to pressure energy conversion can only be moderately efficient. Any changes to the design mass flow and pressure ratio will also result in a loss in diffusion efficiency. For a diffuser with losses, the modified form of Bernoulli's equation is employed, which is equivalent to:

$$\eta d\left(\frac{v^2}{2}\right)+\frac{dp}{\rho} \tag{3.2.22}$$

where η is the diffuser efficiency, can be represented as follows:

1. Polytropic efficiency (η_p)
2. Isentropic efficiency or diffuser effectiveness (η_i)

3.2.3.1 Polytropic Efficiency

In a diffuser, the flow process is supposed to be steady and adiabatic. As shown in Figure 3.2.5, the flow conditions at the diffuser's entrance and exit are marked by 1 and 2, respectively.

The diffuser efficiency in a polytropic process is defined as:

$$\eta_p=\frac{\text{Isentropic enthalpy change}}{\text{Actual enthalpy change}} \tag{3.2.23}$$

The stagnation enthalpy remains constant since the process is adiabatic and steady, which is similar to:

$$h_{01}=h_{02} \tag{3.2.24}$$

$$\Longrightarrow \quad h_2-h_1=\frac{1}{2}\left(v_1^2-v_2^2\right) \tag{3.2.25}$$

The enthalpy change for an equivalent isentropic (reversible and adiabatic) process between 1 and 2s is

$$h_{2s}-h_1=\frac{1}{2}\left(v_1^2-v_{2s}^2\right) \tag{3.2.26}$$

Using Equations (3.2.19) and (3.2.20), the diffuser efficiency can be expressed as:

$$\eta_p=\frac{v_1^2-v_{2s}^2}{v_1^2-v_2^2} \tag{3.2.27}$$

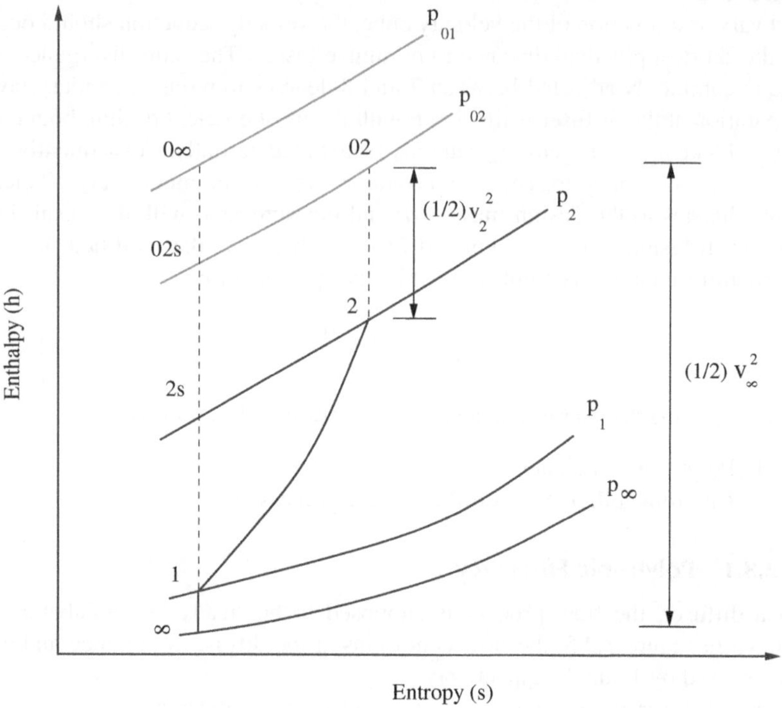

Figure 3.2.5 A Mollier diagram depicting the flow through a divergent diffuser.

equation (3.2.26) can alternatively be represented for an incompressible flow as:

$$h_{2s} - h_1 = \frac{p_2 - p_1}{\rho} \tag{3.2.28}$$

Thus, the expression for diffuser efficiency is:

$$\eta_p = \frac{p_2 - p_1}{\frac{1}{2}\rho\left(v_1^2 - v_2^2\right)} \tag{3.2.29}$$

The conservation of mass (the continuity equation) for a steady and incompressible flow between the diffuser's entry and exit gives:

$$A_1 v_1 = A_2 v_2 \tag{2.7.2}$$

$$\Longrightarrow \quad a_d = \frac{A_2}{A_2} = \frac{v_1}{v_2} \tag{3.2.30}$$

Combining equation (3.2.30) with equation (3.2.29) produces the following alternative form of η_p:

$$\eta_p = \frac{p_2 - p_1}{\frac{1}{2}\rho v_1^2\left(1 - \frac{1}{a_d^2}\right)} \tag{3.2.31}$$

The polytropic efficiency can also be expressed in terms of total head loss, h_L, experienced in the diffuser using the following energy equation:

$$\frac{p_1}{\rho g} + \frac{v_1^2}{2g} = \frac{p_2}{\rho g} + \frac{v_2^2}{2g} + h_L \tag{3.2.32}$$

As a result of this,

$$\eta_p = 1 - \frac{gh_L}{\frac{1}{2}v_1^2\left(1 - \frac{1}{a_d^2}\right)} \tag{3.2.33}$$

Example 21. In a low-speed subsonic wind tunnel, a diffuser with an area ratio of 8 is used. The diffuser is said to be 85% efficient at a test section velocity of 25 m/s. Determine the head loss in the diffuser.

Solution. Given,

$$a_d = 8$$
$$\eta_p = 85\%$$
$$v_1 = 25\ m/s$$

Using equation (3.2.33) for a steady and incompressible flow through the diffuser, we get

$$\eta_p = 1 - \frac{gh_L}{\frac{1}{2}v_1^2\left(1 - \frac{1}{a_d^2}\right)}$$

$$0.85 = 1 - \frac{9.8 \times h_L}{\frac{1}{2} \times 25^2 \times \left(1 - \frac{1}{8^2}\right)}$$

$$\Longrightarrow \quad h_L = 4.7\ \text{m}$$

Thus, the total head loss in the diffuser is 4.7 m.

3.2.3.2 Isentropic Efficiency (Diffuser Effectiveness)

Diffuser performance is frequently measured on the basis of isentropic efficiency, also known as diffuser effectiveness. It is described as:

$$\eta_i = \frac{\text{Actual static pressure rise}}{\text{Ideal static pressure rise}} \tag{3.2.34}$$

The isentropic efficiency compares the actual performance of the diffuser to the theoretical ideal. We shall now construct a mathematical expression for equation (3.2.34). Consider the control volume as shown in Figure 3.2.4. The conservation of mass for a steady and incompressible flow between 1 and 2 yields:

$$v_1A_1 = v_2A_2 \tag{3.2.35}$$

where v_1 and v_2 are the mean axial velocities over cross-sections 1 and 2, respectively.

$$v_{avg} = \frac{1}{A}\int_A u dA \tag{3.2.36}$$

The flow velocity fluctuates between zero at the diffuser walls and highest at the centerline. This velocity distribution is described by the kinetic energy flux factor, which varies from the entrance to the exit of the diffuser. For the same mass flow rate flowing through a given diffuser cross-sectional area A, the kinetic energy flux or correction factor is:

$$\beta_k = \frac{\text{Actual transfer of fluid kinetic energy}}{\text{Least possible transfer of fluid kinetic energy}} \tag{3.2.37}$$

$$= \frac{1}{A}\int_A \left(\frac{v}{v_{avg}}\right)^3 dA \tag{3.2.38}$$

The value of β_k for a uniform flow is one and increases as the flow is accelerated. If the maximum speed at the diffuser exit is minimized, α would also be minimized, thereby reducing the kinetic flow energy at the exit. Thus, the maximum static pressure recovery is attainable when the flow exits the diffuser smoothly. The pressure coefficient for an ideal non-uniform flow is expressed in terms of α and is equivalent to:

$$C_p = \frac{p_2 - p_1}{\frac{1}{2}\rho v_1^2} \tag{3.2.39}$$

$$= \beta_{k1} - \beta k_2 \left(\frac{A_1}{A_2}\right)^2 \tag{3.2.40}$$

where β_{k1} and β_{k2} are the kinetic energy shape factors at the diffuser's entrance and exit, respectively. In the case of an ideal uniform flow, $\beta_{k1} = \beta_{kshock2} = 1$ and $v = v_{avg}$, therefore, equation (3.2.40) becomes:

$$C_p = 1 - \left(\frac{A_1}{A_2}\right)^2 \tag{3.2.41}$$

Therefore, the diffuser effectiveness is:

$$\eta_i = \frac{C_p}{1 - \left(\frac{A_1}{A_2}\right)^2} \tag{3.2.42}$$

3.2.4 DRIVING UNIT

In the absence of losses, a steady and uniform flow could be maintained indefinitely without requiring more energy. However, losses occur due to the dissipation of kinetic energy into heat by turbulence, vortical motion, and other factors. In addition, some kinetic energy is invariably expelled at the exit of the diffuser in open-circuit wind tunnels, which is converted to heat as the tunnel exhaust gases mix with the ambient air. Wind tunnel operation requires an additional power arrangement or drive unit to compensate for these energy losses. A motor and a fan or propeller often drive a low-speed wind tunnel.

Kinetic energy dismissal at the diffuser exit is minimized in closed-circuit wind tunnels. External power, on the other hand, is still required to compensate for losses produced by turbulence, eddying motion, and skin friction on walls and other surfaces. A mechanism ahead of the test section is required to ensure a uniform velocity distribution in the test section, especially with low swirl and eddying motion. It is achieved using guiding vanes at the corners to reduce sharp angle fluctuations. Thus, in an open-circuit wind tunnel, the absolute output power of fans is invested in two things: overcoming circuit losses and imparting kinetic energy to the flow released into the ambient atmosphere.

If the total pressure drop in a subsonic wind tunnel is Δp_T and the measured volume flow rate is $\dot{Q}$, the power required to operate this tunnel is:

$$P_R = \Delta p_T \times \dot{Q} \tag{3.2.43}$$

and the fan efficiency is:

$$\eta_F = \frac{\sum \text{circuit losses}}{P_{input}} \tag{3.2.44}$$

where P_{input} is the power supplied to the fan (or drive unit).

3.2.5 PRESSURE DROPS IN SUBSONIC WIND TUNNELS

The total pressure drop across a wind tunnel is the sum of the pressure drops in its various components. The total pressure drop must equal the pressure rise induced by the fan or driving unit for an efficient tunnel functioning. The following factors cause pressure decrease in a subsonic tunnel circuit:

- Pressure drops in the screens.
- Pressure drops in the honeycomb structures.
- Pressure drops in the contraction cone.
- Pressure drops in the test section.
- Pressure drops in the diffuser due to friction and flow expansion.
- Pressure drops in the guide vanes due to friction and flow expansion.

If the pressure drop in a wind tunnel component is Δp_i, the pressure drop coefficient for this component is expressed as:

$$K_i = \frac{\Delta p_i}{q_i} = \frac{\Delta p_i}{\frac{1}{2}\rho_i v_i^2} \tag{3.2.45}$$

where q_i = dynamic pressure
ρ_i = flow density
v_i = velocity at the component inlet.

3.2.5.1 Pressure Drop in Screens

Screens are deployed in wind tunnels at two places: ahead of the contraction cone to prevent turbulence and just before the fan inlet, wherein velocities are relatively higher, affecting pressure drop significantly. Eckert et al. (1976) [18] provided an empirical relationship for calculating the pressure drop coefficient for such screens and is identical to:

$$K_s = K_m K_{Rn} \sigma_s + \left(\frac{\sigma_s}{\beta_s}\right)^2 \tag{3.2.46}$$

where K_s = pressure drop coefficient for screens
K_m = mesh factor
β_s = screen porosity
$\sigma_s = (1 - \beta_s)$ = screen solidity, and

$$K_{Rn} = \left\{ \begin{array}{lll} 0.785\left(1 - \frac{Re_d}{354}\right) & \text{for} & 0 \le Re_d < 400 \\ 1.0 & \text{for} & Re_d \ge 400 \end{array} \right\} \tag{3.2.47}$$

where Re_d is the flow Reynolds number based on the wire diameter.

3.2.5.2 Pressure Drop in Honeycomb Structures

The pressure drop in honeycomb structures is determined by the porosity of the structure, the Reynolds number based on hydraulic cell diameter, and the ratio of streamwise length to cell hydraulic diameter. To quantify pressure drop coefficient for honeycomb structures, Eckert et al. (1976) [18] presented the following relationship:

$$K_h = \frac{\lambda_h}{\beta_h^2}\left(\frac{L_h}{D_h} + 3\right)\left(\frac{1}{\beta_h} - 1\right)^2 \tag{3.2.48}$$

where K_h = pressure drop coefficient for honeycomb structures
L_h = streamwise length of the honeycomb cell
D_h = hydraulic diameter of honeycomb cell, and

$$\lambda_h = \begin{cases} 0.375\left(\frac{\Delta}{D_h}\right)^{0.4} Re_\Delta^{-0.1} & \text{for} \quad Re_\Delta \le 275 \\ 0.214\left(\frac{\Delta}{D_h}\right)^{0.4} & \text{for} \quad Re_\Delta > 275 \end{cases} \tag{3.2.49}$$

In equation (3.2.49), Δ denotes the material roughness, and Re_Δ signifies the Reynolds number based on the material roughness.

3.2.5.3 Pressure Drop in Contraction Cone

The pressure drop in a contraction cone accounts for about 3-4% of the total pressure drop across the wind tunnel. As a result, errors in determining the loss coefficient are less significant compared to high-speed wind tunnels. Wattendorf (1969) [19] suggested the following relationship to estimate the pressure drop coefficient for a contraction cone:

$$K_c = 0.32 f_{avg} \frac{L_c}{D_h} \tag{3.2.50}$$

where K_c = pressure drop coefficient for a contraction cone
f_{avg} = average friction factor
L_c = contraction cone length
D_h = hydraulic diameter of the settling chamber

The average friction factor, f_{avg}, can be determined using Prandtl's universal law of friction for smooth pipes at high Reynolds numbers, which is similar to:

$$f = \left[2\log_{10}\left(Re_h\sqrt{f}\right) - 0.8\right]^2 \tag{3.2.51}$$

where $Re_h = \frac{\rho v D_h}{\mu}$.

3.2.5.4 Pressure Drop in Test Section

Consider a test section with a constant area and a uniform hydraulic diameter. The pressure drop due to friction in a duct is as follows:

$$\frac{\Delta p}{L} = \frac{f_{avg} L v^2}{2D_h} \tag{3.2.52}$$

where f_{avg} is the average friction factor described by equation (3.2.51), and D_h is the hydraulic diameter of the test section. Combining equation (3.2.45) and equation (3.2.52) yields the pressure drop coefficient for a constant area test section, which is identical to:

$$K_{TS} = \frac{f_{avg} L_{TS}}{D_h} \tag{3.2.53}$$

where K_{TS} = pressure drop coefficient for a test section
L_{TS} = test section length.

3.2.5.5 Pressure Drop in Diffuser

Pressure drop in a diffuser of the subsonic wind tunnel is affected by wall skin friction and flow expansion. Thus, the overall pressure loss coefficient for a subsonic diffuser, K_d, is given by:

$$K_d = K_f + K_{exp} \tag{3.2.54}$$

where K_f = wall skin friction coefficient
K_{exp} = flow expansion coefficient

Consider the subsonic diffuser with a cylindrical cross-section given in Figure 3.2.6. The divergence angle of the wall, α, is kept between 6 and 8 degrees, which is optimal for minimizing pressure drop. The optimum divergence angle, which regulates pressure recovery and gradient, is determined by the diffuser area ratio. If the divergence angle is too large, the unfavorable pressure gradient may cause flow separation on the walls.

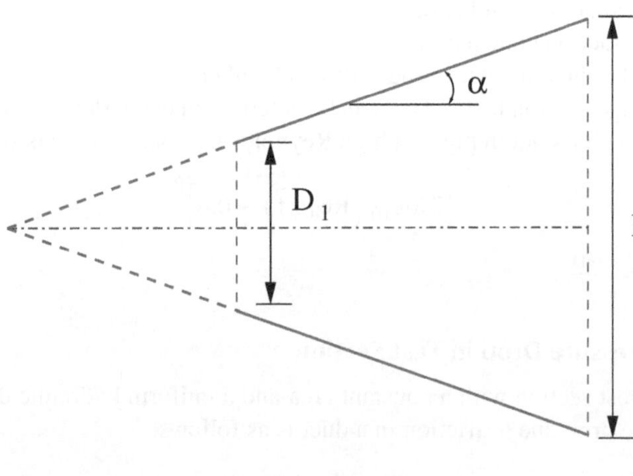

Figure 3.2.6 Schematic diagram of a divergent diffuser.

The pressure drop coefficient owing to skin friction in terms of test section dynamic pressure is:

$$K_f = \frac{f_{avg}}{8\tan\alpha}\left[1-\left(\frac{D_1}{D_2}\right)^4\right]\left(\frac{D_0}{D_1}\right)^4 \tag{3.2.55}$$

where f_{avg} = average friction factor
D_1 = inlet diameter of the round diffuser
D_2 = exit diameter of the round diffuser
D_0 = diameter of the cylindrical test section

The pressure drop coefficient for a gradually expanded flow can be calculated using Fleigner's formula, which is defined as:

$$K_{exp} = \sin 2\alpha\left[1-\left(\frac{D_1}{D_2}\right)^2\right]^2 \tag{3.2.56}$$

Substituting equations (3.2.55) and (3.2.56) into equation (3.2.54) results in

$$K_d = \frac{f_{avg}}{8\tan\alpha}\left[1-\left(\frac{D_1}{D_2}\right)^4\right]\left(\frac{D_0}{D_1}\right)^4 + \sin 2\alpha\left[1-\left(\frac{D_1}{D_2}\right)^2\right]^2 \tag{3.2.57}$$

This is the equation for the overall pressure loss coefficient of a subsonic wind tunnel diffuser. The variation of K_d with α is depicted graphically in Figure 3.2.7.

Example 22. A subsonic wind tunnel round diffuser with a circular cross-section has a divergence angle of 7°. The inlet and outlet areas of the diffuser are 0.125 m^2 and 1.13 m^2, respectively. Determine the total loss coefficient for the diffuser if the wind tunnel contains a cylindrical test section with a diameter of 0.4 m. Assume the average friction coefficient is 0.0309.

Solution. Given,

$$\alpha = 7^o$$

$$A_1 = 0.125\ m^2 \implies D_1 = 0.4\ m$$

$$A_2 = 1.13\ m^2 \implies D_2 = 1.2\ m$$

$$D_0 = 0.4\ m$$

$$f_{avg} = 0.0309$$

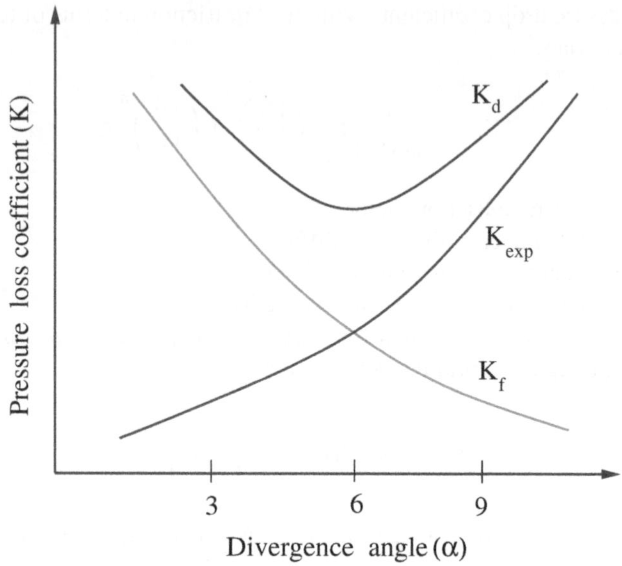

Figure 3.2.7 Variation of K_d with α.

The total loss coefficient of the diffuser is:

$$K_d = \frac{f_{avg}}{8\tan\alpha}\left[1-\left(\frac{D_1}{D_2}\right)^4\right]\left(\frac{D_0}{D_1}\right)^4 + \sin 2\alpha\left[1-\left(\frac{D_1}{D_2}\right)^2\right]^2 \quad (3.2.57)$$

$$= \frac{0.0309}{8\tan 7^o}\left[1-\left(\frac{0.4}{1.2}\right)^4\right] + \sin 14^o\left[1-\left(\frac{0.4}{1.2}\right)^2\right]^2$$

$$\Longrightarrow \quad K_d = 0.22$$

3.2.5.6 Pressure Drop in Corners with Guide Vanes

The corners of an open-circuit wind tunnel are prone to significant losses. They have higher dynamic pressure and are more likely to have flow recirculation or separation. Guide vanes with an efficient blade cross-section and an adequate chord-to-gap ratio are installed at the corners of the closed-circuit tunnels to minimize these losses. Cambered airfoils with straight leading edges are less susceptible to approaching flow angularities than sharp leading edges. In a low-speed wind tunnel, the pressure drop coefficient for corners with guide vanes deployed can be calculated using the same empirical formula as for a flat plate:

$$K_g = 0.1 + \frac{4.55}{(\log_{10} Re_c)^{2.58}} \qquad (3.2.58)$$

where K_g = pressure drop coefficient for the corners with guide vanes
Re_c = Reynolds number based on a guide vane's chord

3.2.6 ENERGY RATIO

The energy ratio of a subsonic wind tunnel is defined as the ratio of flow kinetic energy in the test section to overall energy losses in the tunnel, which is equivalent to:

$$\text{Energy ratio} = \frac{\text{Flow kinetic energy in test section}}{\sum \text{Overall energy losses in tunnel circuit}} \qquad (3.2.59)$$

The energy ratio measures the energy efficiency of a wind tunnel. Multiple definitions of the energy ratio are used depending on the parameter in the denominator. To compensate for circuit losses, the electrical power supplied to a fan or a driving unit, for example, can be used. However, equation (3.2.59) emphasizes the aerodynamic aspects of the energy budget and aids in distinguishing circuit flow parameters from the efficiency of a fan or other driving device.

The energy ratio is always greater than unity in both closed- and open-circuit wind tunnels (excluding open test section walls). It typically ranges between 3 to 7 for closed-circuit tunnels – the higher the energy ratio, the greater the energy efficiency. Since the energy ratio for a free jet configuration is always less than one, an open test section is never used in a large tunnel.

The pressure drop coefficient in a wind tunnel unit is described as:

$$K_i = \frac{\Delta p_i}{q_i} \qquad (3.2.45)$$

$$\Longrightarrow \Delta p_i = K_i q_i \qquad (3.2.60)$$

The various losses in the wind tunnel circuit are commonly expressed in terms of the test section conditions. The pressure loss coefficient in each tunnel component is calculated using the loss coefficient and the dynamic pressure in the test section. We therefore write,

$$K_{TS} = \frac{\Delta p_i}{q_{TS}} = \frac{\Delta p_i}{q_i}\frac{q_i}{q_{TS}} \qquad (3.2.61)$$

$$= K_i \frac{q_i}{q_{TS}} \qquad (3.2.62)$$

$$\Longrightarrow K_i q_i = K_{TS} q_{TS} \qquad (3.2.63)$$

Adding together all of the wind tunnel components,

$$\sum_i K_i q_i = \sum_i K_{TS} q_{TS} \tag{3.2.64}$$

Thus, the energy ratio of a subsonic wind tunnel is:

$$\text{Energy ratio} = \frac{q_{TS}}{\sum_i K_{TS} q_{TS}} = \frac{1}{\sum_i K_{TS}} \tag{3.2.65}$$

3.3 POWER FACTOR

The power factor of a subsonic wind tunnel is the ratio of power supplied to the rate of flow kinetic energy in the test section, which is identical to:

$$\lambda_p = \frac{\text{Power input}}{\text{Rate of flow kinetic energy in the test section}}$$

$$= \frac{P_{input}}{\frac{1}{2}\rho v^3 A} \tag{3.3.1}$$

where λ_p is the power factor, P_{input} is the power input, v is the velocity, and A is the area of the test section.

Combining equation (3.2.44) and equation (3.3.1) yields

$$\lambda_p = \frac{\sum \text{circuit losses}}{\eta_F \frac{1}{2}\rho v^3 A} \tag{3.3.2}$$

where η_F is the fan efficiency. It is worth noting that the reciprocal of the power factor, known as the energy ratio, is another indicator of the wind tunnel's efficiency.

3.3.1 POWER ECONOMY

The power needed to run a wind tunnel is defined as:

$$P_{input} = \lambda_p \frac{1}{2}\rho v^3 A \tag{3.3.1}$$

If c is some characteristic length of the test section, M is the test section Mach number, a is the speed of sound, ρ is the density, μ is the viscosity of air, and γ is the ratio of specific heats, we can define:

$$Re_c = \frac{\rho v c}{\mu} \tag{3.3.3}$$

$$M = \frac{v}{a} \tag{3.3.4}$$

$$a^2 = \gamma RT = \gamma \frac{p}{\rho} \tag{3.3.5}$$

where Re_c is the Reynolds number based on the characteristic length, c.

Using equations (3.3.3)–(3.3.5), the power input relationship is as follows:

$$P_{input} = \lambda_p \frac{1}{2}\frac{A}{c^2} Re_c^2 M \frac{\mu^2 a^3}{\gamma p} \tag{3.3.6}$$

where, $\frac{A}{c^2}$ is called tunnel interference, a constant for a specific tunnel. Equation (3.3.6) reveals that the power factor depends largely on M and Re_c. Thus, for a subsonic wind tunnel operating at given M and Re_c, the power input to the system varies as:

$$P_{input} \propto \frac{\mu^2 a^3}{\gamma p} \tag{3.3.7}$$

which is represented in terms of stagnation properties by:

$$P_{input} \propto \frac{\mu_0^2 a_0^3}{\gamma p_0} \tag{3.3.8}$$

3.3.1.1 Power Economy by Raising Stagnation Pressure

For a certain operating fluid at constant stagnation temperature (i.e., at constant viscosity and constant stagnation speed of sound), equation (3.3.8) reduces to:

$$P_{input} \propto \frac{1}{p_0} \tag{3.3.9}$$

Thus, we can reduce the power required to operate the tunnel by increasing the stagnation pressure. However, the maximum increase in p_0 is restricted by the proportionate increase in aerodynamic forces on the model.

3.3.1.2 Power Economy by Choice of Working Fluid

For a specified stagnation pressure and temperature, μ_0 is constant. Thus, equation (3.3.8) results in

$$P_{input} \propto \frac{\gamma^{1/2}}{\overline{M}_w^{3/2}} \tag{3.3.10}$$

This reveals that the power input can be reduced using a working fluid of a higher specific heat ratio or molecular weight since γ is a narrow range parameter ($1 < \gamma \leqslant 1.67$), the power requirement, in this case, depends mainly on $\overline{M}_w$.

Limitations

1. To achieve complete dynamic similarity, the ratio of specific heats of the working fluid in the wind tunnel must equal that of the actual fluid flow over the prototype.
2. According to thermodynamics, the boiling point of a fluid with a higher molecular weight is high. As a result, the temperature should be high in order to retain the working fluid in a gaseous form. However, it provides significant metallurgical problems to tunnel walls.

3.3.1.3 Power Economy by Lowering Stagnation Temperature

For a certain operating fluid at specified p_0, λ_p, M, and Re, equation (3.3.8) can be written as follows:

$$P_{input} \propto \mu_0^2 a_0^3 \tag{3.3.11}$$

Since $a_0 \propto T_0^{1/2}$ and $\mu_0 \propto \frac{T_0^{3/2}}{(T_0+C)}$ (Sutherland's formula; C is a constant), P_{input} becomes:

$$P_{input} \propto \frac{T_0^{9/2}}{(T_0 + C)^2} \tag{3.3.12}$$

Limitations

1. The lower the stagnation temperature, the lower the flow temperature. Hence, obtaining a power economy by lowering T_0 is not feasible.
2. Severe metallurgical issues may emerge at very low temperatures, limiting the reduction in P_{input}.

3.4 SUMMARY

A wind tunnel is a confined environment with fast-moving air used to study the flow around different objects. Wind tunnels are primarily used to test aircraft and spacecraft scale models, but some are large enough to test full-size vehicles. The airstream moving through an object creates the illusion that it is flying. Wind tunnels are divided into three varieties based on their operational needs: low-speed, high-speed, and special-purpose.

The contraction cone, test section, divergent diffuser, and power unit are the four main components of a general-purpose low-speed wind tunnel.

An effuser is a converging duct located ahead of the test section in which the fluid is accelerated from rest (or extremely low speed) and brought to the appropriate conditions in the test section. As a result, the effuser is also known as the contraction cone. The contraction ratio for an effuser is defined as:

$$a_c = \frac{A_1}{A_2}$$

where A_1 and A_2 are the effuser's inlet and exit cross-sectional areas, respectively.

The test section, also referred to as the working section, is a wind tunnel component with uniform flow properties. It has the same cross-sectional area along its whole length. However, forming a boundary layer over the test section walls reduces net area while raising flow speed. As a result, a horizontal force acts on the model in a downstream direction, known as horizontal buoyancy. Thus, the test section walls must diverge adequately so that the net cross-sectional area remains constant along the length of the test section. The model is placed within the test section, and the relevant measurements and observations are taken. The tunnel is called closed-throat when rigid walls outline the test section.

The diffuser is positioned ahead of the test section and effectively transforms the kinetic energy of the airstream (leaving the test section) into the pressure energy. It is a channel through which a subsonic flow decelerates by increasing the downstream area. The area ratio or the divergence angle, often known as the equivalent cone angle, defines the diffuser. A diffuser's area ratio is given by:

$$a_d = \frac{A_2}{A_1}$$

where A_1 and A_2 are the diffuser's inlet and exit cross-sectional areas, respectively.

In the absence of losses, a steady and uniform flow could be maintained indefinitely without requiring more energy. However, losses occur due to the dissipation of kinetic energy into heat by turbulence, vortical motion, and other factors. In addition, some kinetic energy is invariably expelled at the exit of the diffuser in open-circuit wind tunnels, which is converted to heat as the tunnel exhaust gases mix with the ambient air. Wind tunnel operation requires an additional power arrangement or drive unit to compensate for these energy losses. A motor and a fan or propeller often drive a low-speed wind tunnel.

The total pressure drop across a wind tunnel is the sum of the pressure drops in its various components. The total pressure drop must equal the pressure rise induced by the fan or driving unit for an efficient tunnel functioning. The following factors cause pressure decrease in a subsonic tunnel circuit:

- Pressure drops in the screens.
- Pressure drops in the honeycomb structures.

- Pressure drops in the contraction cone.
- Pressure drops in the test section.
- Pressure drops in the diffuser due to friction and flow expansion.
- Pressure drops in the guide vanes due to friction and flow expansion.
- Due to friction and expansion, pressure losses in the corners (or guide vanes).

The energy ratio of a subsonic wind tunnel is defined as the ratio of flow kinetic energy in the test section to overall energy losses in the tunnel, which is equivalent to:

$$\text{Energy ratio} = \frac{\text{Flow kinetic energy in test section}}{\sum \text{Overall energy losses in tunnel circuit}}$$

The energy ratio is always greater than unity in both closed- and open-circuit wind tunnels (excluding open test section walls). It typically ranges between 3 to 7 for closed-circuit tunnels—the higher the energy ratio, the greater the energy efficiency. Since the energy ratio for a free jet configuration is always less than one, an open test section is never used in a large tunnel.

The power factor of a subsonic wind tunnel is the ratio of power supplied to the rate of flow kinetic energy in the test section, which is identical to:

$$\lambda = \frac{P_{input}}{\frac{1}{2}\rho v^3 A}$$

where λ is the power factor, P_{input} is the power input, v is the velocity, and A is the area of the test section.

The power input to run a wind tunnel is defined as:

$$P_{input} = \lambda \frac{1}{2} \rho v^3 A$$

We can lower the power required to operate the tunnel by increasing the stagnation pressure. However, the maximum increase in stagnation pressure is restricted by the proportionate increase in aerodynamic forces on the model. The power input can also be reduced by utilizing a working fluid with a higher specific heat ratio or molecular weight.

EXERCISE PROBLEMS

Exercise 23. A pitot-static tube measures the dynamic pressure of an incompressible flow. When there is turbulence, the probe produces inaccurate results. Determine the error in pressure readings if a flow contains 5% turbulence. Assume isotropic turbulence.

Example 24. A subsonic wind tunnel has a 0.5 m diameter cylindrical test section. The diffuser is 4.75 m long with 0.5 m and 1.5 m entry and exit diameters, respectively. If the average friction factor is 0.018, calculate the total loss coefficient for the diffuser.

4 High-Speed Wind Tunnels

4.1 INTRODUCTION

A fractional temperature change of 5% or more is considered significant in almost all the fluid flows, necessitating the thermal effects to be accounted for in the analysis. At sea level, more than 5% fractional change in temperature is equivalent to the flow velocity of 650 kmh^{-1} ($M \geq 0.5$). Because compressibility effects prevail at $M \geq 0.3$, flows of $M \geq 0.5$ are known as high-speed flows, and the related wind tunnel is known as the high-speed wind tunnel. When the test section Mach number exceeds one, the corresponding wind tunnel is referred to as a supersonic wind tunnel. The power required to run a low-speed wind tunnel is proportional to the test section speed cube. Although the same proportionality does not exist for a high-speed (or supersonic) wind tunnel, power consumption does increase as the Mach number increases.

4.2 CLASSIFICATION OF HIGH-SPEED WIND TUNNELS

Wind tunnels are categorized in section 1.2.1 depending on the test section Mach number. A high-speed wind tunnel on the basis of operating Mach number range in the test section is divided into transonic, supersonic, and hypersonic tunnels (Table 4.1). A high-speed tunnel may also be categorized as intermittent or continuous operation depending on whether it is utilized for a short or prolonged run (Figure 4.2.1). The open-circuit intermittent tunnels are sub-classified as blowdown and indraft (or induction) tunnels. It is worth emphasizing that the most high-speed wind tunnels are intermittent because of the massive power requirements.

Table 4.1
Wind tunnel classification on the basis of test section Mach number.

Type of tunnel	Mach number working range
Transonic tunnels	$0.8 < M < 1.2$
Supersonic tunnels	$1.2 < M < 5$
Hypersonic tunnels	$M > 5$

DOI: 10.1201/9781003139447-4

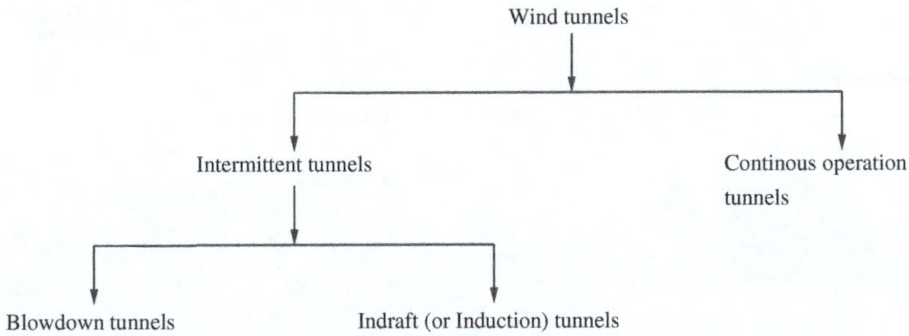

Figure 4.2.1 Wind tunnel classification based on the operational standpoint.

The intermittent blowdown and indraft tunnels generally operates between Mach 0.5 to Mach 5. For the Mach numbers greater than 5, intermittent pressure vacuum tunnels are employed. The continuous operation tunnel, however, has no such Mach number limitation and can be utilized a wide range of speeds. The choice of a specific tunnel for testing depends on the type of measurement and the power requirements. Nevertheless, despite their benefits, both intermittent and continuous operation tunnels have drawbacks. Therefore, a single tunnel is insufficient for all practical applications and flow regimes.

4.2.1 INTERMITTENT BLOWDOWN WIND TUNNELS

An intermittent blowdown wind tunnel can achieve a wide range of test section Mach numbers, from high subsonic to relatively high supersonic. An open-circuit blowdown tunnel obtains supersonic speeds in the test section by creating a pressure difference between a reservoir (through which it inducts the air) and the ambient. A closed blowdown tunnel, on the other hand, includes a high-pressure reservoir upstream and a comparatively low-pressure tank downstream. At the start of a run, the valves in both reservoir and the tank are kept open, and due to the pressure difference between them, air flows toward the tank until an equilibrium is established throughout. The test section is positioned immediately after the supersonic nozzle. Some blowdown tunnels have two throats (one of the supersonic nozzle and the other of the supersonic diffuser) to minimize the pressure drops, primarily due to normal shock. The second throat (also called the diffuser throat) reduces the Mach numbers from supersonic to subsonic before discharging the flow to a low-pressure tank or the ambiance. The schematic layout of an intermittent blowdown wind tunnel is shown in Figure 4.2.2.

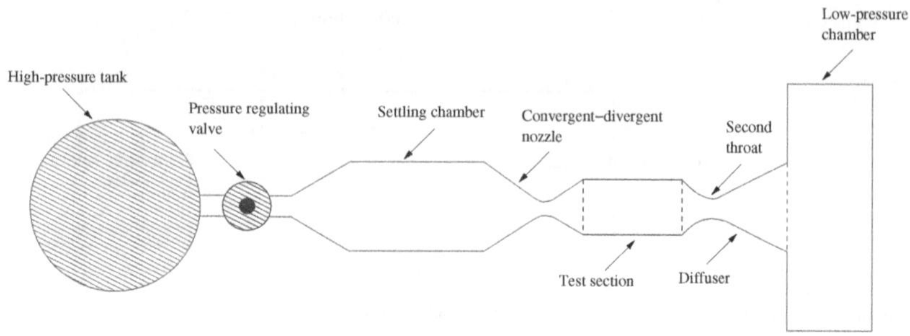

Figure 4.2.2 Schematic layout of an intermittent blowdown wind tunnel.

4.2.1.1 Advantages

The following are the primary benefits of intermittent blowdown tunnels:

- These tunnels are simple to start.
- These tunnels have a simpler layout, which reduces construction cost.
- They provide an excellent design for smoke-flow propulsion and visualization.

4.2.1.2 Disadvantages

The following are the main drawbacks of blowdown tunnels:

- These tunnels have a very limited testing time.
- They necessitate the use of expensive, high-precision measurement equipment.
- Because of the stagnation temperature drop, the Reynolds number in the test section changes during testing.
- The startup load of the tunnel is exceptionally high.

4.2.2 INTERMITTENT INDRAFT WIND TUNNELS

An intermittent indraft wind tunnel achieves supersonic speeds in the test section by producing a vacuum at the downstream end while leaving the upstream end open to the surrounding environment. The opening of the valve eventually pulls air inside the tunnel at a desired vacuum level in the tank, establishing a supersonic flow in the test section. A schematic diagram of an indraft tunnel is shown in Figure 4.2.3.

4.2.2.1 Advantages

The primary benefits of an intermittent indraft tunnel are as follows:

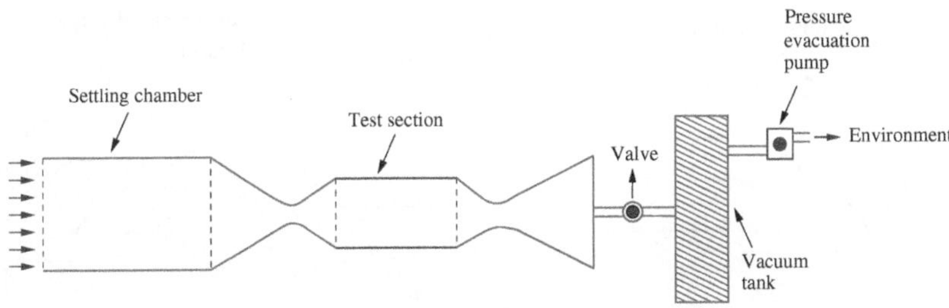

Figure 4.2.3 Schematic layout of an intermittent indraft wind tunnel.

- Unlike intermittent blowdown tunnels, the stagnation temperature in intermittent indraft tunnels remains practically constant during testing.
- The test section flow is pure since the driving unit is at the downstream side.
- These tunnels, unlike blowdown tunnels, can operate at higher Mach numbers prior to actually requiring a heater to eliminate flow liquefaction during expansion.
- Because of the possibility of explosion in a high-pressure reservoir, a vacuum tank is safer.

4.2.2.2 Disadvantages

The primary disadvantages of an intermittent indraft tunnel are as follows:

- These tunnels are far more expensive than blowdown tunnels due to the high-suction demand.
- Because the stagnation pressure is atmospheric, the Reynolds number is quite low.
- Although these tunnels can run without an air drier, they can only reach speeds of up to Mach 1.8 in the absence of condensation.

4.2.3 CONTINUOUS SUPERSONIC WIND TUNNELS

A continuous supersonic wind tunnel, being a closed-circuit assembly, can achieve a wide range of Mach numbers in the test section. The flow in this tunnel does not exhaust the surrounding environment and returns to the test section along a predetermined path. The schematic layout of a continuous supersonic tunnel is shown in Figure 4.2.4.

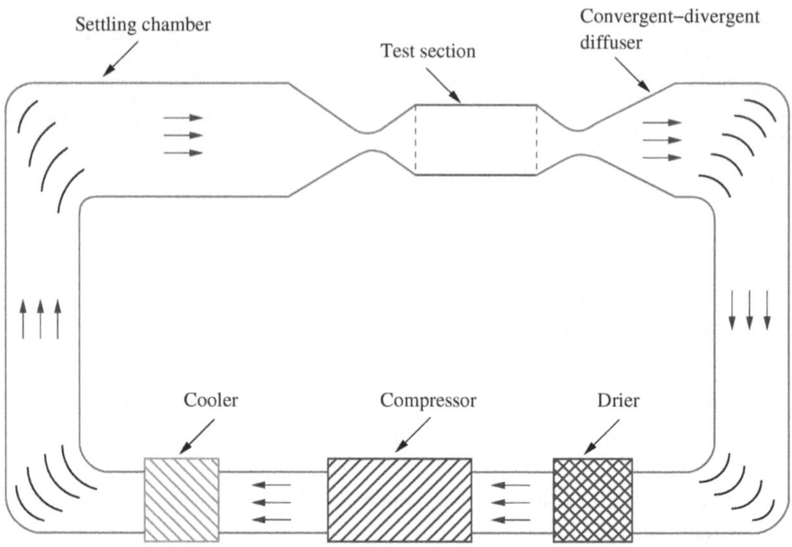

Figure 4.2.4 Schematic layout of a continuous supersonic wind tunnel.

4.2.3.1 Advantages

The main advantages of a continuous supersonic tunnel are as follows:

- Because these tunnels are sealed off from the outside environment, more controlled conditions can be established in the test section.
- The use of guiding vanes in the corners and flow straighteners upstream of the test section ensures that the flow is relatively uniform.
- The test conditions can be kept constant for a prolonged period, and the testing time for each run is longer than for other tunnels.
- The operation of these tunnels is quieter.

4.2.3.2 Disadvantages

The drawbacks of a continuous supersonic tunnel are as follows:

- The power consumption is considerably large due to the extended charging to running time.
- Temperature stabilization necessitates the use of a big cooling system.
- These tunnels are difficult and expensive to build.

4.3 COMPONENTS OF AN INTERMITTENT BLOWDOWN WIND TUNNEL

The following are the major parts of an intermittent blowdown tunnel:

1. Air receiver or storage tank
2. Settling chamber
3. Convergent-divergent (de Laval) or supersonic nozzle
4. Test section
5. Convergent-divergent or supersonic diffuser

4.3.1 AIR RECEIVER

The pressurized air is inducted into the wind tunnel from a large storage tank called an air receiver. The size of a tank depends on the required mass flow rate and the blowdown time. Multiple tanks, mounted horizontally or vertically, store compressed air for a longer operating time. In the author's laboratory, a two-stage reciprocating compressor that can produce 60 cfm of air at a pressure of 40 bar is deployed to fill a vertically mounted 5 m^3 tank (Figure 4.3.1(a)). The compressor is powered by a 150 hp 3-phase induction motor. An inter-cooler consisting of a cooling water circuit is employed to cool the pressurized air. The impurities, such as rust particles and oil droplets in the cooled air, are removed by allowing the air to pass through a pre-filter consisting of porous

(a) Air receiver (b) Settling chamber

Figure 4.3.1 Photographs of the storage tank and the settling chamber.

stone candles – an activated-carbon filter unit follows the pre-filter for finer filtration. The silica gel drier unit finally dries the air before supplying it to the storage tanks.

4.3.2 SETTLING CHAMBER

Pressurized air from tanks is introduced into a settling chamber (Figure 4.3.1(b)) via a short pipe with a gate valve, accompanied by a pressure regulating valve and a 70 mm diameter mixing pipe. The settling chamber has a cylindrical shape with a 300 mm inner diameter and 600 mm length, providing tappings for measuring the pressure and temperature. The mixing pipe is joined with the settling chamber through a bell-shaped diffuser consisting of screens to reduce turbulence at the Laval nozzle entrance.

4.3.3 CONVERGENT-DIVERGENT NOZZLE: SUPERSONIC NOZZLE

The Mach number of a subsonic flow passing through a convergent nozzle increases from inlet to exit; reaching to a maximum of one at the nozzle exit. A divergent duct should be added to the convergent duct to obtain supersonic Mach numbers at the nozzle exit. This design is known as a convergent-divergent or de Laval nozzle after C. G. P. de Laval, who used it in his steam turbines in the late 1800s. Figure 4.3.2 schematically shows a typical convergent-divergent nozzle producing supersonic flow. A carefully designed

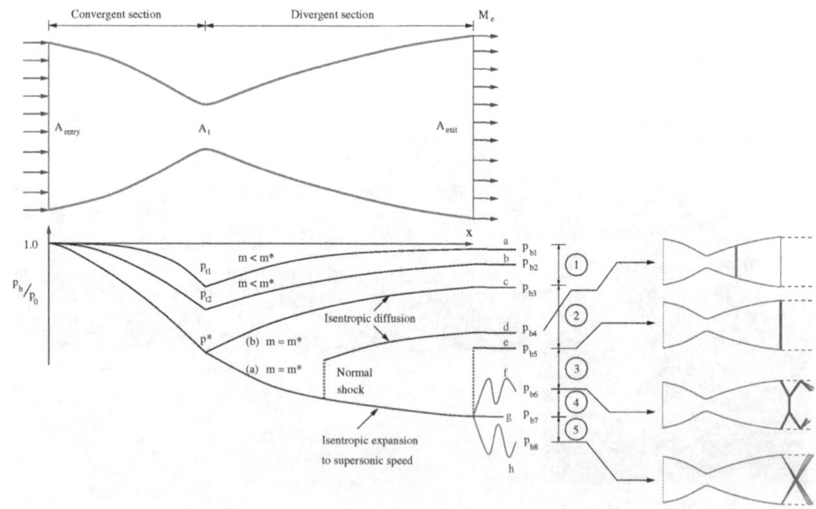

Figure 4.3.2 Convergent-divergent nozzle operation at varied backpressure.

convergent-divergent nozzle ensures the flow parameters are uniform at each cross-section along its length. This article examines the flow characteristics of a convergent-divergent nozzle at varied backpressure p_b while keeping the reservoir pressure p_0 constant.

The Area-Mach number relationship for a convergent-divergent nozzle is[1]:

$$\frac{A}{A^*} = \frac{1}{M}\left[\frac{2}{\gamma+1}\left(1+\frac{\gamma-1}{2}M^2\right)\right]^{\frac{(\gamma+1)}{2(\gamma-1)}} \tag{2.7.17}$$

This equation reveals that the area, A, at any cross-section is always greater than the choked flow throat area, A^*, i.e., $A < A^*$ is physically not feasible in isentropic flows. In addition, a given value of $\frac{A}{A^*}$ corresponds to two different Mach numbers, one subsonic and the other supersonic.

When the backpressure, p_b, equals the reservoir pressure, p_0, there is no flow through the nozzle. When the backpressure is decreased to p_{b1}, a subsonic flow is established across the nozzle, as shown by curve a in Figure 4.3.2. Notice that the Mach number in the convergent section increases with the decreasing area, attaining a maximum at the throat. The static pressure drops in the convergent section, reaching a minimum at the throat. After that, the flow accelerates in the divergent section, and the static pressure progressively increases till it becomes equal to the ambient pressure at the exit. When the back pressure is reduced to p_{b2}, the flow designated by curve b shows that the static pressure drop in the nozzle is more than in the previous case. The flow is accelerated, and the Mach number at each location is comparatively higher. When the back pressure is progressively lowered to p_{b3}, the throat becomes choked ($A_t = A^*$) and attains sonic conditions ($p_t = p^*; M = 1$). In this case, if $p_{b3} > p^*$, the sonic flow entering the divergent section decelerates to subsonic speeds, and the static pressure eventually rises (curve c). On the hand, if $p_{b7} < p^*$, the flow accelerates to supersonic speeds in the divergent section with a progressively decreasing static pressure (curve g). At this point, any decrease in backpressure has no effect on the flow upstream of the throat, and the nozzle is said to be flowing with a maximum mass flow rate, $\dot{m}_{max}$, for a defined p_0 and T_0.

It is interesting to note that the nozzle operation along the curves c and g is fully isentropic. However, between these two limiting cases, a range of backpressures exists where a normal shock forms in the divergent section, thereby making the flow non-isentropic. Also, the flow across the normal shock becomes subsonic. Moreover, in this range of backpressures, the shock gradually moves toward the nozzle exit with a decreasing p_b and stands right at the exit

[1]For more details on this relationship, the interested reader can consult Fundamentals of Gas Dynamics by Mrinal Kaushik, 1st edition, Springer Nature, Singapore, 2022.

when $p_b = p_{b5}$. Consequently, the flow in the nozzle is isentropic throughout and remains supersonic in the divergent section till the nozzle exit. On further reducing the backpressure to p_{b6}, the normal shock moves out of the nozzle, and the flow gets adjusted to the backpressure through oblique shock waves at the exit (curve f). In this situation, the nozzle operation is called overexpanded. Further reduction of the backpressure corresponding to p_{b7} vanishes the oblique shocks, and the nozzle is said to operate at a correctly expanded state (curve g). When the backpressure equals p_{b8}, the flow gets adjusted to the backpressure through expansion waves at the nozzle exit. Under this condition, the nozzle is referred to as underexpanded.

According to the preceding discussion, a convergent-divergent nozzle discharging the maximum mass flow rate at perfect expansion produces a fully isentropic and supersonic flow. At this point, lowering the backpressure does not affect the flow in the nozzle. The jet at the nozzle exit, on the other hand, expands downstream like a jet from a convergent nozzle working at supercritical pressure ratios. Because the supersonic jet is imperfectly expanded, its kinetic energy is lower than it would be if it were perfectly expanded. Consequently, the thrust produced by an underexpanded nozzle is less than that produced by a correctly expanded nozzle. For the same reasons, the thrust produced by a convergent-divergent nozzle when overexpanded is less than that produced when correctly expanded.

Example 25. A pitot probe mounted right at the exit of a Laval nozzle reads 89.5 kPa. The air is inducted in the nozzle from a a reservoir at 202 kPa. Find the area ratio $\frac{A_e}{A^*}$ of the nozzle.

Solution. Given,

$$p_{0e} = 89.5 \text{ kPa}$$
$$p_{01} = 202 \text{ kPa}$$

Note that, in this case a normal shock will be positioned somewhere in the divergent portion of the supersonic nozzle. As a result, the pressure measured by the probe at the nozzle exit will essentially be the stagnation pressure, p_{0e} (downstream of the shock). Thus, the ratio of stagnation pressures across the shock is

$$\frac{p_{0e}}{p_{01}} = \frac{89.5}{202} = 0.443$$

With $\frac{p_{0e}}{p_{01}} = 0.443$, it follows from the normal shock table that

$$M_e = 2.65$$

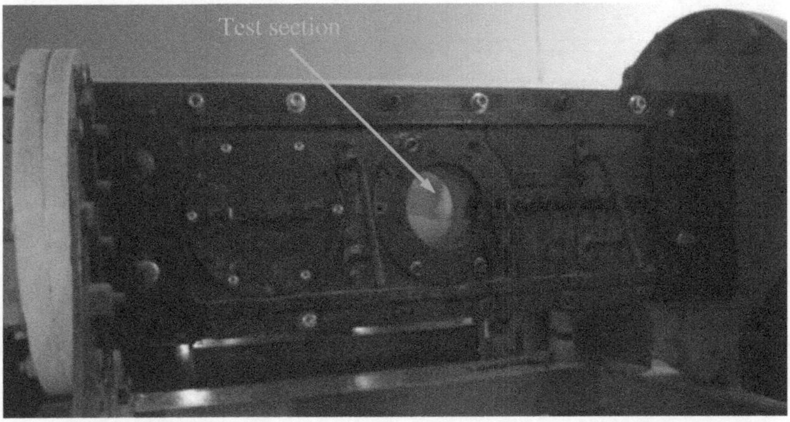

Figure 4.3.3 Test section of a Mach 2 wind tunnel.

which from isentropic table yields

$$\frac{A_e}{A^*} = 3.04$$

4.3.4 TEST SECTION

A test section, also known as the working section, follows the supersonic nozzle for testing the experimental models. The models are mounted on a sting balance deployed inside the test section and have equipment for measuring forces and moments. Figure 4.3.3 depicts a Mach 2 wind tunnel test section in the author's laboratory. The targeted Mach number for the measurements determines the shape and size of the test section. To have a shock-free test section, the cross-section area of the test section must be at least equal to the throat area of the downstream convergent-divergent diffuser.

Interestingly, the size of experimental models in supersonic wind tunnels is decided by a so-called test rhombus formed by the incident and reflected oblique shocks similar to one, as shown in Figure 4.3.4. The model must lie inside the rhombus for precise measurements. However, it is worth emphasizing that the incident and reflected shocks do not subtend equal angles with the test section walls, i.e., shock wave reflection in the test section is not specular. Hence, the streamwise length of the model is selected so that the shock-model interference is avoided.

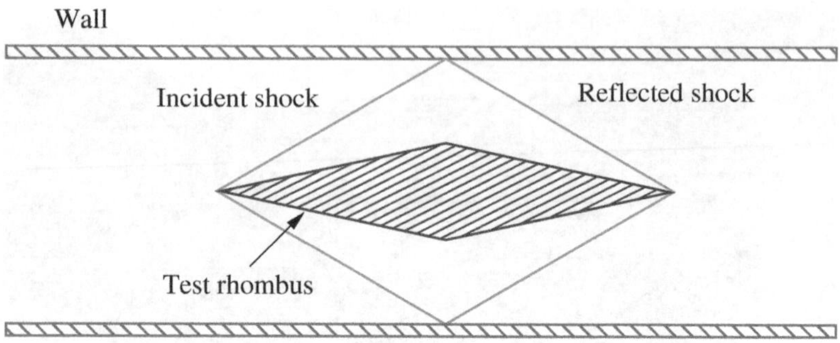

Figure 4.3.4 Model size estimation using test rhombus.

4.3.5 CONVERGENT-DIVERGENT DIFFUSER: SUPERSONIC DIFFUSER

A diffuser is a duct that decreases flow velocity to increase pressure. The diffuser walls should diverge downstream if the entering flow is subsonic. However, if the incoming flow is supersonic, the diffuser walls should converge downstream. The highest retardation obtained at the exit of a convergent diffuser with the supersonic flow at the inlet is Mach 1. A divergent duct should be attached to the exit to decrease the sonic flow further. This type of design is known as a convergent-divergent diffuser. It is also called a reverse nozzle diffuser because it operates in the opposite direction of a convergent-divergent nozzle. A convergent-divergent diffuser (also known as a supersonic diffuser) is attached directly after the test section of a supersonic wind tunnel. Based on the following assumptions, this article investigates the fundamental properties of a fixed-shape convergent and a convergent-divergent diffuser:

- The flow through the diffuser is one-dimensional and steady.
- The flow is isentropic throughout except across an essential normal shock.
- The flow discharges into the ambient with a backpressure, p_b.
- The reservoir conditions p_0 and T_0 are assumed to be constant during the diffuser operation.

4.3.5.1 Normal Shock Recovery

Consider an open-circuit blowdown tunnel with a divergent diffuser, as illustrated in Figure 4.3.5(a). As discussed in section 4.3.3, a normal shock stands right at the nozzle exit if the backpressure equals p_{b8} (Figure 4.3.2). The pressure ratio, σ_1, needed to operate the tunnel is equal to the stagnation pressure ratio, $\frac{p_{01}}{p_{02}}$, across the normal shock formed in the nozzle divergence. This is identical to the following:

$$\sigma_1 = \frac{p_{01}}{p_{02}} = \left[1 + \frac{2\gamma}{\gamma+1}\left(M_1^2 - 1\right)\right]^{\frac{1}{\gamma-1}} \left[\frac{(\gamma+1)M_1^2}{(\gamma-1)M_1^2+2}\right]^{-\frac{\gamma}{\gamma-1}} \tag{2.7.23}$$

A slight decrease in the backpressure at this stage moves the shock to the test section. If the flow exiting the test section is exposed to the surroundings, it encounters tremendous dissipative losses due to high-flow kinetic energy. A divergent diffuser, added soon after the test section, may recover the pressure and reduce the losses (Figure 4.3.5(b)). A substantial decrease in the backpressure from p_{b8} pushes the normal shock in the diffuser, leading to a shock-free test section. This entire process for achieving a shock-free test section is known as normal shock recovery or normal shock swallowing.

Across a normal shock positioned in a divergent diffuser, the supersonic flow becomes subsonic and accelerates isentropically to a new stagnation pressure, $p'_{02} > p_{02}$. The operating pressure ratio across the tunnel, σ_2, is given by:

$$\sigma_2 = \frac{p_{01}}{p'_{02}} = \left[1 + \frac{2\gamma}{\gamma+1}\left(M_1^2 - 1\right)\right]^{\frac{1}{\gamma-1}} \left[\frac{(\gamma+1)M_1^2}{(\gamma-1)M_1^2+2}\right]^{-\frac{\gamma}{\gamma-1}} \tag{4.3.1}$$

where $\sigma_2 < \sigma_1$ since $\frac{p_{01}}{p'_{02}} < \frac{p_{01}}{p_{02}}$. Therefore, adding a divergent diffuser to the test section of a supersonic wind tunnel may reduce the overall pressure ratio for tunnel operation.

4.3.5.2 Consequences of a Supersonic Diffuser Throat

Unlike the previous case of a supersonic wind tunnel with a divergent diffuser, if a convergent-divergent diffuser is employed, the operating pressure ratio of the tunnel can be further reduced. Such an arrangement is examined in this article. Figure 4.3.6 schematically shows a fixed-shape convergent-divergent diffuser following the test section of a supersonic intermittent blowdown tunnel. It is assumed that the flow throughout the wind tunnel is isentropic except across an essential normal shock. The supersonic flow exiting the test section decelerates in the convergent portion of the diffuser. It is eventually compressed to sonic conditions at the diffuser throat (commonly referred to as the second throat). The sonic flow further decelerates in the divergent section attaining a very low speed and, thus, high-static pressure (close to the ambient pressure) at the diffuser exit.

Consider a supersonic intermittent blowdown tunnel with a settling chamber at constant conditions, p_{01} and T_{01}, discharging into ambiance at the backpressure, p_b. For a certain value of p_b, the pressure ratio $\frac{p_b}{p_{01}}$ is sufficient to choke

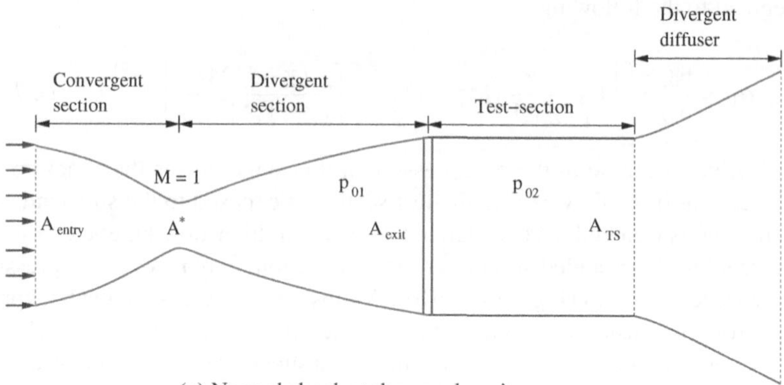

(a) Normal shock at the nozzle exit

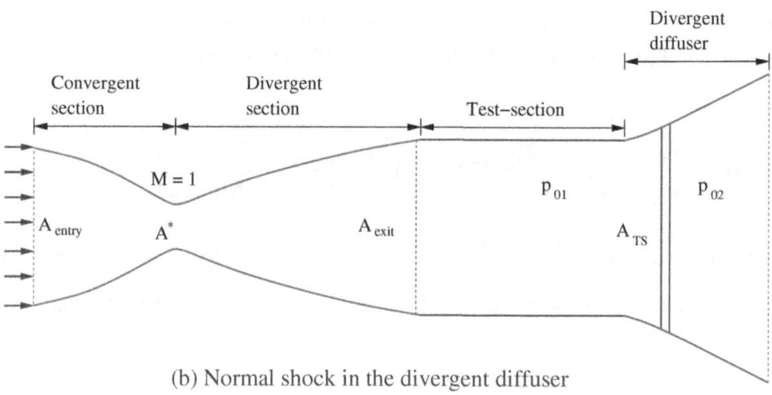

(b) Normal shock in the divergent diffuser

Figure 4.3.5 Supersonic intermittent blowdown tunnel with divergent diffuser.

the nozzle throat with sonic conditions prevailing at the throat. The nozzle discharges a maximum mass flow rate and is said to operate at its first critical condition. In this situation, a progressive decrease in the backpressure causes a normal shock to first appear in the nozzle divergence (section 4.3.3), and then its transportation to the test section. Such a phenomenon of normal shock formation in the test section happens during the starting of a supersonic wind tunnel, which is considered to be the unfavorable (off-design) operation of the tunnel (Figure 4.3.6(a)). If the backpressure is decreased beyond this stage, the diffuser swallows the shock and moves it to its divergence, resulting in a shock-free test section with an isentropic flow throughout. This entire process of normal shock propagation to the diffuser divergence is called normal shock swallowing.

It is worth noting that the test section becomes shock-free, but the tunnel's operational power requirements remain high due to the huge pressure drop across the normal shock in the diffuser. At this point, the backpressure is somewhat raised, which causes the shock to go backward near the second throat. Because the shock strength is low at this position, the power consumption is also low. This is the standard operating condition of a supersonic wind tunnel, often known as the most favorable (design) condition (Figure 4.3.6(b)).

In the discussion to follow, a relationship is derived to quantify the minimum area of the second throat. Consider the following one-dimensional continuity equation:

$$\dot{m} = \rho v A \tag{2.7.2}$$

Substituting the equation of state for a perfect gas ($p = \rho RT$) into above equation and noting $v = M\sqrt{\gamma RT}$, yields

$$\dot{m} = p\sqrt{\frac{\gamma}{RT}}MA \tag{4.3.2}$$

The gas flowing through the nozzle is primarily inducted from the reservoir at conditions p_{01} and T_{01}, and expanded to a point in the nozzle where the local conditions are p and T. The following isentropic relations relate these parameters:

$$\frac{T_{01}}{T_1} = 1 + \frac{\gamma - 1}{2}M_1^2 \tag{2.7.14}$$

$$\frac{p_{01}}{p_1} = \left(1 + \frac{\gamma - 1}{2}M_1^2\right)^{\frac{\gamma}{\gamma-1}} \tag{2.7.15}$$

Eliminating p and T in equation (4.3.2) using above expressions gives

$$\frac{\dot{m}}{A} = \frac{p_{01}\sqrt{\gamma}}{\sqrt{RT_{01}}}M_1\left(1 + \frac{\gamma - 1}{2}M_1^2\right)^{\frac{-(\gamma+1)}{2(\gamma-1)}} \tag{4.3.3}$$

The mass flow rate through a supersonic nozzle with the throat choked ($M = 1$) reaches its maximum and is similar to the following:

$$\frac{\dot{m}_{max}}{A_1^*} = \frac{p_{01}\sqrt{\gamma}}{\sqrt{RT_{01}}}\left(\frac{\gamma + 1}{2}\right)^{-\frac{(\gamma+1)}{2(\gamma-1)}} \tag{4.3.4}$$

where A_1^* = throat area of convergent-divergent nozzle. This equation for air ($\gamma = 1.4$) becomes:

$$\frac{\dot{m}_{max}}{A_1^*} = \frac{0.6847 p_{01}}{\sqrt{RT_{01}}}$$

$$\Longrightarrow \quad \dot{m}_{max} \propto \frac{p_{01}A_1^*}{\sqrt{T_{01}}} \tag{4.3.5}$$

The same mass flow must pass through the diffuser throat, therefore we can also write:

$$\dot{m}_{max} \propto \frac{p_{02}A_2^*}{\sqrt{T_{02}}} \tag{4.3.6}$$

where A_2^* = throat area of convergent-divergent diffuser

p_{02} = stagnation pressure of the ambiance

T_{02} = stagnation temperature of the ambiance

Combining equation (4.3.5) with equation (4.3.6) yields

$$\frac{p_{01}A_1^*}{\sqrt{T_{01}}} = \frac{p_{02}A_2^*}{\sqrt{T_{02}}} \tag{4.3.7}$$

The stagnation temperature upstream and downstream of the shock remains constant, resulting in:

$$\frac{p_{01}}{p_{02}} = \frac{A_2^*}{A_1^*} \tag{4.3.8}$$

which is an important relation for designing a supersonic wind tunnel. Since $p_{01} > p_{02}$, it implies $A_2^* > A_1^*$. This suggests that the second throat (diffuser throat) has to be larger than the first throat (nozzle throat) to accommodate the mass flow rate that the nozzle discharged. For the limiting situation when the normal shock is located in the test section, causing a maximum pressure drop, equation (4.3.8) can be written as follows:

$$\left(\frac{p_{01}}{p_{02}}\right)_{TS} = \frac{A_{2,min}^*}{A_1^*} \tag{4.3.9}$$

where $\left(\frac{p_{01}}{p_{02}}\right)_{TS}$ = stagnation pressure ratio across the normal shock in the test section.

$A_{2,min}^*$ = minimum value of the second throat area at choked flow.

In order to account for errors due to friction, one-dimensional assumption, and so on, the second throat area, A_2^*, should be greater than the theoretical minimum, A_{min}^*, when constructing a supersonic wind tunnel with a fixed-shape convergent-divergent diffuser. Even though this is a simplified analysis based on the isentropic flow assumption, the conclusions can be applied to real flows through a diffuser that eliminates the boundary layer in some way.

Example 26. Air is inducted from a large container at 2 bar and 500 K and passes through a convergent-divergent nozzle (throat area A_t), which is connected to a second large container through a duct of constant area, $A_D = 5A_t$ (Figure 4.3.7). A normal shock is formed in the divergent portion of the

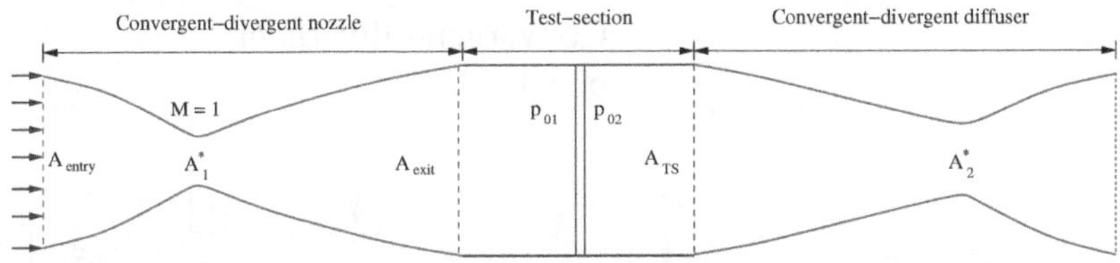

(a) Normal shock in the test–section (startup condition)

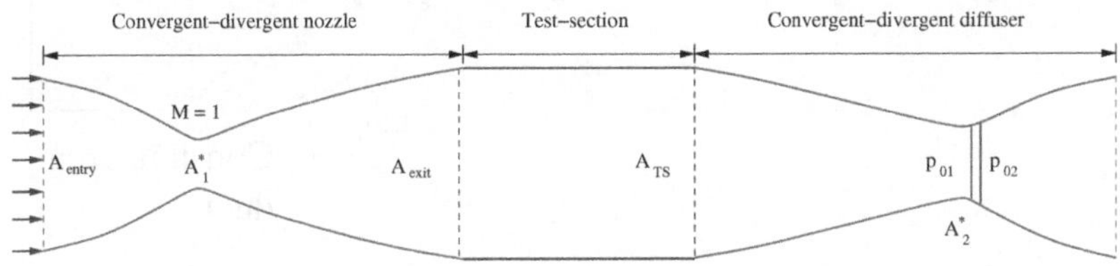

(b) Normal shock at the second throat (running condition)

Figure 4.3.6 Supersonic intermittent blowdown tunnel with a convergent-divergent diffuser.

convergent-divergent nozzle at a location where area, $A_1 = 2A_t$. Except in the second container and within the shock, the flow is assumed to be isentropic throughout. Find:

(a) The Mach numbers M_1, M_2, static pressures p_1, p_2, and temperatures T_1, T_2, upstream and downstream of the shock.

(b) The Mach number M_3, static pressure p_3, and temperature T_3 inside the duct.

(c) The stagnation temperature T_{04}, and pressure p_{04} in the second container.

Solution. Given,

$$p_{01} = 2 \text{ bar}$$
$$T_{01} = 500 \text{ K}$$
$$A_D = 5A_t$$
$$A_1 = 2A_t$$

(a) It is worth noting that when a normal shock forms in the divergent portion of a convergent-divergent nozzle, the throat has to be choked, i.e., $A_t = A^*$.

$$\Longrightarrow \quad \frac{A_1}{A_t} = \frac{A_1}{A^*} = 2$$

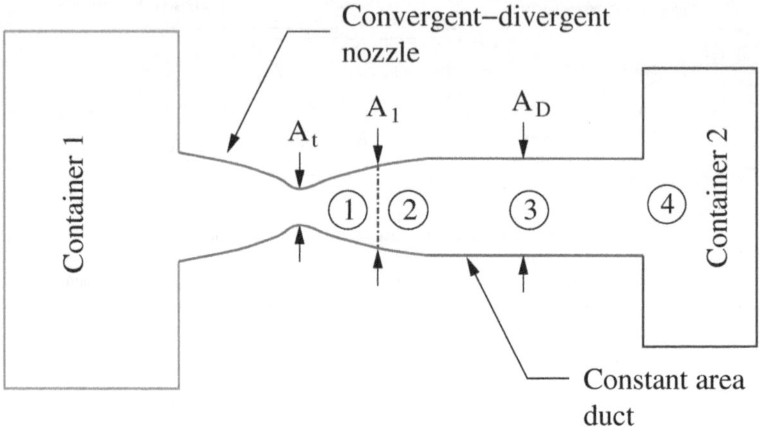

Figure 4.3.7 Normal shock in the divergent section of a convergent-divergent nozzle.

For $\frac{A_1}{A^*} = 2$, the corresponding values from the isentropic table are:

$$M_1 = 2.2$$
$$\frac{p_1}{p_{01}} = 0.0935$$
$$\frac{T_1}{T_{01}} = 0.5081$$

Thus,

$$\begin{aligned} p_1 &= 0.0935 p_{01} \\ &= 0.0935 \times 2 = 0.187 \text{ bar} \\ T_1 &= 0.5081 T_{01} \\ &= 0.5081 \times 500 = 254.05 \text{ K} \end{aligned}$$

Now, for $M_1 = 2.2$, the corresponding values from the normal shock table are:

$$M_2 = 0.5774$$
$$\frac{p_2}{p_1} = 5.48$$
$$\frac{T_2}{T_1} = 1.8569$$
$$\frac{p_{02}}{p_{01}} = 0.6281$$

Hence,

$$\begin{aligned} p_2 &= 5.48p_1 \\ &= 5.48 \times 0.187 = 1.025 \text{ bar} \\ T_2 &= 1.8569T_1 \\ &= 1.8569 \times 254.05 = 471.74 \text{ K} \end{aligned}$$

(b) The constant area duct is attached to the nozzle exit, which implies

$$A_D = A_e$$

where A_e is the nozzle exit area. Moreover, it should be emphasized that the reference throat area at the downstream side of the shock is denoted by A_2^* and the corresponding stagnation pressure and temperature by p_{02} and T_{02}, respectively.

With $M_2 = 0.5774$, it follows from isentropic table that

$$\frac{A_2}{A_2^*} = 1.213$$

Thus, we get

$$\begin{aligned} A_2^* &= \frac{A_2}{1.213} = \frac{A_1}{1.213} \\ &= \frac{A_1}{A_1^*} \times \frac{A_1^*}{1.213} = 2 \times \frac{A_1^*}{1.213} \\ &\Longrightarrow \frac{A_2^*}{A_1^*} = 1.649 \end{aligned}$$

and

$$\begin{aligned} p_{02} &= \frac{p_{02}}{p_{01}} \times p_{01} \\ &= 0.6281 \times 2 = 1.256 \text{ bar} \end{aligned}$$

The stagnation temperature across a shock wave remains constant:

$$T_{02} = T_{01} = 500 \text{ K}$$

The flow downstream of the normal shock is isentropic. Hence,

$$\begin{aligned} p_{0e} &= p_{02} = 1.256 \text{ bar} \\ T_{0e} &= T_{02} = 500 \text{ K} \end{aligned}$$

Further, we have

$$\frac{A_e}{A_2^*} = \frac{A_D}{A_2^*} = \frac{5A_1^*}{A_2^*}$$
$$= \frac{5}{1.649} \approx 3$$

When $\frac{A_e}{A_2^*} = 3$, the isentropic table for subsonic flow in the duct provides

$$M_e = 0.2$$
$$\frac{p_e}{p_{0e}} = 0.9725$$
$$\frac{T_e}{T_{0e}} = 0.9921$$

This gives,

$$p_e = \frac{p_e}{p_{0e}} \times p_{0e}$$
$$= 0.9725 \times 1.256 = 1.222 \text{ bar}$$
$$T_e = \frac{T_e}{T_{0e}} \times T_{0e}$$
$$= 0.9921 \times 500 = 496.05 \text{ K}$$

For the isentropic flow in the duct of constant area,

$$M_3 = M_e = 0.2$$
$$p_3 = p_e = 1.222 \text{ bar}$$
$$T_3 = T_e = 496.05 \text{ K}$$

(c) Because the duct discharges isentropically into the second container, the stagnation temperature is conserved.

$$\Longrightarrow \quad T_{04} = T_{01} = 500 \text{ K}$$

There is subsonic flow within the duct, which is always correctly expanded when discharging into the second container.

$$\Longrightarrow \quad p_4 = p_3 = 1.222 \text{ bar}$$

4.4 PRESSURE DROPS AND POWER REQUIREMENTS

The energy supplied to operate a wind tunnel is used in overcoming the total pressure drops in different sections of the tunnel. A fan or compressor adds power to the tunnel to keep the pressure ratio high enough to overcome the pressure drops in different sections, allowing the test section to run at a constant speed. The total pressure drops in a closed-circuit tunnel is the sum of the individual drops occurring in different components of the tunnel, described as follows:

- Drops due to friction in the contraction cone.
- Drops due to friction in the test section.
- Drops in the diffuser due to flow expansion.
- Drops across shock waves in case of a supersonic diffuser.
- Drops in the corners (with or without guide vanes).
- Drops in the return circuit.
- Drops due to heat exchange in cooling system.
- Drops owing to tunnel balancing drag.

It is interesting to note that pressure drops across a shock and model system contribute approximately 90% of the total pressure drop in a supersonic tunnel. The pressure drop across a shock wave alone accounts for roughly 80% of the total, with the remaining 10% due to tunnel balancing drag. As a result, the total pressure drop in the tunnel was only about 10% due to all other factors. Therefore, while estimating the overall pressure ratio across a supersonic tunnel, only the pressure ratio across the diffuser is taken into account, along with a correction factor, to compensate for pressure drops caused by the remaining factors. This results in the following:

$$\frac{(p_{0c})_{exit}}{(p_{0c})_{inlet}} = \frac{1}{CF}\frac{(p_{0d})_{inlet}}{(p_{0d})_{exit}} \tag{4.4.1}$$

where

$(p_{0c})_{inlet}$ = stagnation pressure at the compressor inlet

$(p_{0c})_{exit}$ = stagnation pressure at the compressor exit

$(p_{0d})_{inlet}$ = stagnation pressure at the diffuser inlet

$(p_{0d})_{exit}$ = stagnation pressure at the diffuser exit

and

$$CF = \text{Correction factor} = \frac{\text{Pressure drop in diffuser}}{\text{Total pressure drops in wind tunnel}}$$

It is worth noting that the correction factor through a supersonic diffuser is mostly determined by the shock structure through which the diffuser

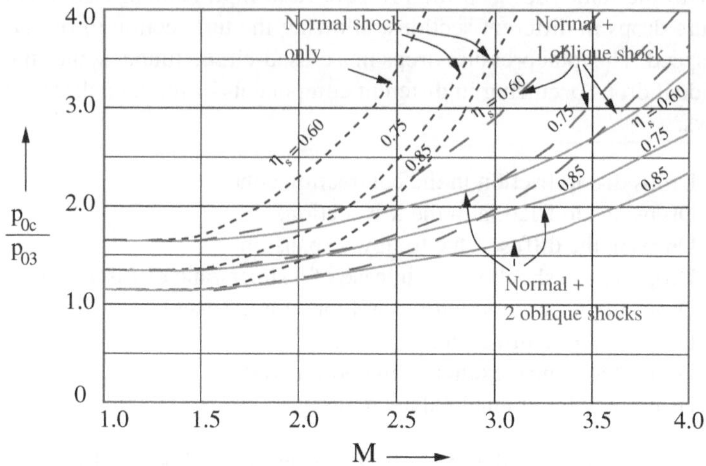

Figure 4.4.1 Variation of compressor pressure ratio with Mach number.

accomplishes pressure recovery and typically ranges from 0.6 to 0.85. In addition, $\frac{(p_{0d})_{inlet}}{(p_{0d})_{exit}}$ is the ratio of stagnation pressures across the diffuser; for a single shock or multiple shock diffuser, its values, together with correction factors, are indicated in Figure 4.4.1.

The minimum power required to compress a gas adiabatically in a multistage reciprocating compressor is

$$P = \frac{N\gamma}{\gamma - 1}\dot{m}RT_{inlet}\left[\left(\frac{(p_{0c})_{exit}}{(p_{0c})_{inlet}}\right)^{\frac{\gamma-1}{N\gamma}} - 1\right] \tag{4.4.2}$$

where P = power required for wind tunnel operation
$\dot{m}$ = mass flow rate of air
N = number of stages
γ = ratio of specific heats
T_{inlet} = static temperature at compressor inlet

Example 27. The test section of a supersonic wind tunnel has the dimensions of 50 cm × 75 cm. The tunnel is desired to operate at Mach 2.5, with a stagnation temperature of 339 K and test section static pressure of 35 kPa. If CF = 0.65, calculate the power required with a five-stage compressor to operate this tunnel. The diffuser is of the simple normal shock type.

Solution. Given,

$$M_{TS} = 2.5$$
$$A_{TS} = 0.5 \times 0.75 = 0.375\ m^2$$
$$T_0 = 339\ K$$
$$p_{TS} = 35\ kPa$$
$$N = 5$$

The stagnation pressure in the test section may be found using equation (2.7.15), where

$$p_0 = p_{TS}\left(1 + \frac{\gamma - 1}{2}M_{TS}^2\right)^{\frac{\gamma}{\gamma-1}}$$
$$= 35\left(1 + 0.2 \times 2.5^2\right)^{3.5}$$
$$= 598\ kPa$$

For Mach 2.5, the pressure ratio across the diffuser is

$$\frac{(p_{0d})_{exit}}{(p_{0d})_{inlet}} = \frac{p_{02}}{p_{01}} = \left(\frac{6M_1^2}{5 + M_1^2}\right)^{3.5}\left(\frac{6}{7M_1^2 - 1}\right)^{2.5} \quad 2.7.24$$
$$= \left(\frac{6 \times 2.5^2}{5 + 2.5^2}\right)^{3.5}\left(\frac{6}{7(2.5^2) - 1}\right)^{2.5}$$
$$= 0.499$$

Thus, the pressure ratio across the compressor is

$$\frac{(p_{0c})_{exit}}{(p_{0c})_{inlet}} = \frac{1}{CF}\frac{(p_{0d})_{inlet}}{(p_{0d})_{exit}}$$
$$= \frac{1}{0.75 \times 0.499}$$
$$= 2.672$$

The mass flow rate may be found using equation (4.3.4), where

$$\dot{m} = \frac{p_0 A^* \sqrt{\gamma}}{\sqrt{RT_0}}\left(\frac{\gamma + 1}{2}\right)^{-\frac{(\gamma+1)}{2(\gamma-1)}}$$
$$= \frac{598 \times 10^3 \times A^*}{\sqrt{339 \times 287}} \times \sqrt{1.4} \times (1.2)^{-3}$$
$$= 1312.7A^*$$

The throat area A^* may be found from the relation,

$$\frac{A}{A^*} = \frac{1}{M}\left[\frac{2}{\gamma+1}\left(1+\frac{\gamma-1}{2}M^2\right)\right]^{\frac{(\gamma+1)}{2(\gamma-1)}} \tag{2.7.17}$$

$$= \frac{1}{2.5}\left[\frac{5+2.5^2}{6}\right]^3 = 2.625$$

or

$$A^* = \frac{A}{2.625} = \frac{0.375}{2.625} = 0.143 \text{ m}^2$$

This gives

$$\dot{m} = 76.83A^*$$
$$= 1312.7 \times 0.143$$
$$= 187.715 \, ^{kg}/_{s}$$

Assuming $T_{inlet} = T_0$, the power required to operate the wind tunnel is

$$P = \frac{N\gamma}{\gamma-1}\dot{m}RT_{inlet}\left[\left(\frac{(p_{0c})_{exit}}{(p_{0c})_{inlet}}\right)^{\frac{\gamma-1}{N\gamma}} - 1\right]$$

$$= \frac{5\times 1.4}{0.4} \times 187.715 \times 287 \times 339\left[(2.672)^{\frac{0.4}{5\times 1.4}} - 1\right]$$
$$= 1846333569 \text{ W}$$
$$= 24749.78 \text{ hp}$$

Remark. The power required decreases depending on the static or stagnation pressure in the test section. Similarly, lowering the stagnation temperature reduces power usage as the square root of the absolute value.

4.5 PROBLEMS IN SUPERSONIC WIND TUNNEL OPERATION

In supersonic wind tunnel operation, condensation and liquefaction of air are two major sources of erroneous results.

4.5.1 CONDENSATION

The quantity of moisture that a unit volume of air can hold rises with temperature but is unaffected by changes in pressure. The temperature drops during testing, and the air may become supercooled[2] as it is isentropically expanded

[2]Cooled to a temperature below dew point temperature.

to higher Mach numbers in the test section. If the moisture content of the air is sufficiently high, condensation happens, resulting in thick fog in the test section. Whether the condensation phenomena would take place or not depends on the following factors:

1. Moisture content in the air.
2. Static temperature of the airstream.
3. Static pressure of the airstream.
4. The duration in which the airstream is at a low temperature.

It is worth emphasizing that the condensation can affect Mach number and other parameters in a flow, leading to inaccurate measurements in a wind tunnel test. The change in Mach number and other properties depends on the amount of heat released (latent heat) during condensation, and can be calculated using the following equations:

$$\frac{dM^2}{M^2} = \frac{1+\gamma M^2}{(1-M^2)}\left[\frac{dQ}{H} - \frac{dA}{A}\right] \tag{4.5.1}$$

$$\frac{dp}{p} = -\frac{\gamma M^2}{(1-M^2)}\left[\frac{dQ}{H} - \frac{dA}{A}\right] \tag{4.5.2}$$

where M = Mach number
γ = ratio of specific heats
dQ = heat released during condensation
H = enthalpy
A = duct area
p = static pressure

In supersonic flows, the Mach number decreases with increasing pressure, as shown by equations 4.5.1 and 4.5.2. The Mach number, on the other hand, increases with pressure in subsonic flows. Interestingly, water vapor in air does not affect the derivation of the temperature ratio, pressure ratio, and Mach number from isentropic relationships.

It is worth mentioning that condensation can be avoided by drying the working fluid before passing it through the test section. Another approach is to utilize stagnation heaters to raise the temperature.

4.5.2 LIQUEFACTION

When the appropriate temperature and pressure conditions are attained, the other components of air gradually liquefy, analogous to moisture condensation in an airstream chilled below its saturation point. It has been observed that when high-pressure air at room temperature is expanded, liquefaction issues can occur as early as Mach 4.

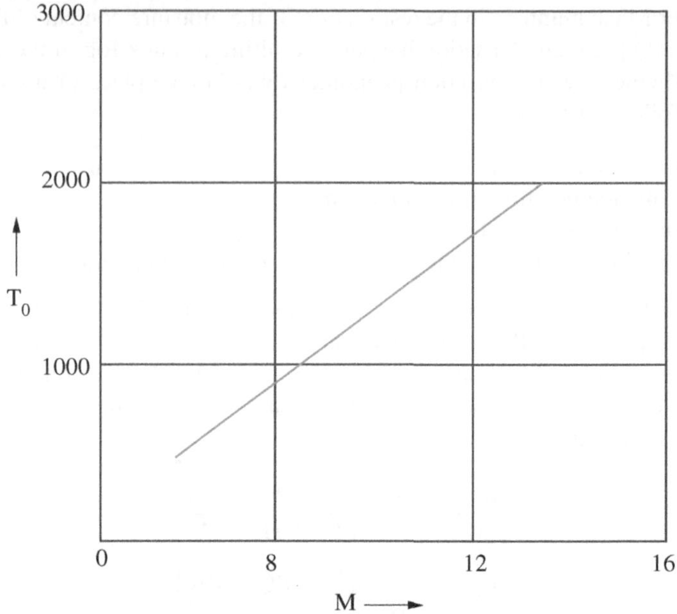

Figure 4.5.1 Necessary stagnation temperatures to prevent liquefaction at varying Mach numbers.

Figure 4.5.1 illustrates the necessary stagnation temperatures to prevent liquefaction at varying Mach numbers. Notice that the lowest temperature needed to prevent liquefaction at Mach 12 and above is roughly 2000 K.

4.6 SUMMARY

A fractional temperature change of 5% or more is considered significant in almost all the fluid flows, necessitating the thermal effects to be accounted for in the analysis. At sea level, more than 5% fractional change in temperature is equivalent to the flow velocity of 650 kmh^{-1} ($M \geq 0.5$). Because compressibility effects prevail at $M \geq 0.3$, flows of $M \geq 0.5$ are known as high-speed flows, and the related wind tunnel is known as the high-speed wind tunnel.

A high-speed wind tunnel on the basis of operating Mach number range in the test section is divided into transonic, supersonic, and hypersonic tunnels. A high-speed tunnel may also be categorized as intermittent or continuous operation depending on whether it is utilized for a short or prolonged run. The open-circuit intermittent tunnels are sub-classified as blowdown and indraft (or induction) tunnels. It is worth emphasizing that the most high-speed wind tunnels are intermittent because of the massive power requirements.

An intermittent blowdown wind tunnel can achieve a wide range of test section Mach numbers, from high subsonic to relatively high supersonic. An open-circuit blowdown tunnel obtains supersonic speeds in the test section by creating a pressure difference between a reservoir (through which it inducts the air) and the ambient. A closed blowdown tunnel, on the other hand, includes a high-pressure reservoir upstream and a comparatively low-pressure tank downstream.

An intermittent indraft wind tunnel achieves supersonic speeds in the test section by producing a vacuum at the downstream end while leaving the upstream end open to the surrounding environment. The opening of the valve eventually pulls air inside the tunnel at a desired vacuum level in the tank, establishing a supersonic flow in the test section.

A continuous supersonic wind tunnel, being a closed-circuit assembly, can achieve a wide range of Mach numbers in the test section. The flow in this tunnel does not exhaust the surrounding environment and returns to the test section along a predetermined path.

The following are the major parts of an intermittent blowdown tunnel:

1. Air receiver or storage tank
2. Settling chamber
3. Convergent-divergent (de Laval) or supersonic nozzle
4. Test section
5. Convergent-divergent or supersonic diffuser

The power required to drive the air through a wind tunnel is only required to overcome the total energy losses in the tunnel's various sections. Power is added to the system through fans or compressors, the purpose of which is to keep the pressure ratio sufficiently high to overcome energy losses in the system or resistance to flow in the tunnel at any speed and thus to maintain a given speed in the test section.

The energy supplied to operate a wind tunnel is used in overcoming the total pressure drops in different sections of the tunnel. A fan or compressor adds power to the tunnel to keep the pressure ratio high enough to overcome the pressure drops in different sections, allowing the test section to run at a constant speed. The total pressure drops in a closed-circuit tunnel is the sum of the individual drops occurring in different components of the tunnel, described as follows:

- Drops due to friction in the contraction cone.
- Drops due to friction in the test section.
- Drops in the diffuser due to flow expansion.
- Drops across shock waves in case of a supersonic diffuser.
- Drops in the corners (with or without guide vanes).
- Drops in the return circuit.

- Drops due to heat exchange in cooling system.
- Drops owing to tunnel balancing drag.

In supersonic wind tunnel operation, condensation and liquefaction of air are two major sources of erroneous results. The quantity of moisture that a unit volume of air can hold rises with temperature but is unaffected by changes in pressure. The temperature drops during testing, and the air may become supercooled as it is isentropically expanded to higher Mach numbers in the test section. If the moisture content of the air is sufficiently high, condensation happens, resulting in thick fog in the test section. Condensation can be avoided by drying the working fluid before passing it through the test section. Another approach is to utilize stagnation heaters to raise the temperature.

When the appropriate temperature and pressure conditions are attained, the other components of air gradually liquefy, analogous to moisture condensation in an airstream chilled below its saturation point. It has been observed that when high-pressure air at room temperature is expanded, liquefaction issues can occur as early as Mach 4.

EXERCISE PROBLEMS

Example 28. A supersonic jet flying at Mach 2.5 at 16000 m above sea level. How long after passing directly above a ground observer will the jet noise be audible to him?

Exercise 29. The airstream passes at a rate of 2 $^{kg}/_{s}$ through a 200 mm diameter frictionless duct. The static pressure in one point of the duct is 50 kPa. Determine the stagnation pressure, Mach number, and velocity at this point if the stagnation temperature in the duct is 350 K.

Example 30. The airstream flows steadily and isentropically through a Laval nozzle with throat conditions of 150 kPa and 80°C. The cross-section area of the throat is 400 cm^2. The pressure is 90 kPa at a certain point in the diverging portion of the nozzle. Find the area, Mach number, and velocity at this point.

Example 31. Air flows isentropically via a supersonic nozzle with a 10 cm^2 exit area. At 1 MPa and 300 K, the nozzle draws air from a huge reservoir. The nozzle exit static pressure is 965 kPa, and the Mach number at the throat is 0.75. Determine (a) the pressure and temperature at the throat, and (b) the Mach number at the nozzle exit. What is the nozzle exit Mach number and mass flow rate if the nozzle is intended for a backpressure of 75 kPa but is being operated at 45 kPa?

Example 32. A 2.5 m × 2.5 m supersonic wind tunnel is supposed to operate at Mach 2.2. Find the throat area of the nozzle. If all other losses are ignored except that occurs due to normal shock formation, what should be the throat area of the nozzle to start the tunnel? Air is inducted in the tunnel from a large tank at 101 kPa and 300 K. Calculate the pressure drop across the tunnel.

Example 33. It is intended to build a supersonic tunnel with a 40 cm × 40 cm test section that will operate at Mach 1.5 with a stagnation temperature of 325 K and a stagnation pressure of 175 kPa. Find (a) the horsepower required from a six-stage compressor to operate the tunnel if CF = 0.65, and (b) the pressure in the test section. The diffuser is of the normal shock type.

5 Hypersonic and Hypervelocity Facilities

5.1 HYPERSONIC FLOW – SPECIAL CHARACTERISTICS

At hypersonic Mach numbers, the kinetic energy of a flow is extraordinarily large in comparison to its thermal energy. Indeed, in hypersonic flows, the ratio of kinetic energy to thermal energy is directly proportional to the square of the Mach number and is roughly equal to $\frac{1}{2}\gamma(\gamma-1)\,M^2$ for a perfect gas. Interestingly, for an airstream traveling at Mach 10, the kinetic energy is thirty times more than the thermal energy. Consequently, when a hypersonic flow is decelerated, it generates a large amount of thermal energy, inducing a huge temperature rise and ultimately modifying the thermodynamic properties significantly. Moreover, the intense shock waves in the flow field create a significant increase in temperature, which further contributes to the change in thermodynamic characteristics. The air is considered a perfect gas at standard temperature and pressure, primarily composed of oxygen and nitrogen. However, it becomes imperfect at very high temperatures due to the excitation of vibrational modes and the dissociation of nitrogen and oxygen molecules. The molecular dissociation, being an endothermic process, causes a tremendous decrease in the temperature of an imperfect gas and changes the surface pressure distribution considerably.

The following are some of the basic characteristics of a hypersonic flow:

- Thin shock layer
- Entropy layer
- Viscous interactions
- High-temperature effects
- Low-density flow

5.1.1 THIN SHOCK LAYER

In supersonic flows, the shock angle (wave angle) decreases as the Mach number increases for a given flow turning angle (wedge angle). As a result, an attached oblique shock wave approaches the body surface with increasing Mach numbers. The shock layer is the flow zone enclosed by a shock wave at one end and the body surface at the other. A typical shock layer at Mach 28 is depicted schematically in Figure 5.1.1. A thin shock layer formed at very high Mach numbers increases the possibility of the shock wave interacting with the

DOI: 10.1201/9781003139447-5

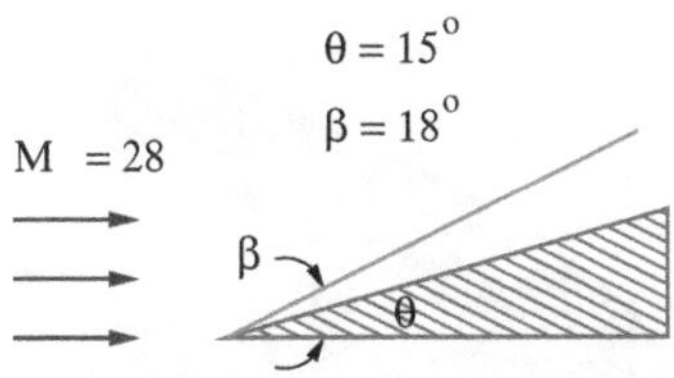

Figure 5.1.1 Schematic diagram of a typical shock layer at Mach 28.

boundary layer, a phenomenon known as shock-wave/boundary-layer interaction (SBLI). It is more prominent at low Reynolds numbers when the boundary layer is comparatively wider. Nonetheless, even for an inviscid flow without a boundary layer or a high Reynolds number flow with a very thin boundary layer, a flow domain with a thin shock layer is easily examined by the Newtonian flow model, which is often employed in hypersonic aerodynamics for approximate computations.

5.1.2 ENTROPY LAYER

SBLI can be reduced by restricting the development of a thin shock layer closer to the body surface in hypersonic flows. This is usually accomplished by changing the nose section of the body from a pointed to a blunt shape. Such a blunt-nosed body causes a detached bow shock formation ahead of the nose, as shown in Figure 5.1.2. A bow shock comprises a normal shock in the center and multiple oblique shocks around it, degenerating into Mach waves near theoretical infinity. The entropy of a flow through a shock wave increases, and this entropy increase intensifies with shock strength. As a result, the entropy increase along the central streamline passing through the normal shock is greater than those along the nearby streamlines traveling through the peripheral and relatively weaker oblique shocks. An entropy gradient exists downstream of the bow shock, resulting in an entropy layer with vorticity all along the body surface. Because the boundary layer grows beneath the entropy layer, it must interact with the vorticity in the latter.

5.1.3 VISCOUS INTERACTIONS

When a fluid flows past a body, a thin viscous layer, also known as the boundary layer, forms on the surface. The viscous action of the boundary layer slows a hypersonic flow with enormous kinetic energy. The decrease in kinetic energy increases the flow's internal energy, eventually raising the boundary layer temperature. This process, known as viscous dissipation, significantly affects the boundary layer characteristics. The viscosity of the gaseous fluid in the

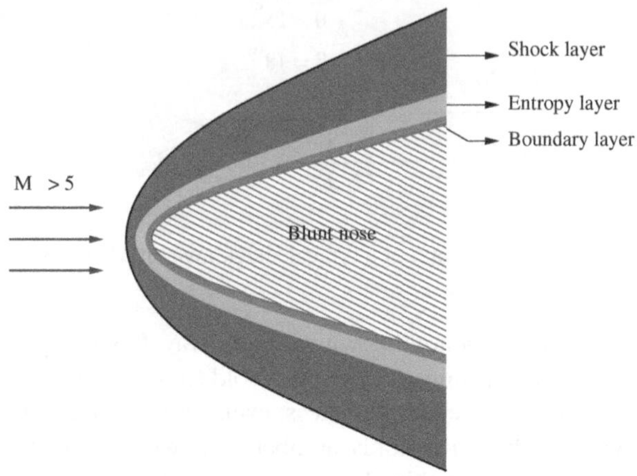

Figure 5.1.2 Schematic representation of an entropy layer (full of vorticity) in a hypersonic flow.

boundary layer grows with temperature, increasing the thickness of the boundary layer. Furthermore, as the temperature rises with constant pressure in the boundary layer, the density increases according to the equation of state. Consequently, the boundary layer thickness increases to maintain the same mass flow rate at a lower density.

The boundary layer theory for incompressible flows assumes the viscous effects to be confined within a thin layer on the surface and that the outer inviscid flow stays untouched by this layer. In other words, the boundary layer in an incompressible flow does not interact with the outer inviscid flow. On the other hand, due to its larger displacement effects, the thick boundary layer in a hypersonic flow considerably affects the outer inviscid flow properties. The changes in outer flow characteristics further change the boundary layer characteristics, and their interaction continues, creating a feedback loop between them. This mutual interaction, usually called the viscous-inviscid interaction, is of two types. The first instance, where an extremely thick boundary layer interacts with the outer flow, is called the pressure interaction. In the second instance, a shock impinges upon the boundary layer and imposes an unfavorable pressure gradient on it, causing either retardation or separation of the boundary layer. This interaction is known as shock-wave/boundary-layer interaction. A viscous-inviscid interaction modifies the surface pressure distribution and, consequently, the lift and drag on a body. It also alters the frictional and heat transport properties of a body.

Figure 5.1.3 depicts a boundary layer beneath the shock layer over a thin flat plate in a hypersonic flow. At a distance x from the leading edge, the boundary

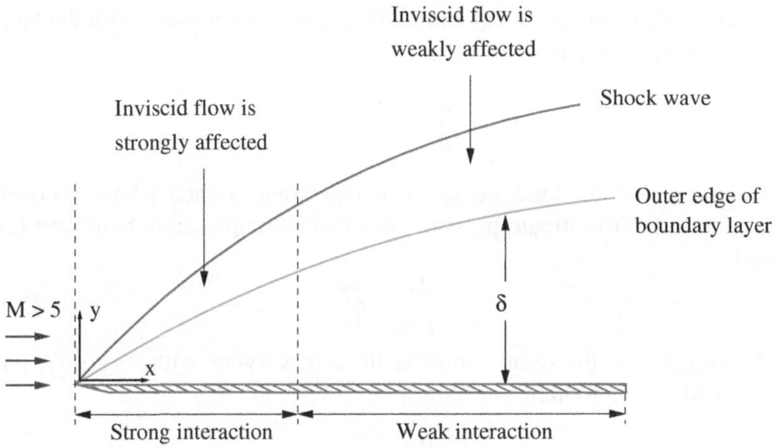

Figure 5.1.3 Schematic diagram of a boundary layer above a thin flat plate in hypersonic flow.

layer thickness, δ, is given by:

$$\delta \propto \frac{x}{\sqrt{Re_x}} \tag{5.1.1}$$

where $Re_x \left(= \frac{\rho v x}{\mu}\right)$ is the local Reynolds number at a distance x. Based on the wall temperature, T_w, the local Reynolds number can also be specified as:

$$Re_x = \frac{\rho_w u_w x}{\mu_w} \tag{5.1.2}$$

where ρ_w = flow density at T_w

μ_w = coefficient of viscosity at T_w

u_w = flow velocity in the boundary layer at T_w. From equation (5.1.1) and equation (5.1.2), we get

$$\delta \propto \frac{x}{\sqrt{u_w x}}\sqrt{\frac{\mu_w}{\rho_w}} = \frac{x}{\sqrt{\frac{\rho_\infty u_\infty x}{\mu_\infty}}}\sqrt{\frac{\rho_\infty}{\rho_w}}\sqrt{\frac{\mu_w}{\mu_\infty}} \tag{5.1.3}$$

$$\Longrightarrow \quad \delta \propto \frac{x}{\sqrt{Re_\infty}}\sqrt{\frac{\rho_\infty}{\rho_w}}\sqrt{\frac{\mu_w}{\mu_\infty}} \tag{5.1.4}$$

where $Re_\infty \left(= \frac{\rho_\infty u_\infty x}{\mu_\infty}\right)$ is the freestream Reynolds number at a distance x

ρ_∞ = flow density at freestream temperature T_∞

μ_∞ = coefficient of viscosity at T_∞

The density ratio, $\frac{\rho_\infty}{\rho_w}$, in equation (5.1.4) can be expressed, with the help of perfect gas equation, as follows:

$$\frac{\rho_\infty}{\rho_w} = \frac{p_\infty}{p_w}\frac{T_w}{T_\infty} \tag{5.1.5}$$

Assuming that the static pressure in a hypersonic boundary layer is constant and equal to the freestream pressure, as in an incompressible boundary layer, we get:

$$\frac{\rho_\infty}{\rho_w} = \frac{T_w}{T_\infty} \tag{5.1.6}$$

Assuming that the temperature is linearly varying with viscosity, equation (5.1.6) can be written as follows:

$$\frac{\mu_w}{\mu_\infty} = \frac{T_w}{T_\infty} \tag{5.1.7}$$

Substituting equations (5.1.6) and (5.1.7) into equation (5.1.4) results in

$$\delta \propto \frac{x}{\sqrt{Re_\infty}}\frac{T_w}{T_\infty} \tag{5.1.8}$$

The wall temperature, T_w, for a no-slip wall under adiabatic conditions is almost the same as the freestream stagnation temperature. Hence, the temperature ratio, $\frac{T_w}{T_\infty}$, can be represented using isentropic relations as follows:

$$\frac{T_w}{T_\infty} = 1 + \frac{\gamma - 1}{2}M_\infty^2 \tag{5.1.9}$$

Therefore, equation (5.1.8) implies

$$\delta \propto \frac{M_\infty^2}{\sqrt{Re_\infty}}x \tag{5.1.10}$$

From equation (5.1.10), it is clear that the thickness of the boundary layer is directly proportional to the square of the freestream Mach number. This means that at hypersonic Mach numbers, the boundary layer thickens significantly.

5.1.4 HIGH-TEMPERATURE EFFECTS

Another feature of hypersonic flow is a significant increase in temperature. The excessive kinetic energy dissipation of a high-speed flow over a surface generates a large amount of frictional heat. As a result, the vibrational mode of molecular motion is activated at extremely high temperatures caused by viscous dissipation in a hypersonic boundary layer, which may result in

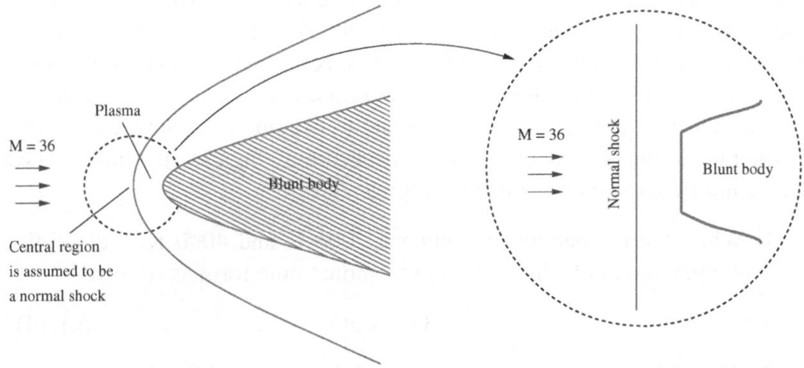

Figure 5.1.4 Atmospheric reentry of a spacecraft.

gas molecule dissociation or ionization. This results in a chemically reacting boundary layer which soaks the surface of a hypersonic vehicle.

Consider a spacecraft reentering into the atmosphere at Mach 36 at the sea-level altitude of 60 km (Figure 5.1.4). As stated previously, the central region of a bow shock behaves similarly to a normal shock, which gives the static temperature ratio across it at Mach 36 as follows:

$$\frac{T_2}{T_1} = 253 \tag{5.1.11}$$

The static temperature at 60 km above sea level is

$$T_1 = -15^{\circ}C + 273 = 258\ K \tag{5.1.12}$$

Thus, the static temperature after the shock is

$$T_2 = 253 \times 258 = 65274\ K \tag{5.1.13}$$

In essence, long before this temperature is reached, the air molecules dissociate and ionize. The shock layer transforms into a partially ionized plasma in which the specific heat of the fluid becomes a strong function of both pressure and temperature. As a result, the previous assumption of a calorically perfect gas must be revised. However, if the chemically reacting flow is computed correctly, the temperature behind the shock wave would be roughly 11000 K – still a high figure at which the assumption of a perfect gas is incorrect.

There are two thermodynamic properties that are primarily responsible for air to become imperfect at high temperatures. The oxygen and nitrogen gases in the air usually stay diatomic at room temperature. However, as the temperature rises well above room temperature, the vibrational motion of molecules becomes active, absorbing a fraction of the energy that would otherwise be used for translational and rotational motions. As a result of the activation of the vibrational motion of molecules, the specific heat of the air becomes a function of temperature. When the temperature rises sufficiently, the diatomic oxygen and nitrogen molecules dissociate to become their monatomic forms. These monatomic gases get ionized at exceedingly high temperatures. The air molecules dissociate and ionize as follows:

1. When the temperature is between 2000 K and 4000 K, a diatomic oxygen molecule dissociates to its monatomic form as follows:

$$O_2 \rightarrow 2O \tag{5.1.14}$$

2. When the temperature is between 4000 K and 9000 K, a diatomic nitrogen molecule dissociates to its monatomic form as follows:

$$N_2 \rightarrow 2N \tag{5.1.15}$$

   It is important to note that molecular dissociation is an endothermic process that absorbs energy. The temperature range denotes that a gas (oxygen or nitrogen) begins to dissociate at the lower limit and completely dissociates at the upper limit. The air eventually becomes a mixture of monatomic and diatomic gases over the full temperature range. Nevertheless, the composition of monatomic particles increases as temperature rises.
3. When the temperature is above 9000 K, the monatomic oxygen and nitrogen ionize as follows:

$$O \rightarrow O^+ + e^- \tag{5.1.16}$$

$$N \rightarrow N^+ + e^- \tag{5.1.17}$$

   Likewise a molecular dissociation, the ionization of gases is also an endothermic process that absorbs energy.
4. When the temperature exceeds 9000 K, air contains both ionized and unionized gas particles. However, the content of ionized particles increases as temperature rises. At higher temperatures, some more chemical reactions may occur, such as the formation of nitrous oxide from the reaction of monatomic nitrogen and oxygen particles.

The preceding explanation reveals that at high Mach numbers, the temperature rise behind the shock wave can be significant enough to cause changes in c_p and $c_\forall$ due to gas dissociation or even ionization. The specific heats c_p and $c_\forall$, as well as their ratio γ, are no longer constants and have become temperature dependent. At temperatures of about 800 K, variations in γ must be considered. Because the air downstream of the bow shock is in the plasma state, normal shock relationships are inappropriate for flow analysis.

5.1.5 LOW-DENSITY FLOW

The mean free path for sea level air molecules is only about 0.1 μm. Thus, if an astronaut walks through the air on the earth's surface, the air itself would give him the impression of a continuous medium. A hypersonic vehicle, on the other hand, normally flies in the outer atmosphere, where air density is substantially lower. As a result, the mean free path is significantly greater than at sea level. The mean free path for air molecules, for example, is nearly 0.1 m at an altitude of around 100 km. If an astronaut walks through the air at this altitude, he will be able to feel the individual collisions of molecules, and the air will no longer appear to him as a continuous medium. The air molecules are widely spaced and begin to preserve their identities. The governing equations, aerodynamic principles, and results based on the continuum hypothesis become invalid, and the flow analysis is performed employing gas kinetics concepts. The associated flow domain is known as low-density flow.

The similarity parameter that characterizes a low-density flow is the Knudsen number (given as $\mathrm{Kn} = \frac{\lambda}{\ell}$, where ℓ is a characteristic dimension of a body). Based on the Knudsen number, flow regimes are classified into three types:

1. Continuum ($\mathrm{Kn} < 0.1$)
2. Transitional flow ($0.01 < \mathrm{Kn} < 1.0$)
3. Free molecular flow ($1.0 < \mathrm{Kn} < \infty$).

5.2 HYPERSONIC WIND TUNNEL WITH AIR HEATER

A hypersonic wind tunnel is built to obtain flow speeds of Mach 5 and higher in its test section (Figure 5.2.1). Although the layout and functioning of a hypersonic tunnel are analogous to those of a supersonic tunnel, the speed in the test section is substantially higher than that of a supersonic tunnel, which typically ranges between Mach 5 and 15. To store high-pressure air in tanks, a multistage compressor is used. The number of compressor stages is determined by the required stagnation pressure at the compressor exhaust. The pressurized air is then sent to the storage tanks after passing through the dryer.

The dry air is expanded in a specially designed hypersonic nozzle[1], which causes it to liquefy[2]. Therefore, it is pre-heated to a high enough temperature to prevent flow condensation in the test section. In a hypersonic tunnel,

---

[1]Appropriately shaped axisymmetric convergent-divergent nozzles achieve Mach 5 and higher flow speeds in the test section of hypersonic tunnels. The nozzles are bent properly to ensure a uniform flow at the exit. A nozzle used to achieve speeds of Mach 10 or more has an extremely narrow throat area, making fabrication difficult. The throat is routinely cooled with water to tolerate severe temperatures. To strengthen the throat while retaining good heat conductivity, alloys of beryllium, copper, titanium, zirconium, and molybdenum are extensively employed as lining materials.

[2]It should be noted that whenever the static temperature drops below 90 K, air tends to liquefy.

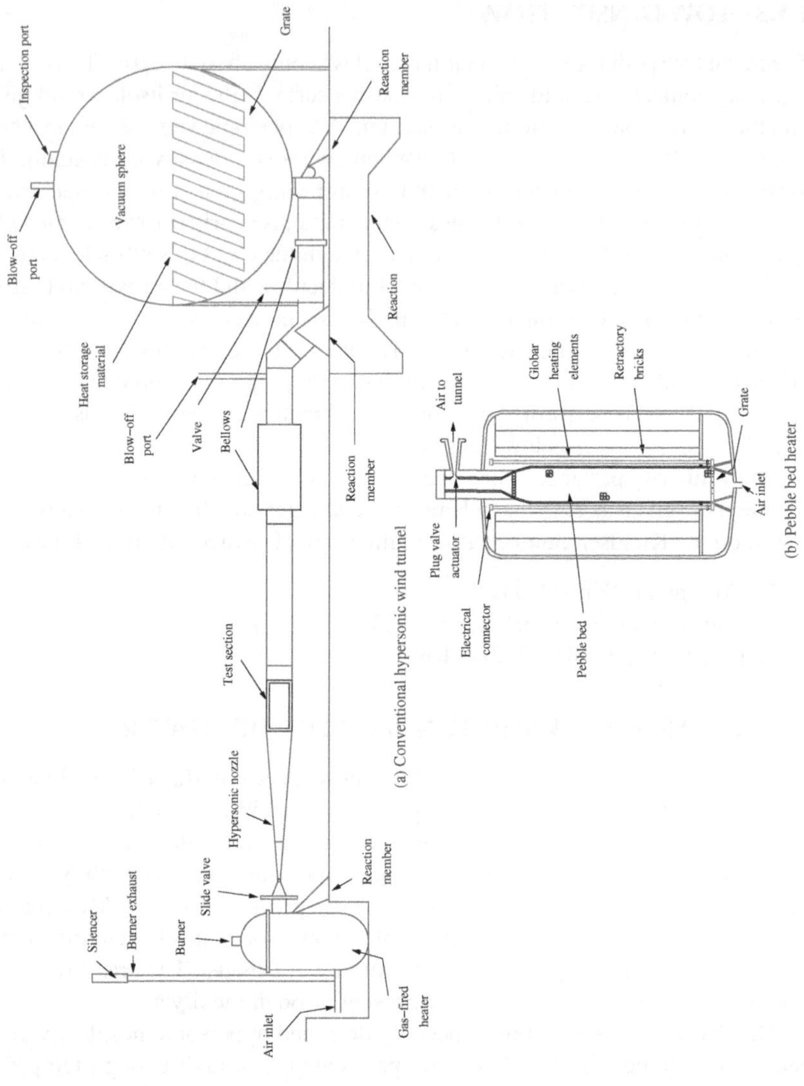

Figure 5.2.1 Schematic diagram of a hypersonic wind tunnel with air heater.

several heating mechanisms, such as combustors, electric resistance, and arc jets, are typically used for this purpose. Air pre-heating not only minimizes the probability of condensation but also cuts electricity consumption. Using lower boiling point fluids instead of air in some tunnels also aids in achieving condensation-free hypersonic flow.

5.3 HYPERVELOCITY TUNNELS

The previous articles discuss wind tunnels that can generate a range of flow speeds from low to moderately high levels. In many engineering applications, a flow speed much higher than these conventional tunnels is desired to test the models of interest. Such high speeds are achieved in specifically designed tunnels called hypervelocity tunnels. Although these non-conventional facilities obtain very high testing speeds, the other corresponding flow parameters, such as run time, are proportionately low. Some of the extensively used hypervelocity tunnels, like, the Ludwieg tube, hotshot tunnel, plasma arc tunnel, shock tube, shock tunnel, and gun tunnel, are briefly discussed in the following articles.

5.3.1 LUDWIEG TUBE

A Ludwieg tube, named after its inventor Hubert Ludwieg who first proposed its concept in 1955, is an affordable and effective method of generating supersonic flow. It is a wind tunnel in which flow speeds of up to Mach 4 can be easily achieved without the use of extra heating systems. A conventional Ludwig tube is schematically represented in Figure 5.3.1. The figure shows a long cylindrical tube connected to a hypersonic nozzle through a pressure-regulating valve. The tube has a cross-sectional area much larger than the nozzle throat area. The nozzle is followed by a test section, separated from a large dump tank by a thin membrane. The tube-nozzle assembly is filled with air at high-pressure. As the diaphragm ruptures, a shock wave travels toward the dump tank, and the expansion fans travel opposite toward the nozzle and the tube. The unsteady movement of expansion fans in the tube produces a steady

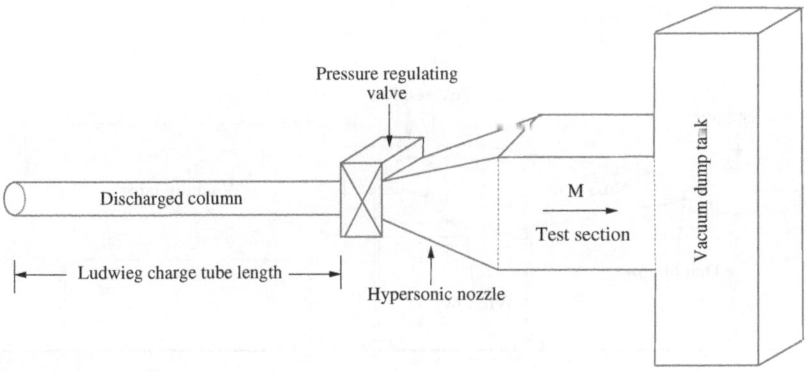

Figure 5.3.1 Schematic diagram of a conventional Ludwieg tube.

subsonic flow toward the nozzle and gets expanded to supersonic speeds. Between the opposite traveling shock and expansion waves, a steady supersonic flow is produced for a short duration, limited by the reflection of expansion fans from the tube's upstream end and arriving at the nozzle again. In a conventional Ludwieg tube, a testing time of around 100 milliseconds is normally achieved, which is sufficient enough for testing.

5.3.2 HOTSHOT TUNNEL

A hotshot tunnel is a test facility for producing high-velocity flows at extremely high pressures and temperatures for a brief period of time. A conventional hotshot tunnel is schematically shown in Figure 5.3.2. The high-pressure and -temperature conditions in the test section are produced by introducing a large-amount of electrical energy to a small mass of air in the arc chamber, which expands through the nozzle and enters the test section at high velocities. The arc chamber is generally kept at a maximum pressure of roughly 2800 bar, while the rest of the tunnel circuit is kept at very low pressure or vacuum. A thin metal plate separates the high- and low-pressure sections of the tunnel. Electric energy is discharged for a few milliseconds from a capacitance or inductance-based system, raising pressure and temperature in the arc chamber. This causes the diaphragm to rupture, allowing high-pressure, high-temperature air to expand through the nozzle and deliver high-speed flow in the test section. The high-speed flow typically lasts 10 to 100 milliseconds and is unsteady due to the arc chamber's progressive decline in pressure and temperature. A high-speed flow stops when a shock wave propagates through the tunnel toward the vacuum tank and is reflected by expansion fans traveling in the opposite direction to reach the model in the test section. It is important to note that data collecting in a hotshot tunnel is challenging due to the short run time.

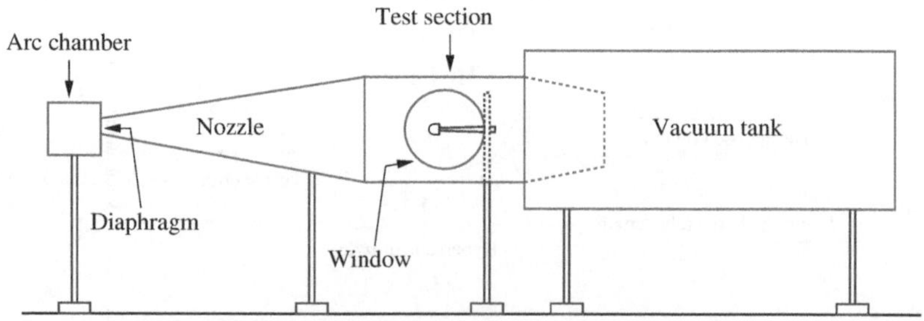

Figure 5.3.2 Schematic diagram of a conventional hotshot tunnel.

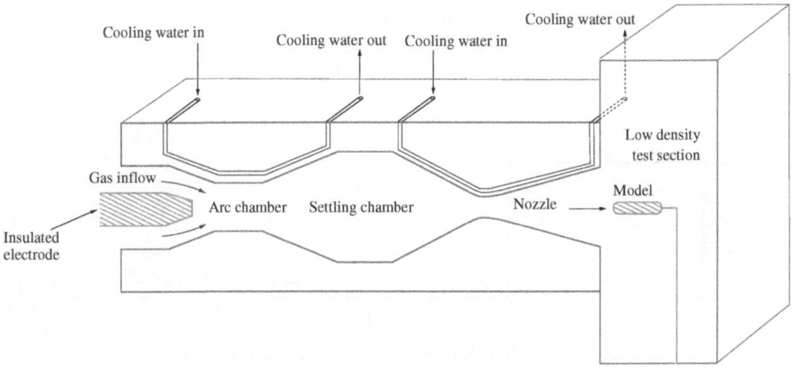

Figure 5.3.3 Schematic diagram of a conventional plasma arc tunnel.

5.3.3 PLASMA ARC TUNNEL

Another hypervelocity facility for producing supersonic flow in the test section is a plasma arc or wind tunnel. As shown in Figure 5.3.3, a conventional plasma arc tunnel consists of an arc chamber, a settling chamber, a convergent-divergent nozzle, a test section, and a low-pressure or -vacuum tank. The tunnel uses a strong electric current in the arc chamber to heat the air to an extremely high temperature of roughly 12000 K, yielding plasma. It is then fetched to a high-pressure settling chamber before expanding through the nozzle to deliver a supersonic flow in the test section. Using both direct and alternating currents, a testing time of several minutes is easily produced.

Argon is frequently utilized in place of air to achieve a higher degree of ionization. A plasma arc tunnel is especially effective for studying reentry vehicles that entail significant heat transfer rates. Furthermore, the surface ablation test can only be performed in this tunnel, which is nearly impossible in low-temperature or high-temperature short-duration tunnels.

5.3.4 SHOCK TUBE

A shock tube and its various developments have gained prominence for a long time as a device for the study of hypersonic flow. This instrument has an advantage over the wind tunnel in that it can replicate both the Mach number and the temperature. The downside is that overall testing time is very short, on the order of one millisecond, necessitating specialized instrumentation to measure the flow parameters. However, methodologies for physical measurements have been developed in a short period, with relatively acceptable results.

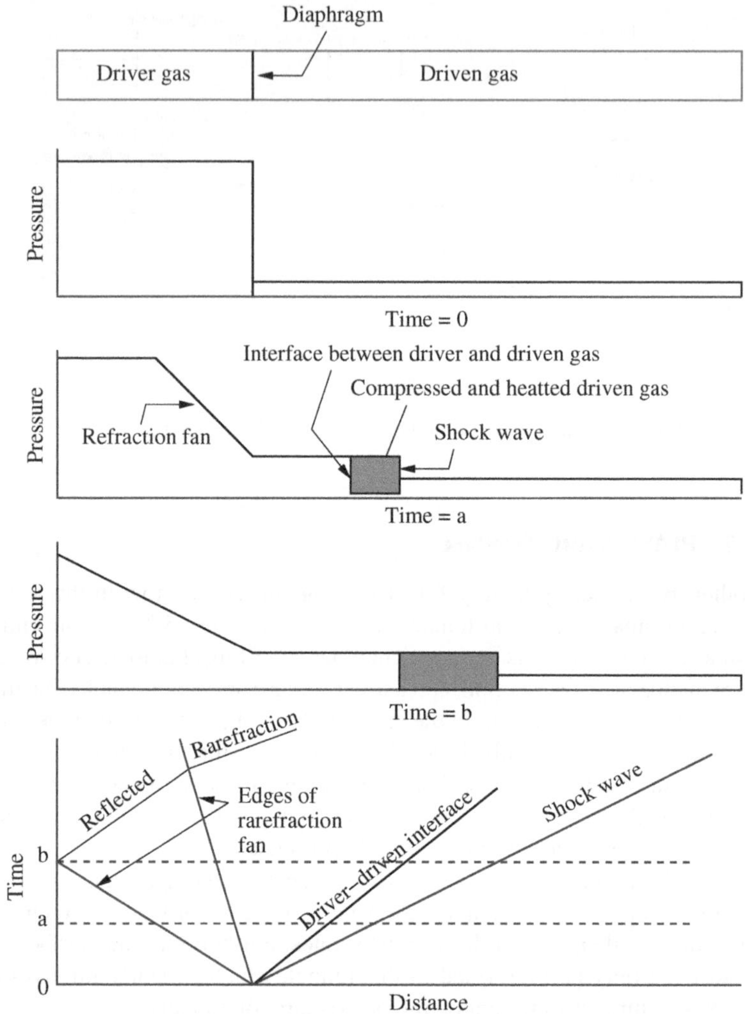

Figure 5.3.4 Pressure variation and shock wave propagation in a shock tube.

A shock tube is made up of a long straight tube with a uniform cross-section divided by a thin diaphragm that splits the tube into two compartments and creates a pressure ratio between them. The high-pressure chamber or driver section is one compartment, and the low-pressure chamber or driven section is the other. A conventional shock tube is schematically shown in Figure 5.3.4. The driving section is generally pressurized, whilst the driven section may be evacuated or at ambient pressure. Normally, the gases in both sections are in thermal equilibrium. The wave system shown in Figure 5.3.4 is established upon rupturing the diaphragm. In theory, when the diaphragm is removed, a shock wave propagates into the low-pressure section, and the high-pressure gas expands into the driven zone through a rarefaction wave centered at the origin. In practice, this is not true since the rarefaction wave is not centered at the origin, and the shock wave takes a finite amount of time to emerge from a succession of compression waves formed when the diaphragm is withdrawn. However, these variations from the ideal theory are minor for practical reasons, and the shock is assumed to proceed at a constant velocity after it has originated.

The shock wave compresses and heats the gas in the driven section, while the rarefaction wave expands and cools the driver gas. The two bodies of gas in the shock tube are brought to the same pressure and have the same particle velocity, however, their temperature, density, and entropy are different due to separate formative processes. The interface or contact surface, which is more of an area than a plane surface, separates these two states. This contact surface, also known as the contact front, moves at a constant velocity called particle velocity. As a result, there is an area of steady flow at high pressure and temperature behind the shock, while the flow behind the contact surface is steady with the same pressure but at a lower temperature. Because of the steady-state characteristics of these flow regimes, several approaches for using them in aerodynamic testing have evolved. However, the duration of the steady nature of these flow zones is short, resulting in a limited testing time of only a few milliseconds in shock tubes of adequate length.

5.3.5 SHOCK TUNNEL

Even though a shock tube can produce the required temperature, the highest Mach number is too low for hypersonic experiments. This problem is solved by creating a shock tunnel, in which a convergent-divergent nozzle is added to the downstream end of a conventional shock tube to increase the flow behind the shock wave to a higher Mach number. Thus, selecting the appropriate shock strength and nozzle area ratio makes it possible to replicate both temperature and Mach number. As shown in Figure 5.3.5, a conventional hypersonic shock tunnel consists of pressurized gas cylinders, a shock tube, a convergent-divergent or expansion nozzle, a test section, and a dump tank/vacuum system assembly.

As discussed in section 5.3.4, a shock tube works on the principle of shock-induced compression of the test gas by generating moving shocks when a high-pressure gas is suddenly exposed to the low-pressure section. A metal diaphragm separates the high-pressure driver section and the low-pressure driven section. In the driver section, a gas (normally helium) is filled at sufficiently

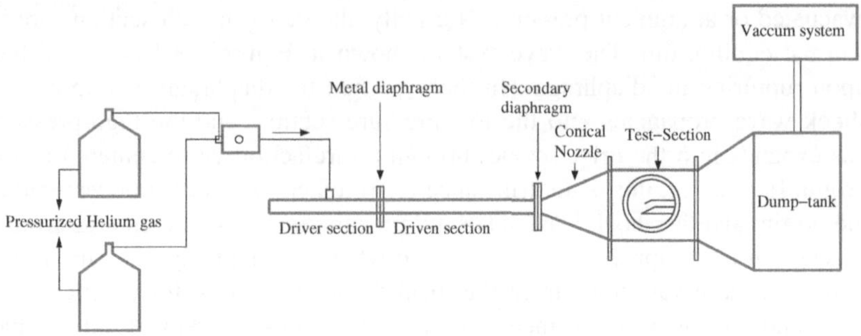

Figure 5.3.5 Schematic view of a hypersonic shock tunnel.

high pressure at which the diaphragm ruptures. Besides, the driven section consists of test gas at a pressure lower than the atmospheric pressure. One end of the shock tube is closed, and the other is connected to a conical convergent-divergent nozzle. Normally, the diverging section of the nozzle is longer than the converging section to keep the divergence angle small. This essentially helps reduce or mitigate the flow separation at the inner wall of the diverging section. Depending upon a high or low Mach number range, the nozzle would operate with the throat insert (reflected mode) or without the throat insert (straight-through mode). Initially, the driven section of the shock tube is separated from the nozzle utilizing a secondary diaphragm. The secondary diaphragm is normally made of tracing paper; hence, it is known as the paper diaphragm. The sole purpose of the secondary diaphragm is to separate the test gas from other sections (nozzle, test section, and dump tank assembly) that operate under a vacuum. The secondary diaphragm ruptures when the pressure in the driven section is slightly higher than the atmospheric pressure (soon after the rupture in the primary diaphragm). High-vacuum pressure is maintained in the nozzle, test section, and dump tank assembly to achieve a high-pressure difference between the nozzle inlet and exit sections. Generally, the vacuum pressure level in the dump tank is 10^{-8} bar.

The Helium gas from the pressurized tank is discharged into the driver section at a predetermined pressure, controlled by the primary diaphragm. At a certain pressure, the primary diaphragm ruptures, and the high-pressure driver gas comes into contact with the driven section, which is nearly vacuum. The sudden expansion of the high-pressure gas into a vacuum creates shock and expansion waves. The shock wave propagates toward the end of the driven section and gets reflected. The primary and the reflected shock waves increase the temperature and pressure of the test gas (air is used in the present study). This high-pressure and -temperature test gas are expanded through a convergent-divergent nozzle to achieve the required hypersonic Mach number (Figure 8.3.2). Due to the high-temperature of the driver gas, a shock tunnel can stimulate a much higher enthalpy than that produced in the wind tunnel. The enthalpy can be altered by varying the driver gas pressure.

5.3.6 GUN TUNNEL

The operation of a gun tunnel is quite similar to that of a shock tunnel. It can reach a Mach number of 25 or higher in the test section for a few milliseconds. The operating fluid is air, which is heated and pressurized in the tunnel. A gun tunnel, as shown in Figure 5.3.6, comprises three sections: the driver section, the driven section, and the test section. There is a high-pressure membrane between the diver and driven sections and a low-pressure membrane between the driven and test sections. As the high-pressure membrane ruptures, a lightweight piston deployed in the driven tube accelerates at supersonic speeds and compresses the fluid in the tube. This causes a shock wave to propagate toward the driven tube's downstream end, raising the fluid temperature. The shock wave is reflected from the low-pressure diaphragm and travels in the opposite direction of the piston, heating the fluid and increasing its pressure even

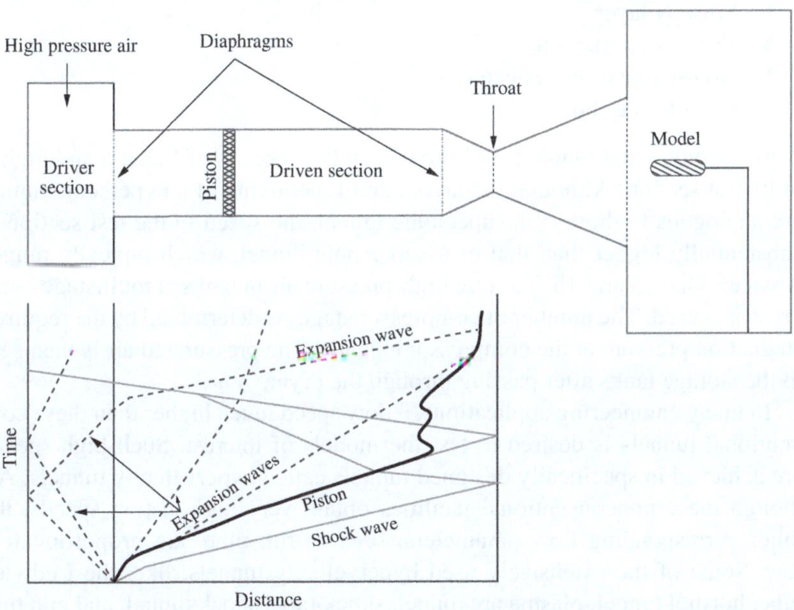

Figure 5.3.6 Schematic layout of a gun tunnel together with wave diagram.

more. The high-temperature pressurized fluid punctures the low-pressure diaphragm and enters the nozzle, resulting in supersonic flow in the test section. A common gun tunnel provides enough test time for all practical applications; nevertheless, its utility is limited due to the temperature rise achieved with a given piston geometry. A typical gun tunnel reaches a maximum temperature of around 2000 K.

5.4 SUMMARY

At hypersonic Mach numbers, the kinetic energy of a flow is extraordinarily large in comparison to its thermal energy. Indeed, in hypersonic flows, the ratio of kinetic energy to thermal energy is directly proportional to the square of the Mach number and is roughly equal to $\frac{1}{2}\gamma(\gamma-1)\,M^2$ for a perfect gas. Interestingly, for an airstream traveling at Mach 10, the kinetic energy is thirty times more than the thermal energy. Consequently, when a hypersonic flow is decelerated, it generates a large amount of thermal energy, inducing a huge temperature rise and ultimately modifying the thermodynamic properties significantly.

The following are some of the basic characteristics of a hypersonic flow:

- Thin shock layer
- Entropy layer
- Viscous interactions
- High-temperature effects
- Low-density flow

A hypersonic wind tunnel is built to obtain flow speeds of Mach 5 and higher in its test section. Although the layout and functioning of a hypersonic tunnel are analogous to those of a supersonic tunnel, the speed in the test section is substantially higher than that of a supersonic tunnel, which typically ranges between Mach 5 and 15. To store high-pressure air in tanks, a multistage compressor is used. The number of compressor stages is determined by the required stagnation pressure at the compressor exhaust. The pressurized air is then sent to the storage tanks after passing through the dryer.

In many engineering applications, a flow speed much higher than these conventional tunnels is desired to test the models of interest. Such high speeds are achieved in specifically designed tunnels called hypervelocity tunnels. Although these nonconventional facilities obtain very high testing speeds, the other corresponding flow parameters, such as run time, are proportionately low. Some of the extensively used hypervelocity tunnels, like, the Ludwieg tube, hotshot tunnel, plasma arc tunnel, shock tube, shock tunnel, and gun tunnel, are briefly discussed in this chapter.

A Ludwieg tube, named after its inventor Hubert Ludwieg who first proposed its concept in 1955, is an affordable and effective method of generating supersonic flow. It is a wind tunnel in which flow speeds of up to Mach 4 can be easily achieved without the use of extra heating systems.

A hotshot tunnel is a test facility for producing high-velocity flows at extremely high pressures and temperatures for a brief period of time. The high-pressure and -temperature conditions in the test section are produced by introducing a large amount of electrical energy to a small mass of air, which expands through the nozzle and enters the test section at high velocities.

Another hypervelocity facility for producing supersonic flow in the test section is a plasma arc or wind tunnel. A conventional plasma arc tunnel consists of an arc chamber, a settling chamber, a convergent-divergent nozzle, a test section, and a low-pressure or -vacuum tank. The tunnel uses a strong electric current in the arc chamber to heat the air to an extremely high temperature of roughly 12000 K, producing plasma. It is then fetched to a high-pressure settling chamber before expanding through the nozzle to deliver a supersonic flow in the test section. Using both direct and alternating currents, a testing time of several minutes is easily produced.

A shock tube and its various developments have gained prominence for a long time as a device for the study of hypersonic flow. This instrument has an advantage over the wind tunnel in that it can replicate both the Mach number and the temperature. The downside is that overall testing time is very short, on the order of one millisecond, necessitating specialized instrumentation to measure the flow parameters. However, methodologies for physical measurements have been developed in a short period, with relatively acceptable results.

Even though a shock tube can produce the required temperature, the highest Mach number is too low for hypersonic experiments. This problem is solved by creating a shock tunnel, in which a convergent-divergent nozzle is added to the downstream end of a conventional shock tube to increase the flow behind the shock wave to a higher Mach number. Thus, selecting the appropriate shock strength and nozzle area ratio makes it possible to replicate both temperature and Mach number. A conventional hypersonic shock tunnel consists of pressurized gas cylinders, a shock tube, a convergent-divergent or expansion nozzle, a test section, and a dump tank/vacuum system assembly.

The operation of a gun tunnel is quite similar to that of a shock tunnel. It can reach a Mach number of 25 or higher in the test section for a few milliseconds. The operating fluid is air, which is heated and pressurized in the tunnel. A gun tunnel comprises three sections: the driver section, the driven section, and the test section. There is a high-pressure membrane between the diver and driven sections and a low-pressure membrane between the driven and test sections. As the high-pressure membrane ruptures, a lightweight piston deployed in the driven tube accelerates at supersonic speeds and compresses the fluid in the tube. This causes a shock wave to propagate toward the driven tube's downstream end, raising the fluid temperature. The shock wave is reflected from the low-pressure diaphragm and travels in the opposite direction of the piston, heating the fluid and increasing its pressure even more. The high-temperature pressurized fluid punctures the low-pressure diaphragm and enters the nozzle, resulting in supersonic flow in the test section.

6 Supersonic Open Jet Facility at IIT Kharagpur

6.1 INTRODUCTION

In wind tunnels, models are often tested in a test section surrounded by walls. It is often preferable to test the model in an open environment with no diffusers ahead of the test section and an open investigating zone. An open test section wind tunnel is typical of the free jet design, with air at the desired initial pressure delivered through a three-dimensional de Laval nozzle that exhausts the atmosphere. Test models can be placed in the supersonic stream at the nozzle exit. The supersonic open jet facility is a configuration with an open test section for producing supersonic flow. The open jet can, however, be enclosed by an anechoic chamber to investigate the jet noise characteristics. In these cases, the jet leaving the nozzle exit is surrounded by the stagnant air in the laboratory. As a result, the turbulent mixing zone and the jet boundary are developed right from the nozzle exit. The open jet facility has numerous advantages, such as:

- We can examine the large models for the same nozzle exit area compared to a solid-walled test section.
- There is an ease in the mounting and unmounting of the models.
- We can minimize window heating issues, particularly at high-temperature operations.
- No additional pressure requirement since the pressure ratio for starting and running the facility is nearly the same.

Besides the above advantages, the open jet facility has some disadvantages too. For example, an open jet facility necessitates a greater compression ratio, which leads to a lot of noise. This facility has poor flow quality compared to the wind tunnels with the closed test section. In the following articles, author's indigenously developed supersonic open jet facility in the Department of Aerospace Engineering at the Indian Institute of Technology Kharagpur is discussed at length. Figure 6.1.1 exhibits the photographic view of the installed experimental setup.

DOI: 10.1201/9781003139447-6

Figure 6.1.1 Photographic view of the supersonic open jet facility.

6.2 COMPONENTS OF SUPERSONIC OPEN JET FACILITY

The supersonic open jet facility, as illustrated in Figure 6.2.1, consists of a compressor, air receiver (storage tank), oil filter, air drier, wide angle diffuser and settling chamber assembly, pressure regulator, and gate valves. Each of these components is separately discussed below.

6.2.1 COMPRESSOR

The compressor pumps atmospheric air into a storage tank. Because of their extensive availability and low cost, reciprocating compressors are commonly used among various compressors. Depending on the requirements, they can deliver single, double, or triple-stage compression. A single-stage compressor

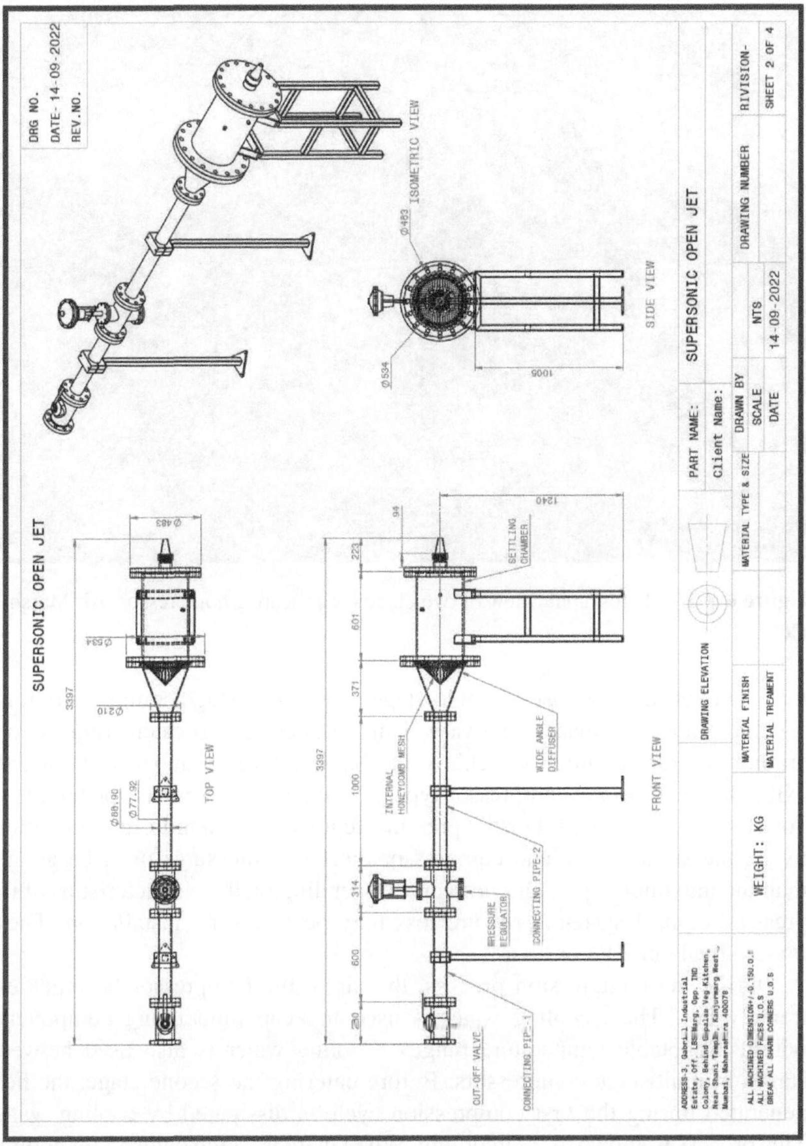

Figure 6.2.1 Design of supersonic open jet facility.

can produce up to 10 bar of discharge pressure, a two-stage compressor up to 20 bar, and a three-stage compressor with significantly higher pressures. Figure 6.2.2 shows a two-stage reciprocating compressor installed in the author's laboratory, and Table 6.1 displays its technical specifications. It is important

Figure 6.2.2 Photographic view of two-stage reciprocating compressor (Air Marshal: GC-65TH).

to remember that the cost of single-stage and double-stage compressors for a specific pumping capacity can vary. Thus, it is critical to calculating the operating pressure required to achieve the highest Mach number at the nozzle exit before selecting a compressor type. A suitable compressor type (single or double-stage) is chosen. For the pressure-regulating system to function effectively, the storage tank and compressor discharge pressures must be greater than the maximum operating pressure. Depending on the characteristics of the pressure control system, pressure loss may occur during installation. These losses are about 20% or less.

During the compression process, the air in the compressor becomes extremely hot. Thus, cooling water is used to keep functioning components within acceptable temperature ranges. Cooling water is also used between stages in multistage compressors. Before entering the second stage, the heat generated during the first compression cycle is dissipated by cooling water. This process is known as intercooling. When there is a high demand for cooling water, a cooling tower, as well as piping, valves, and pumps, may be erected. Because of the cooling tower, water is reused rather than wasted. Large compressors typically include a number of safety precautions. For example, when the lubricating oil level falls too low, the cooling water stops flowing. When the discharge pressure becomes too high, the controls automatically shut down the compressor. The compressor should immediately stop when the storage

Table 6.1
Technical specifications for a two-stage reciprocating compressor installed at the author's laboratory.

Compressor technology	Reciprocating
Compressor type	Air compressor
Mounting configuration	Base mounted
Free air delivery at maximum operating pressure	60 cfm
Maximum working pressure	40 bar
Working fluid	Air
Number of compression stages	2
Operating ambient Temperature	45°C
Operating humidity	Up to 90% non-condensing
Horse Power (HP)	20 hp
Filters	Pre- and post-filters
Lubrication style	Oil flooded
Cooling method	Water cooled
Power supply	3-phase, 230 V, 50 Hz

tank reaches its design pressure level. The compressor should restart when the storage tank pressure falls below the prescribed lower limit. A separate motor control unit with substantial electrical power is usually necessary to turn them on or off the compressor.

Compressors are typically rated based on their ability to deliver a certain number of cubic feet of sea level air per minute. The time it takes them to pump a tank from an initial pressure of p_i to a final pressure of p_f may be calculated using the following relationship [20]:

$$t_p = \frac{\forall_T}{14.7\dot{Q}}(p_f - p_i) \tag{6.2.1}$$

where t_p = pumping time, minutes

$\dot{Q}$ = compressor rating, cfm (air delivery rate in cubic feet per minute at sea level)

p_f = final pressure, psia (per square inch, absolute)

p_i = initial pressure, psia (per square inch, absolute)

$\forall_T$ = volume of high-pressure tank, ft^3 (cubic feet)

In reality, p_f represents the run start pressure and p_i represents the run end pressure. It is worth noting that the number 14.7 in equation (6.2.1) implies the atmospheric static pressure at sea level (absolute).

Example 34. Calculate the pump time required to fill a 177 ft^3 tank from 14.7 psia to 435 psia with a 60 cfm reciprocating compressor.

Solution. Given,

$$\forall_T = 177 \text{ ft}^3$$
$$p_i = 14.7 \text{ psia}$$
$$p_f = 435 \text{ psia}$$
$$\dot{Q} = 60 \text{ cfm}$$

The time taken by the compressor to fill the tank is:

$$t_p = \frac{\forall_T}{14.7\dot{Q}}(p_f - p_i) \tag{6.2.1}$$
$$= \frac{177}{14.7 \times 60}(435 - 14.7)$$
$$\approx 84 \text{ minutes}$$

6.2.2 AIR RECEIVER OR STORAGE TANK

The compressor draws in air from its surroundings and pumps it into a storage tank. Depending on the requirement, the compressed air is stored in a cylindrical or spherical tank. Typically, spherical tanks are used for higher storage pressures of around 300 bar, while cylindrical tanks are preferred for lower pressures of around 20 to 30 bar. A cylindrical tank can be mounted horizontally or vertically, depending on the space available.

The costs are about the same whether it is a high- or low-pressure tank; however, a higher pressure tank must be stronger than a lower pressure tank. The tank should be equipped with a drain pipe to avoid air blast when draining at higher pressures. The tank also has a safety disc that is activated once the tank pressure reaches its maximum. In the event of a malfunction, the safety disc allows the tank to be emptied if the pressure becomes dangerously high. As the air is sucked out, the remaining air in the tank expands with a stagnant pressure drop. This expansion process is polytropic since the heat is transferred from the walls to the air with a temperature drop. The temperature variation in the reservoir strongly affects the Mach number and the Reynolds number at the nozzle exit, so we need to install a heater downstream of the tank to prevent the temperature drop.

The storage tank can take various forms, such as; cylindrical, spherical, or conical. However, making a tank of unusual shapes can have dangerous consequences. Although a spherical tank is best, the manufacturing difficulties involved and the requirements of the large tank diameter make this design impractical. Therefore, to take advantage of cylindrical and spherical shapes, a

Figure 6.2.3 Photographic view of high-pressure air receiver.

cylindrical tank is generally made with hemispherical top and bottom covers (or domed end covers, called heads). In the author's laboratory, a vertically placed thin-walled cylindrical air receiver with a hemispherical cover at the top is designed and installed (Figure 6.2.3). The design of the air receiver is illustrated in Figure 6.2.4. Because it is usually understood that the thickness of a thin-walled cylinder should be less than one-tenth of its inner radius, the wall thickness of the storage tank is kept at 30 mm because the inner radius of the tank is 685 mm. The storage tank has a volume[1] of 5 m^3 and can be charged up to a maximum pressure of 30 bar. The tank is made of mild steel

[1] The size of the storage tank is determined by two factors: mass flow rate and test time required.

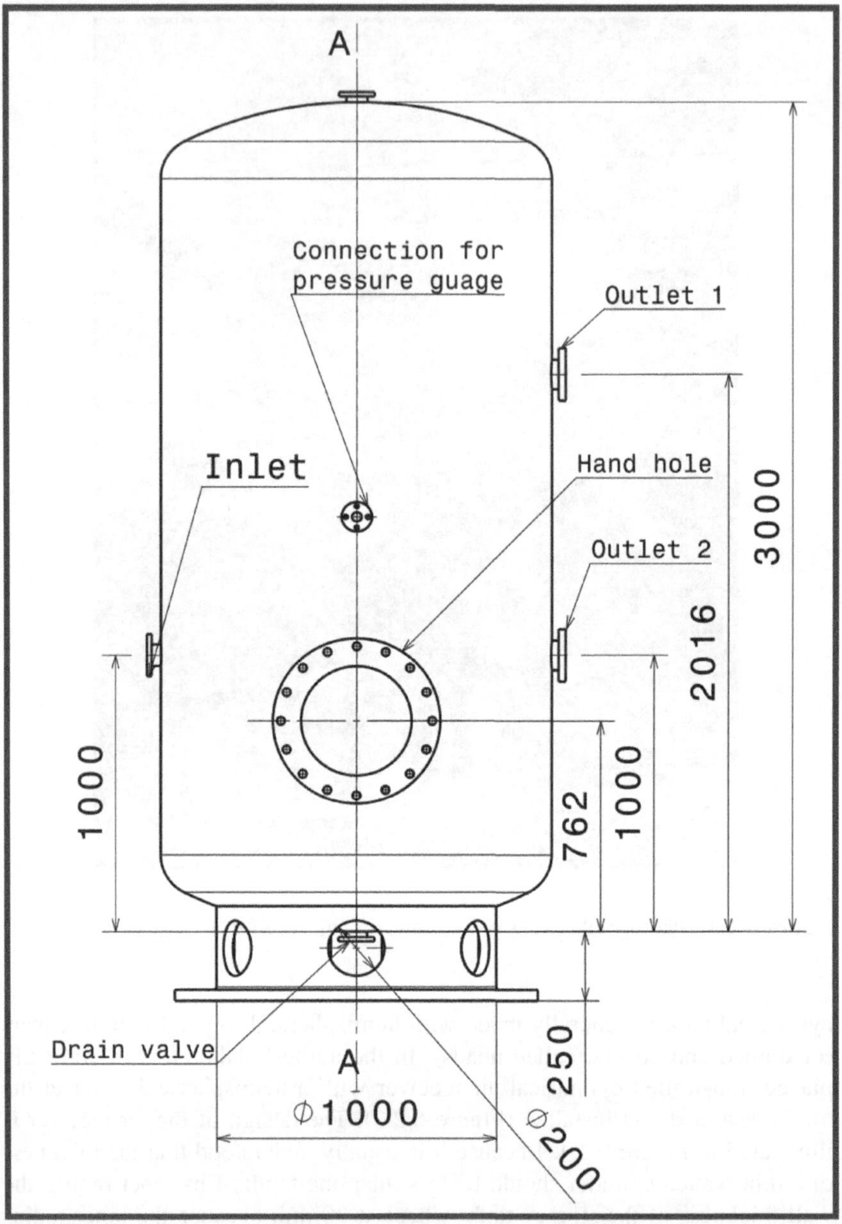

Figure 6.2.4 Design of high-pressure air receiver.

(MS IS-2062), whose ultimate strength is 342.5 MPa. A factor of safety of 5 is considered in the tank fabrication.

According to the polytropic process, the stagnation temperature in the storage tank reduces during expansion. For a polytropic expansion of air,

$$\frac{\rho_{0f}}{\rho_{0i}} = \left(\frac{p_{0f}}{p_{0i}}\right)^{\frac{1}{n}} \tag{6.2.2}$$

where, ρ_{0i} = initial air density, kg/m^3
ρ_{0f} = final air density, kg/m^3
p_{0i} = initial tank pressure (absolute), Pa
p_{0f} = final tank pressure (absolute), Pa
n = polytropic exponent

The polytropic exponent is equal to 1.4 for short runs with high-mass flow, and it approaches 1.0 for long runs with heat sink material in the tank. For more spherical tanks, the value of n is closer to 1.4.

6.2.3 WIDE ANGLE DIFFUSER

In order to obtain uniform flow in the nozzle, a large cylindrical tube with a low-velocity zone, the so-called settling chamber, is used on which the nozzle is mounted. A diffuser is inserted between the smaller diameter tube with the high-velocity flow and the larger diameter settling chamber with the near stagnant flow. The diffuser is used to restore dynamic pressure in the pipe by reducing the flow rate. Still, a simple diffuser cannot have efficient pressure recovery because the flow is very turbulent and exits the pressure regulator asymmetrically. Therefore, a wide-angle diffuser connecting the tube to the settling chamber is generally used. The divergence angle of the diffuser wall is usually kept in the range of 45 to 90 degrees to minimize losses due to boundary layer separation. In addition, since the flow at the diffuser inlet is non-uniform and turbulent, a series of perforated plates are installed inside the diffuser, resulting in uniform flow distribution.

6.2.4 SETTLING CHAMBER

The settling chamber is a cylindrical duct of a few diameters in length, located downstream of the wide-angle diffuser in the jet facility. The settling chamber receives air from the wide-angle diffuser and settles it down to provide a steady and uniform flow. The perforated sheets inside the chamber reduce the flow irregularities and turbulence levels; thus, uniform flow is achieved ahead of the nozzle inlet. At low speeds, the reduction in turbulence level in the settling chamber is caused by successive pressure drops through a series of perforated sheets. It is always preferred to obtain a low-pressure drop with multiple sheets rather than a high-pressure drop with a single sheet. A single sheet creates additional turbulence, whereas multiple sheets continuously reduce the initial turbulence level. The interior of the settling chamber should be easily accessible for the maintenance of perforated sheets. The settling chamber has provisions for measuring pressure and temperature at some downstream locations from the perforated sheets. A pressure regulating valve (PRV) is used to adjust the pressure level in the settling chamber.

The settling chamber thickness is determined similarly to the storage tank; it is also assumed to be a thin cylinder. Once again, circumferential stress is the desired criterion for determining the thickness of the settling chamber wall. The relationship between the circumferential stress and the wall thickness is given by

$$t_c = \frac{p_{0s} \times D_{ic}}{2 \times \sigma_{cc}} \tag{6.2.3}$$

where p_{0s} is the maximum allowable pressure in the settling chamber, σ_{cc} is the circumferential stress inside the settling chamber, D_{ic} is the inner diameter and t_c is the wall thickness of the settling chamber. The circumferential stress is 18.7 MPa, so the chamber thickness is 12 mm for the chamber's inner diameter of 300 mm at the maximum storage pressure of 15 bar.

6.2.5 OIL FILTERS

The effectiveness of air filters decreases when oil particles from the compressor contaminate the air. If these particles stick to the walls of the test section, the quality of the photos taken during flow visualization degrades. In the worst case, air with oil particles can cause explosions when the ambient temperature is very high. Therefore, it is unavoidable to use oil filters in the compressors to remove oil particles from the air. An oil filter is a mechanical device with a large area for oil vapor to condense. The air is usually passed through

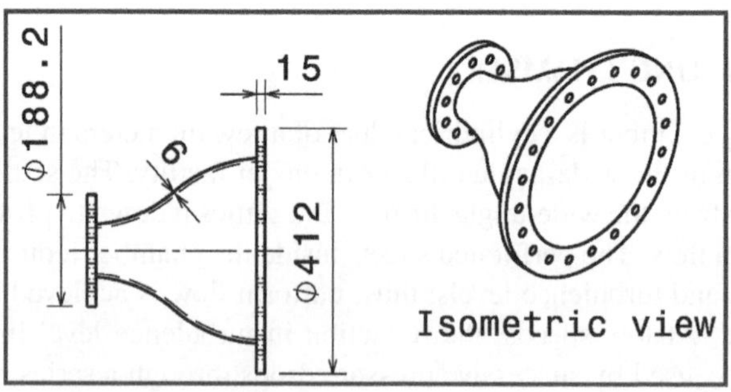

Figure 6.2.5 Design of wide angle diffuser.

desiccant granules for drying. Thus, the desiccant in the oil filter should be cleaned regularly; otherwise the oil particles adhering to the granules will reduce the effectiveness of the filter. However, the frequency of filter cleaning depends on the amount of oil present in the incoming air from the storage tank. The oil filter also acts as a good moisture separator, collecting the water droplets condensed by the aftercooler. The cavity at the bottom of the filter collects and drains the water.

6.2.6 AIR DRIER

The air drier used in the jet test facility is a high-pressure device between the storage tank and the settling chamber. Since the volumetric flow through the air drier is less at high pressure, the size and cost of the air drier are reduced. The adsorption drying method is used in this type of setup. This method collects moisture in a condensed form on the surface of the desiccant. We can utilize desiccants in the form of granules with extremely porous structures, such as silica gel, activated alumina, or zirconia. Moisture condensation occurs on the granule's outer surface and is subsequently fed into the pores of the granules by capillary action. In addition, the moisture on top of the desiccant is eliminated by heating the granules to around 310 K.

6.2.7 PIPING AND VALVES

A pipe with a smaller diameter is always desirable since the cost increases with diameter for a given working pressure. Also, we should choose an optimum pipe size to avoid the whistling and pressure losses when the mass flow is maximum. It is essential to determine the pressure drop between the storage tank and the pressure regulator and the piping and valves to ensure a higher storage tank capacity. When the pressure in a storage tank drops, the run time decreases. Generally, the jet facility includes a gate or quick-opening valve, followed by a pressure-regulating valve. A gate valve consists of a plate that shuts off the flow in the pipe. When the gate valve is closed, a pressure difference is created between the plate and the sealing surfaces of the valve body, holding the sealing surface together. Hence, the gate valve cannot function as a quick-opening valve.

6.2.8 PRESSURE REGULATOR

The pressure regulating valve maintains constant pressure in the settling chamber during the experiment. A pressure regulating valve has an opening through which air flows uniformly with respect to the valve position, whether fully or partially open. When a valve is fully open, the area of airflow through the valve is roughly equal to the pipe area. However, when the valve is partially open,

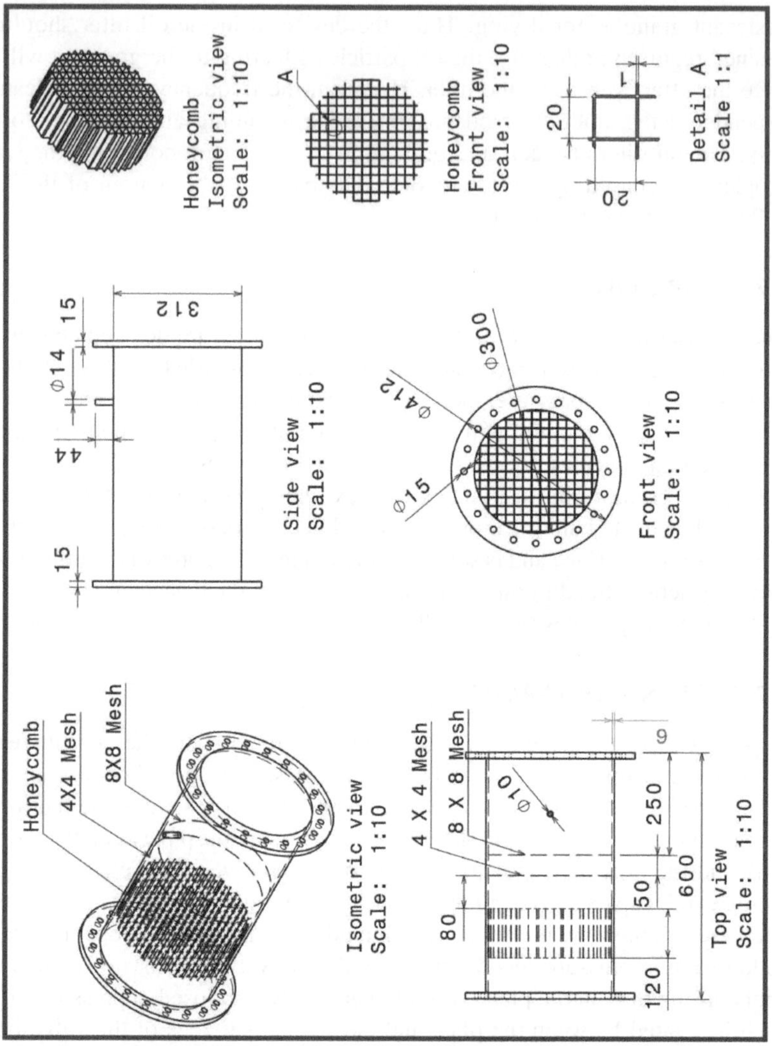

Figure 6.2.6 Designs of settling chamber, honeycomb structures, and screens.

the required pressure in the storage tank to maintain the same Mach number is greater, and the running time is reduced. The pressure regulator can be employed in different ways in the operation of a jet-test facility. In some setups, the pressure regulating valve is suddenly opened and adjusted manually or automatically to the desired position to keep the settling chamber at a constant stagnation pressure. The alternative system includes a quick opening valve followed by a pressure regulating valve. The pressure regulating valve is fixed at the required position before operating the quick opening valve so that the pressure regulating valve takes over control once the quick opening valve is released. Another option is to pre-program the function of the pressure controlling valve, which allows the valve to be adjusted automatically to deliver the required pressure to begin the operation. The valve is then held in that position for 2 or 3 s to allow the setup operation to begin, after which it is closed to provide the minimum operating pressure ratio. Finally, the valve can operate at this pressure ratio during the operation. The pressure regulating valve works with the gate valve and regulators. This arrangement can be operated quickly, either manually or automatically, to provide good control. Choosing a single pressure regulator for a wide range of operations can be difficult. For instance, if the Mach number range is broad, so are the mass flow and operating pressure ranges; selecting a single pressure regulator for a wide range of operations is challenging.

6.3 PERFORMANCE PARAMETERS

6.3.1 MASS FLOW RATE

The air passing through the convergent-divergent nozzle is introduced from the settling chamber at conditions p_{0s} and T_{0s}. It is then expanded to local conditions p and T in the nozzle. The mass flow rate through the nozzle is determined by:

$$\frac{\dot{m}}{A} = \frac{p_{0s}\sqrt{\gamma}}{\sqrt{RT_{0s}}} M \left(1 + \frac{\gamma - 1}{2} M^2\right)^{\frac{-(\gamma+1)}{2(\gamma-1)}} \quad 4.3.3$$

For air ($\gamma = 1.4$), this is reduced to

$$\frac{\dot{m}}{A} = \frac{p_{0s}}{\sqrt{T_{0s}}} \frac{0.07M}{(1 + 0.2M^2)^3} \tag{6.3.1}$$

Example 35. A Mach 1.8 nozzle is mounted at the outlet of the settling chamber with air at 8 bar and 30°C in a supersonic open jet facility as shown in Figure 6.1.1. The nozzle has an exit area of 132.66 mm^2. Calculate the mass flow rate of air passing through the nozzle if it discharges under correct expansion. Also, calculate the Reynolds number for this flow based on the nozzle exit diameter.

Solution. Given,

$$M_D = 1.8$$
$$A_e = 132.66 \text{ mm}^2$$
$$p_{0s} = 8 \text{ bar}$$
$$T_{0s} = 30^\circ C + 273 = 303 \text{ K}$$

When a supersonic nozzle operates under correct expansion, the Mach number at the exit is equal to the design Mach number, i.e., $M_e = M_D = 1.8$. Thus, the mass flow rate of air passing through the nozzle is:

$$\frac{\dot{m}}{A_e} = \frac{p_{0s}}{\sqrt{T_{0s}}} \frac{0.07M_e}{(1+0.2M_e^2)^3} \tag{6.3.1}$$

$$\Longrightarrow \quad \dot{m} = 132.66 \times 10^{-6} \times \frac{8 \times 101325}{\sqrt{303}} \times \frac{0.07 \times 1.8}{1+0.2 \times (1.8)^3}$$

$$= 0.36 \ \frac{kg}{s}$$

For Mach 1.8, from isentropic table,

$$\frac{p_e}{p_{0s}} = 0.1740$$

$$\Longrightarrow \quad p_e = 0.1740 \times 8 \times 101325 = 141.04 \text{ kPa}$$

$$\frac{T}{T_{0s}} = 0.6068$$

$$\Longrightarrow \quad T_e = 0.6068 \times 303 = 183.86 \text{ K}$$

The density of air at the nozzle exit is:

$$\rho_e = \frac{p_e}{RT_e} = \frac{141.04 \times 10^3}{287 \times 183.86}$$

$$= 2.67 \ {}^{kg}/{}_{m^3}$$

The flow velocity at the nozzle exit is:

$$V_e = M_e a = 1.8 \times \sqrt{1.4 \times 287 \times 183.86}$$

$$= 489.24 \ {}^{m}/{}_{s}$$

The nozzle exit diameter is:

$$d_e = \sqrt{\frac{4A_e}{\pi}} = \sqrt{\frac{4 \times 132.66 \times 10^{-6}}{3.14}}$$

$$= 0.013 \text{ m}$$

From Sutherland's law, the viscosity of air at 303 K is:

$$\mu_e = 1.46 \times 10^{-6} \left(\frac{T_e^{3/2}}{T_e + 111} \right) \tag{2.3.12}$$

$$= 1.46 \times 10^{-6} \times \frac{(183.86)^{1.5}}{183.86 + 111} = 1.23 \times 10^{-5} \ {}^{kg}/{}_{(m.s)}$$

The Reynolds number based on the nozzle exit diameter is:

$$\mathrm{Re_d} = \frac{\rho_e v_e d_e}{\mu}$$
$$= \frac{2.67 \times 489.24 \times 0.013}{1.23 \times 10^{-5}}$$
$$\Longrightarrow \quad \mathrm{Re_d} = 1.38 \times 10^6 = 1.38 \text{ million}$$

6.3.2 RUN TIME

The expression to compute the run time of an open jet test facility with a constant mass flow rate of air through a supersonic nozzle positioned at the settling chamber exit is as follows [20]:

$$t = 0.0862\left(\frac{\forall}{A^*}\right)\left(\frac{\sqrt{T_{0s}}}{T_0}\right)\left(\frac{p_{0i}}{p_{0s}}\right)\left[1-\left(\frac{p_{0f}}{p_{0i}}\right)^{\frac{1}{n}}\right] \tag{6.3.2}$$

where t = run time, seconds

$\forall$ = tank volume, m^3

A^* = Laval nozzle throat area, m^2

T_{0s} = settling chamber temperature, K

p_{0s} = settling chamber pressure (absolute), Pa

p_{0i} = initial tank pressure (absolute), Pa

p_{0f} = final tank pressure (absolute), Pa

n = polytropic exponent

It is worth noting that p_{0f} is the minimum pressure that should be maintained in the tank to produce the required Mach number at the nozzle exit. However, due to losses in the duct and the pressure regulating valve (PRV), the pressure in the tank may drop by up to 50% of its initial value. Thus, it should be considered in the final tank pressure computations. If p_{0s} is the settling chamber pressure and Δp_0 is the stagnation pressure loss, the final tank pressure is:

$$p_{0f} = p_{0s} + \Delta p_0 \tag{6.3.3}$$

As discussed in section §6.2, a free jet test facility is designed and installed in the author's laboratory. Using a convergent or convergent-divergent nozzle mounted at the end of the settling chamber, this facility can produce subsonic or supersonic jets. The facility run time can be computed using equation (6.3.2) for a supersonic nozzle with a design Mach number and a specific nozzle pressure ratio (NPR).

Example 36. A supersonic open jet facility, such as the one depicted in Figure 6.1.1, is used to examine the aerodynamic characteristics of a Mach 1.8

jet. The facility is supplied by a high-pressure air receiver (storage tank) with a volume of 5 m^3 that is initially charged with air at 20 bar and 30°C. The pressure in the tank may drop by up to 50% due to losses in the duct and pressure regulating valve (PRV). The nozzle pressure ratio at the facility is 8.0. The Laval nozzle mounted at the settling chamber exit has a 13 mm exit diameter. Determine the jet facility run time. The storage tank contains heat absorbing material.

Solution. Given,

$$\forall = 5 \text{ m}^3$$

$$p_{0i} = 20 \text{ bar}$$

$$\text{NPR} = \frac{p_{0s}}{p_a} = 8 \implies p_{0s} = 8 \text{ bar}$$

$$T_{0s} = 30^{\circ}\text{C} + 273 = 303 \text{ K}$$

$$D_e = 13 \text{ mm} \implies A_e = 132.67 \text{ mm}^2$$

The storage tank contains heat-absorbing material; hence the exponent, n, equals 1. Furthermore, with Mach 1.8, the stagnation pressure loss induced by the duct and pressure regulating valve might be 50%, resulting in

$$p_{0f} = 1.5 \times p_{0s}$$
$$= 1.2 \times 8 = 12 \text{ bar}$$

The isentropic table for Mach 1.8 gives:

$$\frac{A_e}{A*} = 1.439$$

$$\implies A^* = \frac{132.67}{1.439}$$

$$= 92.19 \text{ mm}^2 = 92.19 \times 10^{-6} \text{ m}^2$$

For a constant mass flow rate operation, the run time of the jet facility is:

$$t = 0.0862 \left(\frac{\forall}{A^*}\right)\left(\frac{\sqrt{T_{0s}}}{T_0}\right)\left(\frac{p_{0i}}{p_{0s}}\right)\left[1 - \left(\frac{p_{0f}}{p_{0i}}\right)^{\frac{1}{n}}\right] \tag{6.3.2}$$

$$= 0.0862\left(\frac{5}{92.19 \times 10^{-6}}\right)\left(\frac{\sqrt{303}}{303}\right)\left(\frac{20}{8}\right)\left[1 - \left(\frac{12}{20}\right)^{\frac{1}{1.0}}\right]$$

$$\approx 268.57 \text{ seconds} = 4.48 \text{ minutes}$$

6.4 SUMMARY

The installation of the author's indigenously developed supersonic open jet facility at the Department of Aerospace Engineering, Indian Institute of Technology Kharagpur, is extensively covered in this chapter. In contrast to wind tunnels, which normally test the models in a test section with closed walls, an open jet facility examines the models in an open environment with no diffuser positioned ahead of the test section.

A supersonic open jet facility is made up of the following components:

1. Compressor
2. Air receiver or storage tank
3. Wide angle diffuser
4. Settling chamber
5. Oil filters
6. Air drier
7. Piping and valves
8. Pressure regulator

The expression to compute the run time of an open jet test facility with a constant mass flow rate of air through a supersonic nozzle positioned at the settling chamber exit is as follows:

$$t = 0.0862\left(\frac{\forall}{A^*}\right)\left(\frac{\sqrt{T_{0s}}}{T_0}\right)\left(\frac{p_{0i}}{p_{0s}}\right)\left[1-\left(\frac{p_{0f}}{p_{0i}}\right)^{\frac{1}{n}}\right]$$

where t = run time, seconds

$\forall$ = tank volume, m^3

A^* = Laval nozzle throat area, m^2

T_{0s} = settling chamber temperature, K

p_{0s} = settling chamber pressure (absolute), Pa

p_{0i} = initial tank pressure (absolute), Pa

p_{0f} = final tank pressure (absolute), Pa

n = polytropic exponent

The mass flow rate through a convergent-divergent duct inducted from the settling chamber with conditions p_{0s} and T_{0s} is given by

$$\frac{\dot{m}}{A} = \frac{p_{0s}\sqrt{\gamma}}{\sqrt{RT_{0s}}}M\left(1+\frac{\gamma-1}{2}M^2\right)^{\frac{-(\gamma+1)}{2(\gamma-1)}}$$

where M is the Mach number at a particular point in the nozzle with area A. For air, this results in

$$\frac{\dot{m}}{A} = \frac{p_{0s}}{\sqrt{T_{0s}}}\frac{0.07M}{(1+0.2M^2)^3}$$

Section II

Measurements in Compressible Flows

7 Measurement Techniques in Compressible Flows

7.1 INTRODUCTION

This chapter examines briefly various strategies for measuring different variables in compressible flows. However, only conventional techniques from the standpoint of a wind tunnel observer are presented. Nevertheless, many of these techniques are equally applicable to the problem of flight measurement.

7.2 STATIC PRESSURE MEASUREMENT

Static pressure is measured on an aerodynamic surface such as a wind tunnel wall or airfoil using a small opening (hole) drilled perpendicular to the wall and attached to a manometer or other pressure gauge (Figure 7.2.1(a)). The orifices, commonly referred to as static pressure orifices, should be tiny and free from contaminants such as burrs.

Apart from wall, the orifices are also constructed along the periphery of a narrow tube to measure the static pressure at a point inside the flow. Such a device is called the static pressure probe, schematically shown in Figure 7.2.1(b). In order to minimize the obstruction to the flow, the probe must be slender and aligned parallel to the freestream direction. Moreover, to minimize the nose effects on the flow downstream, the static holes are drilled at a distance of 8D to 10D, where D stands for the tube diameter.

The sensitivity to yaw my be reduced by by constructing several holes around the circumference so that the mean pressure is measured. Despite of this provision, the yaw should never be more than 5° to keep the measurement error below 1%.

The shape of the probe nose may be either blunt or conical in compressible flows. However, a conical nose shape is usually preferred in supersonic flows. Because a blunt shape experiences a detached bow shock ahead of the probe nose, on the other hand, a conical shape will lead to a comparatively weak oblique shock attached to the probe nose.

7.3 STAGNATION PRESSURE MEASUREMENT

The total or stagnation pressure in a flow is defined as the pressure when the flow is isentropically brought to rest. When the flow itself is isentropic (i.e., no change in the entropy of a fluid), the stagnation pressure remains the same

DOI: 10.1201/9781003139447-7

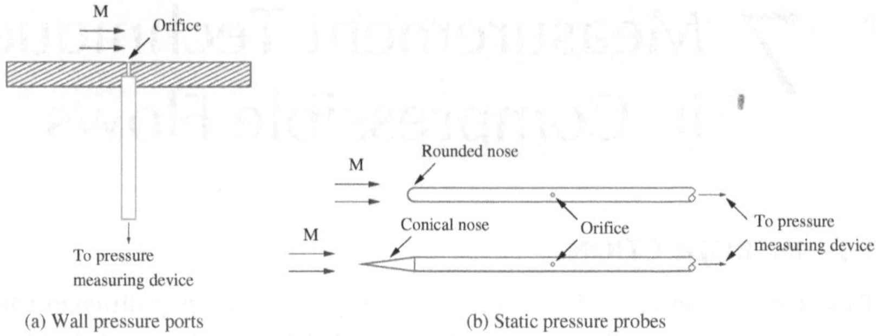

Figure 7.2.1 Measurement of static pressure.

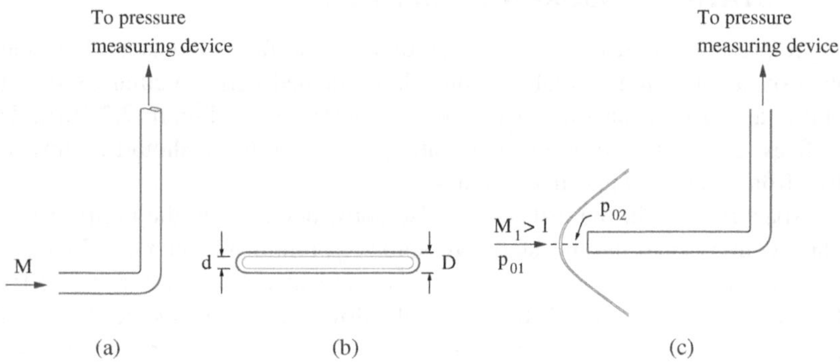

Figure 7.3.1 Measurement of stagnation pressure.

as the reservoir pressure at any point in the flow. On the other hand, if the flow experiences an entropy change (i.e., the flow is not isentropic), the total pressure at a point will be different from the reservoir pressure.

For measuring the stagnation pressure, a pitot probe, a simple narrow tube with an orifice at the nose tip, is aligned parallel to flow, as shown in Figure 7.3.1. The other end of the tube is connected to a pressure measuring device. The fluid inside the tube is at rest. Extensive experimentation has shown that for the Reynolds number of more than 500, the viscous effects on pressure measurements can be ignored.

A pitot probe is less sensitive to flow alignment than a static probe. A simple open-ended probe, as shown in Figure 7.3.1, gives 1% accuracy up to yaw angles of 20°. The probes with rounded noses and small orifices (high outer-to-inner diameter ratio) are more sensitive to flow alignment.

It is worth noting that in supersonic flows, a pitot probe does not measure the local stagnation pressure because a bow shock develops ahead of the probe nose, therefore it measures the stagnation pressure downstream of the shock wave. By definition, the portion of bow shock at the center, normal to the stagnation streamline, is assumed to be a normal shock. Thus, the ratio of actual stagnation pressure to the measured pitot pressure may be written using equation (2.7.23), which may be slightly rewritten as

$$\frac{p_{01}}{p_{02}} = \left[\frac{2\gamma}{\gamma+1}M_1^2 - \frac{\gamma-1}{\gamma+1}\right]^{\frac{1}{\gamma-1}} \left[\frac{\frac{\gamma-1}{2}M_1^2+1}{\frac{\gamma+1}{2}M_1^2}\right]^{\frac{\gamma}{\gamma-1}} \tag{7.3.1}$$

Therefore, if M_1 is known, the true stagnation pressure p_{01} can be computed from equation (7.3.1). In case M_1 is not known beforehand, an additional measurement is required.

Example 37. A supersonic wind tunnel operates at a test section Mach number of 1.8. The operating conditions in the test section are 249 kPa (gauge) and −85°C. A pitot probe placed inside the test section experience a pressure loss of 18.7%. Calculate (a) the local stagnation conditions in the test section, and (b) the pressure sensed by the probe. Consider the atmospheric pressure as 101 kPa.

Solution. Given,

$$\begin{aligned} M_{TS} &= 1.8 \\ p_{TS} &= 249 \text{ kPa (gauge)} \\ &= 249 + 101 = 350 \text{ kPa (absolute)} \\ T_{TS} &= -85^{\circ}\text{C} + 273 = 193 \text{ K} \end{aligned}$$

(a) Using the isentropic relation,

$$\frac{T_0}{T_{TS}} = 1 + \frac{\gamma-1}{2}M^2 \tag{2.7.14}$$

$$= 1 + \frac{1.4-1}{2}(1.8)^2 = 1.648$$

i.e.,

$$\begin{aligned} T_0 &= 1.648 T_{TS} \\ &= 1.648 \times 193 = 318.06 \text{ K} \end{aligned}$$

Now,

$$\frac{p_0}{p} = \left(1+\frac{\gamma-1}{2}M^2\right)^{\frac{\gamma}{\gamma-1}} \tag{2.7.15}$$

$$= \left(\frac{T_0}{T_{TS}}\right)^{\frac{\gamma}{\gamma-1}}$$

$$= \left(\frac{318.06}{193}\right)^{3.5} = 5.746$$

Thus,

$$p_0 = 5.746 p_{TS}$$
$$= 5.746 \times 350$$
$$= 2011.02 \text{ kPa} = 2.01 \text{ MPa}$$

(b) Because the pitot probe is placed inside the test section, a bow shock forms ahead of the probe nose, resulting in a loss of total pressure. As a result, the pressure sensed by the probe is essentially downstream of the shock, as indicated by

$$p_{02} = \left(1-\frac{18.7}{100}\right) p_{01}$$
$$= 0.813 \times 2.01 = 1.634 \text{ MPa}$$

7.4 MACH NUMBER FROM PRESSURE MEASUREMENTS

In compressible flows, the Mach number is one of the most important parameters that may be calculated from numerous equations in which it appears, given that other values, such as pressures, are measured. If the flow field at the measurement point is isentropic, the stagnation pressure will stay equal to the reservoir pressure, p_0. A static pressure measurement will then allow computing Mach number using the following relation (applicable for both subsonic and supersonic flows):

$$\frac{p_0}{p} = \left(1+\frac{\gamma-1}{2}M^2\right)^{\frac{\gamma}{\gamma-1}} \tag{2.7.15}$$

The static pressure p_1 at the wall is determined using static pressure ports and, for the flow field, using a static probe. In actual flight operations and non-isentropic flows, the reservoir pressure is unknown; hence, both static and stagnation pressures must be measured to calculate the Mach number. In subsonic flows, the pitot pressure is the true total pressure, and the Mach number is calculated from equation (2.7.15), provided the static pressure is measured through a static probe. On the other hand, if the flow is supersonic, the pressure indicated by the pitot probe is p_{02}, the stagnation pressure downstream of the normal shock. In this case, since actual stagnation pressure p_{01} cannot be measured, the Mach number may be obtained by eliminating p_{01} in equation (7.3.1) using equation (2.7.15). This results in

$$\frac{p_1}{p_{02}} = \frac{\left[\frac{2\gamma}{\gamma+1}M_1^2 - \frac{\gamma-1}{\gamma+1}\right]^{\frac{1}{\gamma-1}}}{\left[\frac{\gamma+1}{2}M_1^2\right]^{\frac{\gamma}{\gamma-1}}} \tag{7.4.1}$$

which is referred as the Rayleigh-pitot formula.

Example 38. The dynamic pressure in a Mach 1.8 wind tunnel test section is 200 kPa. Calculate the stagnation pressure in the test section.

Solution. Given,

$$M_{TS} = 1.8$$
$$q_{TS} = 200 \text{ kPa}$$

The dynamic pressure can be described in terms of static pressure and Mach number as follows:

$$q_{TS} = \frac{1}{2}\gamma p_{TS} M_{TS}^2 \tag{2.7.13}$$

$$200 \times 10^3 = \frac{1}{2} \times 1.4 \times p_{TS} \times (1.8)^2$$

$$\Longrightarrow \quad p_{TS} = 88.18 \text{ kPa}$$

Thus, the stagnation pressure is:

$$\frac{p_0}{p_{TS}} = \left[1 + \frac{\gamma-1}{2}M_{TS}^2\right]^{\frac{\gamma}{\gamma-1}} \tag{2.7.15}$$

$$\frac{p_0}{88.18 \times 10^3} = \left[1 + 0.2 \times (1.8)^2\right]^{3.5}$$

$$\Longrightarrow \quad p_0 = 506.66 \text{ kPa}$$

7.5 VELOCITY MEASUREMENTS

The following equation can be used to compute the velocity of a perfect gas:

$$v = M\sqrt{\gamma RT} \tag{2.7.7}$$

Furthermore, the isentropic relations can be used to calculate the Mach number using static and stagnation pressure measurements:

$$\frac{p_0}{p} = \left[1 + \frac{\gamma - 1}{2} M^2\right]^{\frac{\gamma}{\gamma - 1}} \tag{2.7.15}$$

or

$$M = \sqrt{\frac{2}{\gamma - 1}\left[\left(\frac{p_0}{p}\right)^{\frac{\gamma - 1}{\gamma}} - 1\right]} \tag{7.5.1}$$

and using static and stagnation temperature:

$$\frac{T_0}{T} = 1 + \frac{\gamma - 1}{2} M^2 \tag{2.7.14}$$

or

$$T = \frac{T_0}{1 + \frac{\gamma - 1}{2} M^2} \tag{7.5.2}$$

In supersonic flows, the inclination of shock waves is also used to compute the Mach number. The inclination of Mach waves, called the Mach angle, is given by:

$$\mu_M = \sin^{-1}\left(\frac{1}{M}\right) \tag{2.4.3}$$

Although the shock angle β is somewhat greater than the Mach angle μ_M, the former can be assumed equal to the latter for the propagation of weak shock waves. This results in,

$$\beta = \sin^{-1}\left(\frac{1}{M}\right) \tag{7.5.3}$$

Thus, the shock angle created by an object, such as a wedge in supersonic flow, can be determined optically (photographically). However, the obtained Mach number is slightly less than its true value because the weak shock waves travel faster than the Mach waves, resulting in a shock angle that is less than the Mach angle.

Example 39. An aircraft is flying at an altitude where the air temperature is 15°C. At the leading edge of the wing where the relative air velocity is zero, the temperature is found to be 51°C. Determine the Mach number and speed of the aircraft.

Solution. Given,

$$T_\infty = 15^\circ C + 273 = 288\ K$$
$$T_0 = 51^\circ C + 273 = 324\ K$$

From isentropic relation,

$$\frac{T_0}{T_\infty} = 1 + \frac{\gamma - 1}{2} M_\infty^2 \qquad (2.7.14)$$

or

$$M_\infty = \sqrt{\frac{2}{\gamma - 1}\left(\frac{T_0}{T_\infty} - 1\right)}$$
$$= \sqrt{\frac{2}{1.4 - 1}\left(\frac{324}{288} - 1\right)}$$
$$= 0.85$$

Thus, the speed of the aircraft is

$$v_\infty = M_\infty \sqrt{\gamma R T_\infty} \qquad (2.7.7)$$
$$= 0.85 \times \sqrt{1.4 \times 287 \times 288}$$
$$= 289.15\ {}^{m}/_{s}$$

Example 40. The Shadowgraphic flow visualization of a supersonic jet at overexpanded conditions is shown in Figure 7.5.1. The inclination of the oblique shock wave at the nozzle exit is 24°. Determine the Mach number at the nozzle exit.

Solution. Given,

$$\beta_s = 24^\circ$$

From equation (7.5.3),

$$M = \frac{1}{\sin \beta_s} = \frac{1}{\sin 24^\circ} = 2.46$$

The Mach number at the nozzle exit is thus 2.46.

Remark. The Shadowgraphic flow visualization image in Figure 7.5.1 depicts a supersonic jet emitted by a convergent-divergent nozzle with a design Mach number (M_D) of 2.5. It is interesting to note that M_D is more than the value indicated by the Mach-angle relation for the same reasons explained in the previous article.

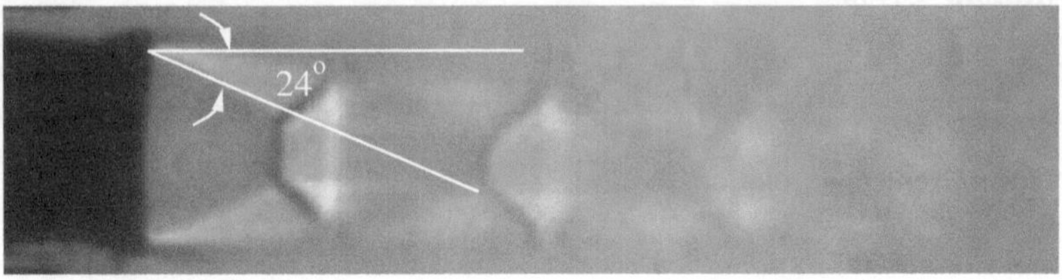

Figure 7.5.1 Supersonic jet at moderate overexpanded conditions.

7.5.1 HOT WIRE ANEMOMETRY (HWA)

A hot wire anemometer, also called the hot wire probe, is a thermal transducer widely used to measure the flow velocity and direction. The sensing element of this instrument is an electrically heated thin metal wire with a diameter of a few micrometers and a length of a few millimeters. The wire is exposed to freestream and oriented to face the incoming flow directly. The sensing wire is connected to a Wheatstone bridge, and an electric current is passed to raise the wire temperature above the freestream temperature. When a heated wire is exposed to a flow, it experiences cooling due to convective heat transfer. The heat transfer from the wire to the flow depends on the wire's characteristics, flow properties, and flow velocity. Since the electrical resistance of most metals varies with temperature, a correlation between the wire resistance and the airspeed can be established.

Let the resistance of the sensing element at any reference temperature T_0 be denoted by R_0. The resistance of the wire R at any temperature T is given by

$$R = R_0 [1 + \alpha (T - T_0)] = R_0 [1 + \alpha \Delta T] \tag{7.5.4}$$

whereα is the temperature coefficient for electrical resistance. At thermal equilibrium, the electric energy supplied to the wire equals the convective heat transfer to the flow. Thus,

$$Q = i^2 R = hA_s (T_w - T_\infty) = \frac{V^2}{R} \tag{7.5.5}$$

where, Q = heat loss due to convection
i = electric current through the wire
R = electric resistance of the wire
V = voltage across the wire
h = convective heat transfer coefficient
A_s = wire surface area exposed to the flow
T_w = wire temperature
T_∞ = freestream temperature

The amount of heat energy transferred by the sensing element can be correlated to the freestream velocity using convective heat transfer equations. However, the complexity of the resulting equations makes its application impracticable. As a result, the relation between heat transfer and flow velocity is frequently estimated empirically using anemometer calibration experiments.

The relation between the heat loss Q from the wire and the flow velocity v at any wire temperature T_w is given by

$$Q = i^2 R = \left(c_1 + c_2\sqrt{v}\right)(T_w - T_\infty) \tag{7.5.6}$$

equation (7.5.6) is also known as Kings law, where c_1 and c_2 are constants for a given wire and are estimated by calibrating the wire in a given flow. The theoretical values of c_1 and c_2 can be calculated using the heat transfer correlations for the flow past the heated wire. For incompressible flows, the convective heat transfer correlation is

$$Nu = A + B\sqrt{Re} \tag{7.5.7}$$

where Nu is the Nusselt number and Re is the Reynolds number. The constants A and B depend on the wire parameters such as length, diameter, and material.

For compressible flows, the constants are additionally affected by other parameters such as the Mach number, Prandtl number, and ratio of specific heats. Thus, the convective heat transfer correlation for compressible flows can be expressed as:

$$Nu = f(Re, M, Pr, \gamma, \ell/d) \tag{7.5.8}$$

Hot wire anemometers are extremely effective for monitoring rapid velocity fluctuations, and as a result, they are the primary instruments used to measure turbulence, employing probes with high resolution and great frequency response characteristics. Platinum, tungsten, and occasionally platinum-rhodium or iridium are used to make the sensing wire. The wire is 1 to 5 microns in diameter and has a length-to-diameter ratio of at least 100. Using an adequate electric circuit, the hot wire can also be utilized to detect flow velocity.

Hot wire anemometers are classified into two types: constant temperature anemometers (CTA) and constant current anemometers (CCA). Following Ohm's law, the voltage produced by these anemometers is thus the consequence of some form of a circuit within the device attempting to maintain the constant of the specified variable (temperature or current).

7.5.1.1 Constant Temperature Anemometer (CTA)

Figure 7.5.2(a) depicts a typical electronic circuit for a constant temperature hot wire anemometer, also known as constant resistance hot wire anemometer.

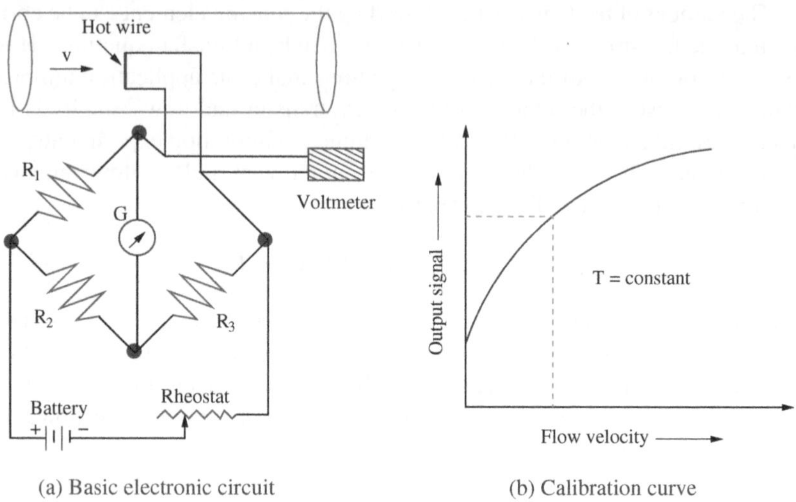

Figure 7.5.2 Schematic diagram of CTA.

Because the resistance of a wire is proportional to its temperature (equation (7.5.4)), maintaining the probe resistance R_W constant necessitates maintaining its temperature constant. The voltage applied to the top of a Wheatstone bridge measures heat transfer from the probe, which eventually becomes a measure of the fluid parameter under consideration at that time.

Assume the bridge is originally balanced under particular conditions. A decrease in R_W due to increased heat transfer will result in a temperature drop, causing the bridge to become imbalanced. The wire resistance is restored to its original value by altering the variable resistance and rebalancing the bridge. This is accomplished through the employment of an electrical servo system. It should be noted that for each value of flow velocity, there is a corresponding value of current across the hot wire, which may be measured using the ammeter in the circuit. At any given time, the ammeter measurement serves as a measure of flow velocity.

Figure 7.5.2(b) displays the calibration curve of a typical constant temperature hot wire anemometer. Because the sensitivity of the hot wire increases with temperature, it is generally heated to extremely high temperatures (sometimes as high as 950 K). Furthermore, because the amplifier has a high gain and the wire is small, the anemometer can detect rapid changes in velocity. In fact, hot wire probes have cutoff frequencies that exceed several hundred kHz, and in some cases, they can reach above 1 MHz. and in certain situations, they can exceed 1 MHz.

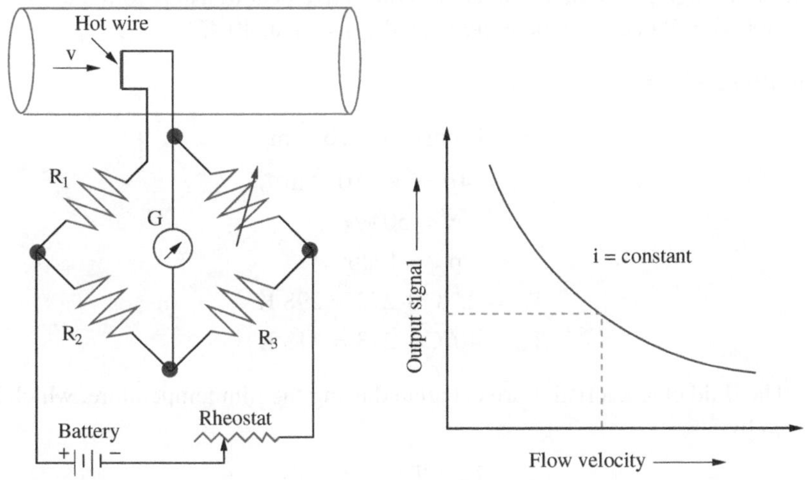

(a) Basic electronic circuit (b) Calibration curve

Figure 7.5.3 Schematic diagram of CCA.

7.5.1.2 Constant Current Anemometer (CCA)

In the constant current hot wire anemometer, as shown in Figure 7.5.3(a), a high-impedance current source is utilized, and the voltage across the hot wire is used to determine heat transfer and flow velocity. Because the current flowing through the heated wire remains constant, the temperature of the wire, and hence its resistance, varies with flow velocity. Variations in resistance produce variations in voltage across the wire when the current is constant. Thus, at any given instant, the voltage across the wire is a measure of the instantaneous flow velocity. Figure 7.5.3(b) depicts a typical calibration curve for a constant current anemometer.

7.5.1.3 Disadvantages

- The response of CCA is limited to about 1 kHz due to the thermal inertia of the sensing wire.
- The current is typically set to a level that heats the wire well above the freestream temperature. If the mass flow rate or flow velocity drops, the wire may be damaged by burning out since there is no convective heat transfer from it.

Example 41. In an airstream, a constant temperature hot wire anemometer is kept at 50 m/s, 1 atm, and 25°C. The wire filament is 4 micron in diameter and

1 mm in length. Calculate the rate of convective heat transfer from the wire if the hot wire filament temperature is held constant at 40°C.

Solution. Given,

$$\ell = 1\text{ mm} = 1\times10^{-3}\text{ m}$$
$$d = 4\mu = 4\times10^{-6}\text{ m}$$
$$v = 50\ {}^{m}/_{s}$$
$$p_\infty = 1\text{ atm}$$
$$T_\infty = 25^\circ C + 273 = 298\text{ K}$$
$$T_w = 40^\circ C + 273 = 313\text{ K}$$

The fluid characteristics are calculated using the film temperature, which is given by

$$T_f = \frac{T_\infty + T_w}{2}$$
$$= \frac{298+313}{2} = 305.5\text{ K}$$

The viscosity of air at the film temperature is

$$\mu = 1.46\times10^{-6}\left(\frac{T_f^{3/2}}{T_f+111}\right) \tag{2.3.12}$$
$$= 1.46\times10^{-6}\times\frac{(305.5)^{1.5}}{305.5+111}$$
$$= 1.871\times10^{-5}\ {}^{kg}/_{(m.s)}$$

The density of air is calculated using the equation of state for a perfect gas:

$$\rho_\infty = \frac{p_\infty}{RT_f}$$
$$= \frac{101.325\times10^3}{287\times305.5} = 1.16\ {}^{kg}/_{m^3}$$

Based on the wire diameter, the Reynolds number is

$$Re = \frac{\rho_\infty v d}{\mu}$$
$$= \frac{1.16\times50\times4\times10^{-6}}{1.871\times10^{-5}} = 12.4$$

The coefficient of thermal conductivity for air is $k = 0.026\ {}^{W}/_{(m^\circ C)}$ at $T_f = 305.5$ K, and the Prandtl number is $Pr = 0.7$.

The empirical heat transfer correlation for the flow past an infinite rod under forced convection is

$$\begin{aligned}\mathrm{Nu} &= 0.42\mathrm{Pr}^{0.2} + 0.57\mathrm{Pr}^{0.33}\mathrm{Re}^{0.5} \\ &= 0.42\,(0.7)^{0.2} + 0.57\,(0.7)^{0.33}\,(12.4)^{0.5} \\ &= 2.175\end{aligned}$$

The Nusselt number is defined as

$$\mathrm{Nu} = \frac{\mathrm{hd}}{\mathrm{k}} = 2.175$$

This results in the heat transfer coefficient,

$$\mathrm{h} = \frac{2.175 \times 0.026}{4 \times 10^{-6}} = 14137.5\ ^{W}/(\mathrm{m}^2.\mathrm{K})$$

The rate of convective heat transfer from the wire is

$$\begin{aligned}\mathrm{Q} &= \mathrm{hA}\,(\mathrm{T_w} - \mathrm{T_\infty}) \\ &= 14137.5 \times 3.14 \times 4 \times 10^{-6} \times 1 \times 10^{-3}\,(313 - 298) \\ &= 2.66 \times 10^{-3}\ \mathrm{W} = 2.66\ \mathrm{mW}\end{aligned}$$

7.5.2 PARTICLE IMAGE VELOCIMETRY (PIV)

Particle Image Velocimetry (PIV) is a non-intrusive flow visualization technology that allows viewing two or three-dimensional flow fields[1]. It is a whole-flow-field technique for obtaining instantaneous velocity vectors over a flow cross-section. Because this technique is non-invasive, it can be used to investigate supersonic and hypersonic flows, and boundary layer fluid flows. As shown in Figure 7.5.4, a typical PIV setup consists of a high-speed camera, a high-power multiphased laser, an optical arrangement to convert laser output to a light sheet to illuminate the region of interest, tracer particles, a synchronizer to control the synchronization between the laser and the camera, a fiber optic cable or liquid guide to connecting the laser to the lens setup, and a dedicated computer system (with PIV software installed) for saving the particle images and conduct image processing. Some of these elements are briefly explored in the following articles.

[1] In contrast to other velocity measurement techniques, such as Laser Doppler Velocimetry (LDV) and Hot Wire Anemometry (HWA), which detect fluid velocity at a single point, PIV produces two or even three-dimensional velocity fields.

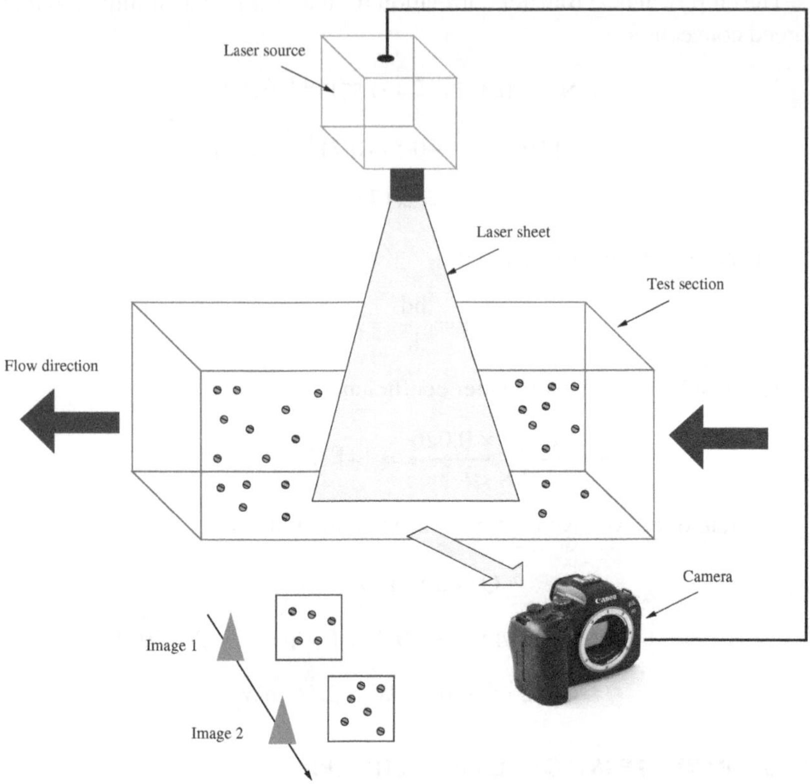

Figure 7.5.4 Schematic illustration of Particle Image Velocimetry (PIV) setup.

7.5.2.1 Seeding of Tracer Particles

In a PIV study, the fluid in the interest region is injected with small tracer particles without disturbing the flow. Seeding is the term given to this particle injection procedure. The seeded particles are expected to be neutrally buoyant and move at the same velocity as the local flow velocity[2]. Additionally, these particles should be large enough to efficiently scatter the illumination light rays. Tables 7.1 and 7.2 show a variety of particles that can be used for flow visualization in PIV for liquid and gaseous flows, respectively. Tracer particles are believed to follow fluid dynamics reliably at sufficiently small sizes. The seeding method entails putting microscopic tracer particles into the flow in a position where the flow of interest will not be interrupted. The particles must

[2]The PIV technique is based on this important assumption that the recorded particle velocity equals the local flow velocity!

Table 7.1
Seeding particles for liquid flows.

Type of tracer particle	Material	Mean diameter (μm)
Solid	Polystyrene	10 – 100
	Aluminum flakes	2 – 7
	Hollow glass spheres	10 – 100
	Granules for synthetic coatings	10 – 500
Liquid	Different oils	50 – 500
Gaseous	Oxygen bubbles	50 – 1000

Table 7.2
Seeding particles for gaseous flows.

Type of tracer particle	Material	Mean diameter (μm)
Solid	Polystyrene	0.5 – 10
	Aluminum flakes	0.2 – 5
	Hollow glass spheres	0.2 – 3
	Granules for synthetic coatings	10 – 50
Liquid	Different oils	0.5 – 3
Gaseous	Different polypropylene glycols	0.5 – 1.5
	Glycerine-water mixture	0.5 – 2
	Helium-filled soap bubbles	200 – 3000

have the same fluid flow structure as turbulent flow and be large enough to be detected by scanning methods. Because an uneven region may cause entrained particles to move at different speeds, all particles should be the same size. With different stages of fluid flow, numerous types of seeding are feasible, as is the intricacy of seeding with different fluid media.

Seeding is simpler in a liquid flow than in a gaseous flow. Seeding is a simple procedure for most liquid flows that involve floating solid particles in the fluid and mixing them to generate a homogeneous distribution. The greater the density difference between the particles and the gaseous bulk fluid, the greater the velocity lag, making seeding in gaseous flows much more challenging. Because the experimenters were able to breathe in seeded air, as in wind tunnels with open test sections, health issues are even more pressing.

Stokes number represents the degree to which these particles follow the flow. A particle with a low Stokes number[3] follows the fluid streamlines, while a particle with a high Stokes number is dominated by its inertia and hence follows its initial trajectory. The fluid with tracer particles is illuminated to make them visible. The motion of seeding particles is used to calculate the instantaneous velocity and direction of the flow. Furthermore, the particle concentration in a PIV measurement is such that it is possible to identify individual particles in a picture but not with the confidence to monitor them between different photographs. If the concentration is low enough, it is feasible to track an individual particle; such a visualization approach is known as Particle Tracking Velocimetry (PTV). However, if the concentration is exceedingly high, following individual particles becomes extremely challenging, and the examination is carried out using another technique known as Laser Speckle Velocimetry (LSV).

7.5.2.2 Illumination System

In order to make the tracer particles visible, the planar or voluminous region of the flow has to be illuminated in a short time. A high-power light source, such as a high-density laser, is used in gaseous flows to illuminate tracer particles and expose the video sensor by scattered light well. In PIV, homogeneous illumination of the region of interest is crucial for choosing a light source. Technically, the light beam profile should have no holes or gaps and closely resemble a top hat or Gaussian shape.

A light source and several lenses[4] make up the typical PIV illumination system. Because they emit monochromatic light with a high energy density that can be easily bundled into a thin light sheet for illuminating and recording tracer particles without chromatic aberrations, lasers like the Argon-ion laser and the standard double pulse Nd: YAG laser are frequently used as light sources. A continuous laser beam is split into light pulses by an electro-valve, which are then bent by a cylindrical lens to create a light plane. The light sheet used to highlight the particles involves different lens configurations.

---

[3]The Stokes number, given after George Gabriel Stokes, is a non-dimensional number that describes the behavior of particles suspended in a flow. It is defined as the ratio of the characteristic or relaxation time of a particle to the characteristic time of the flow:

$$\mathrm{Stk} = \frac{v_f t_c}{\ell_c}$$

where v_f denotes the fluid velocity sufficiently away from the tracer particle, t_c is the characteristic time of the particle, and ℓ_c is the characteristic dimension of the particle (usually its diameter).

[4]An arrangement of cylindrical lenses and mirrors to shape the light beam into a planar sheet in order to illuminate the flow field.

7.5.2.3 Camera and Recording

Two laser light exposures onto the camera from the flow are necessary in order to accomplish PIV analysis on the flow. Since conventional cameras (photographic film-based cameras) cannot record multiple frames at high speeds, both exposures are generally recorded on a single frame to calculate the flow velocity. This analysis utilizes an auto-correlation or cross-correlation technique. However, due to cross-correlation, it is difficult to relate which particle spot came from the first pulse and which from the second pulse, making it difficult to determine the direction of the flow. The other drawback of conventional cameras is that they can only take two shots at high speed. Since each pair of images must be uploaded to the computer before another set is captured, only two shots can be taken per second with these cameras.

Since then, high-speed digital cameras that can take two frames swiftly with a few hundred nanosecond difference between the frames have been developed using charge-coupled device (CCD) or complementary metal-oxide-semiconductor (CMOS) circuits. This has made it possible to isolate each exposure on a separate frame, allowing for a more precise cross-correlation study. In addition, the conventional auto-correlation technique coupled with a special framing technique in CCD or CMOS cameras allows for measuring higher flow velocities.

7.5.2.4 Synchronizer

A Synchronizer externally triggers both the laser and the camera. The timing of each frame in the camera's sequence in relation to the laser's discharge can be accurately controlled by the Synchronizer, which is operated by a host computer. As a result, the timing of the camera allows for exact control of the distance between each laser pulse and the positioning of the laser beam. It is crucial to comprehend this timing because only this allows to calculate the flow velocity in a PIV analysis

7.5.2.5 Host Computer System

The primary purpose of the host computer during PIV is to communicate the timing control parameters to the Synchronizer; it directs the Synchronizer at what particular time intervals the laser and CCD camera should operate to acquire the images. The host computer is also utilized to store particle images and perform image processing. During PIV, massive amounts of data (often in terabytes) are generated, necessitating the use of a computer with a big storage capacity and significant processing capability.

Data Acquisition

The laser and CCD camera should be triggered at regular intervals while inspecting a flow field. A Synchronizer accomplishes this by monitoring particles in two subsequent frames using the time step provided by the host computer. The flow speed usually determines the time step; the higher the flow speed, the smaller the time step. In addition, the laser and CCD camera operate at the time specified on the host computer. At predetermined time intervals, it takes several snapshots of the flow field, which are stored on the host computer. Two subsequent images can be used to compute the velocity field at that point.

7.5.2.6 PIV in High-Speed Flows

In order to eliminate particle lag, smaller particles must be used in high-speed flows. However, because these small particles scatter less light, a high-energy-density laser is frequently required to compensate. Laser and camera setups can sometimes be advantageously configured to collect scattered light. Blow-down facilities are far more difficult to seed than closed-return supersonic and hypersonic tunnels in order to achieve a sufficient homogeneity of particle dispersion in time and space. Seeding in high-speed flows is done under pressurized environments and transferred to the settling chamber. Although many researchers throughout the world have used PIV in supersonic and hypersonic flow regimes, they have met various problems. The most challenging part in these regimes is to effectively seed the flow. Due to the fact that particles cannot accelerate or decelerate as quickly as the flow, particularly flows with shock waves, challenges arise in the behavior of tracer particles in the presence of considerable velocity gradients across the shocks. Because of the short reaction time, high-speed flows are also difficult to assess. As a result, incredibly modern technology, such as a fast image processing computer and a CCD camera with high spatial and temporal resolution, is required. Furthermore, while implementing PIV, researchers made several alterations to the setup based on the nature of the difficulties, such as selecting the seeding particle, laser type, high-speed camera resolution, and so on.

7.5.3 LASER DOPPLER VELOCIMETRY (LDV)

LDV (Laser Doppler Velocimetry) is a technique for determining the instantaneous velocity of a flow field. This method, like PIV, is non-intrusive and can evaluate all three velocity components. The laser Doppler velocimeter casts a monochromatic laser beam onto the target and detects the reflected radiation (Figure 7.5.5). According to the Doppler effect, the change in wavelength of reflected radiation is a function of the relative velocity of the target object. Thus, the velocity of the object can be determined by detecting the change in wavelength of the reflected laser light, which is accomplished by producing an interference fringe pattern (i.e., superimposing the original and reflected signals). This framework serves as the foundation for LDV. A flow is injected with small, neutrally buoyant particles that scatter light. To illuminate the

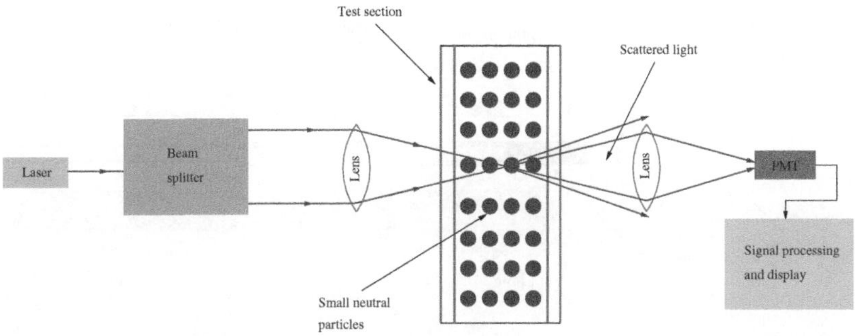

Figure 7.5.5 Schematic view of Laser Doppler Velocimetry (LDV) setup.

particles, a known frequency of laser light is used. A photomultiplier tube (PMT) captures the dispersed light and generates and amplifies a current proportional to the absorbed photon energy. The frequency difference between incident and scattered light is denoted as the Doppler shift. The Doppler equivalent frequency of the laser light scattered in the flow can be used to calculate the local fluid velocity.

7.6 TEMPERATURE MEASUREMENTS

The static temperature can be computed using the equation of state from the measured pressure and density; if the flow field is isentropic, one of these is sufficient. However, no direct method exists for determining static temperature in fluid flows. Because of the boundary layer on its surface, the temperature detected by a device, such as a thermometer, is higher than T. The static temperature rises from T at the boundary layer edge to T_r at the surface (also known as the recovery temperature). Because T_r varies across the surfaces of a measuring device, the thermometer will display a mean recovery temperature. The recovery temperature depends on the following factors:

1. Local Mach number (or static temperature) at the outer edge of the boundary layer.
2. Dissipative kinetic energy due to friction in the boundary layer.
3. Rate of heat transfer.

In theory, measuring stagnation or total temperature $T_0$is straightforward. Because the flow inside a pitot probe comes to rest at equilibrium, the temperature inside the probe in both subsonic and supersonic flows should be the stagnation temperature. As a result, the temperature can be detected by inserting a thermometer into the probe. Because heat is conducted or radiated from the

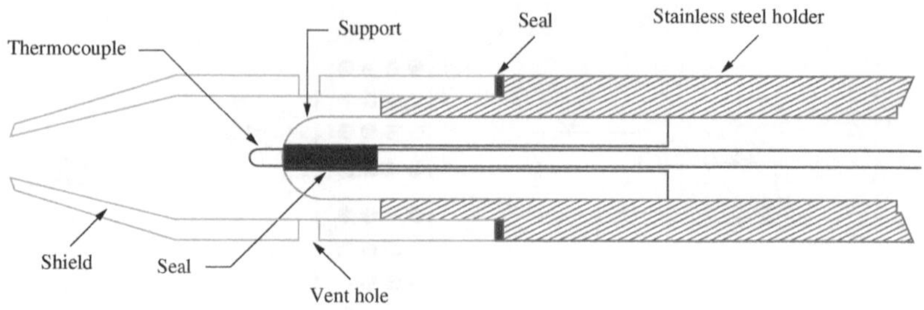

Figure 7.6.1 Stagnation temperature probe.

thermometer surface or probe wall, the equilibrium condition does not exist in reality. Consequently, the stagnation temperature decreases from T_0 to a lower value, T_r.

A total or stagnation temperature probe, consisting of a thermocouple for a sensing element, is schematically shown in Figure 7.6.1. The insulation and supports are provided in the probe design to minimize the heat losses due to conduction and radiation. However, a small flow through the probe is permitted by employing a vent hole to recover some of the lost energy. The performance of such a probe is usually defined by a coefficient of thermal recovery (recovery factor) as follows:

$$r = \frac{T_r - T_\infty}{T_0 - T_\infty} \tag{7.6.1}$$

The primary objective is to keep r constant over a wide range of conditions, since the probe is a calibrated instrument. equation (7.6.1) essentially shows that the difference between recovery and static temperatures is a fraction of the adiabatic temperature rise. The recovery factor represents the fraction of the kinetic energy of the medium that is recovered as heat. It depends on body shape, Mach number M, Reynolds number Re, Prandtl number Pr, and specific heat ratio γ. For a poorly streamlined bodies the recovery factor varies between 0.6 and 0.7, and for well streamlined bodies it is between 0.8 and 0.9. For a given gas, Pr and γ are constant over a wide range of temperatures (for air, $Pr = 0.72$ and $\gamma = 1.4$), and so r is a function of M and Re only.

The relationship between recovery temperature and stagnation temperature depends on the Mach number and may be derived from equation (7.6.1) and equation (2.7.14):

$$\frac{T_r}{T_0} = 1 - \frac{\frac{\gamma-1}{2}M^2}{1+\frac{\gamma-1}{2}M^2}(1-r) \tag{7.6.2}$$

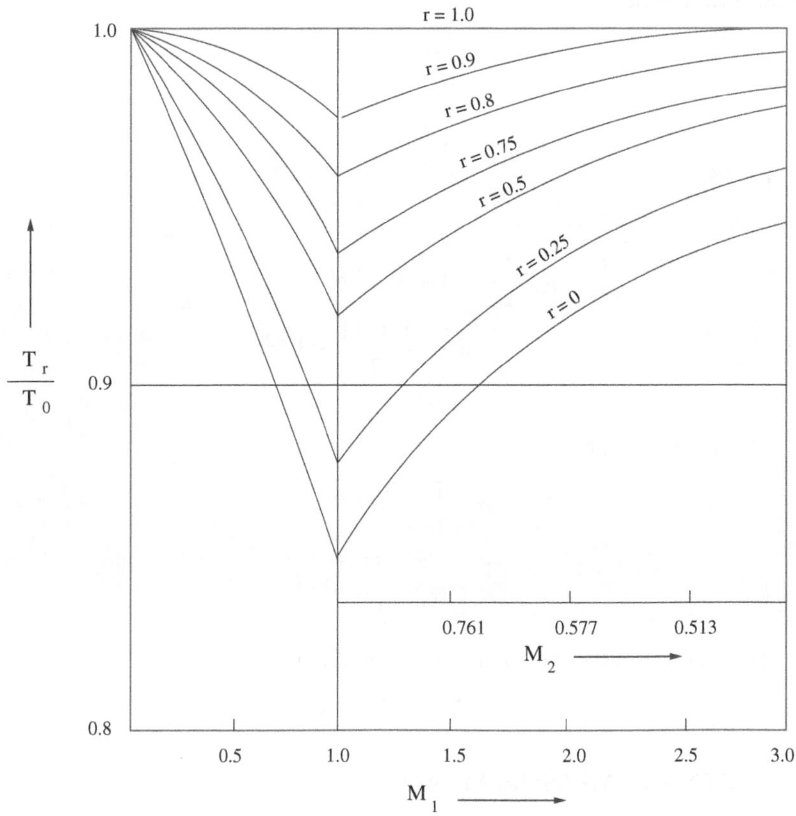

Figure 7.6.2 The ratio, $\frac{T_r}{T_0}$, as a function of M.

The ratio, $\frac{T_r}{T_0}$, as a function of Mach number is schematically illustrated in Figure 7.6.2. The figure shows that when the Mach number exceeds one, a shock occurs that increases in magnitude as the incident Mach number increases. With no heat transfer, a thermometer on the wall of a pipe inserted into a flow would show a recovery temperature that depends only on the flow properties in the boundary layer around the pipe. When, $r = 1$, $T_r = T_0$.

Example 42. A stagnation temperature probe reads 530 K for a Mach 1.5 airstream. Determine the static temperature of the airstream if the probe has a recovery factor of 0.98.

Solution. Given,

$$T_r = 530\ \text{K}$$
$$M_\infty = 1.5$$
$$r = 0.98$$

For $M_\infty = 1.5$,

$$\frac{T_0}{T_\infty} = 1 + \frac{\gamma - 1}{2} M^2 \tag{2.7.14}$$
$$= 1 + \frac{1.4 - 1}{2} \times 1.5^2$$
$$= 1.45$$

The recovery factor is

$$r = \frac{T_r - T_\infty}{T_0 - T_\infty} \tag{7.6.1}$$
$$= \frac{\frac{530}{T_\infty} - 1}{\frac{T_0}{T_\infty} - 1} = 0.98$$

This gives the static temperature of airstream as

$$T_\infty = 367.8\ \text{K}$$

7.7 DENSITY MEASUREMENTS

In many applications, flow density must be determined in addition to pressure. All other variables can be easily determined after measuring pressure and density. For example, the speed of sound can be calculated by plugging the measured static pressure and density into the equation:

$$a = \sqrt{\frac{\gamma p}{\rho}} \tag{7.7.1}$$

The velocity is then calculated using the following equation:

$$v = aM \tag{7.7.2}$$

where, the Mach number, M, is obtained using equation (2.7.15). The following formulae are used to calculate the local values of static and stagnation temperature:

$$p = \rho RT \tag{2.7.1}$$
$$\frac{T_0}{T} = 1 + \frac{\gamma - 1}{2} M^2 \tag{2.7.14}$$

In the case above, density measurements accompany pressure measurements, but density measurements alone are occasionally required. For instance, in isentropic flows with known p_0 and T_0, a measurement of density yields the pressure and Mach number using the following expressions:

$$\frac{p}{p_0} = \left(\frac{\rho}{\rho_0}\right)^{\gamma} \tag{7.7.3}$$

$$\frac{\rho_0}{\rho} = \left[1 + \frac{\gamma - 1}{2} M^2\right]^{\frac{1}{\gamma - 1}} \tag{2.7.16}$$

The methods used to measure or visualize a changing density field nearly invariably rely on the effects of fluid density on some electromagnetic radiation. The techniques can be refractive index dependent (optical techniques), absorption, absorption, and emission. Optical techniques are by far the most advanced and widely used. They are briefly explored in the sections that follow.

7.8 FLOW VISUALIZATION

We can observe flow patterns in gas streams using optical flow visualization techniques sensitive to gas density variations. At high velocities, the density changes can be large enough to cause comparable changes in the refractive index of the gas. Optical techniques provide an important tool for the subsonic and supersonic flow of gases around bodies without interference from probes. These techniques are particularly useful when the gas flow is accompanied by heat transfer and shock waves.

The three principal optical methods, namely, Shadowgraph, Schlieren, and Interferometer depend on the speed of light varies with the density of the medium through which it is passing. We know that the speed of light in any medium, c, is related to the speed of light in vacuum, c_0, by the refractive index,

$$n = \frac{c_0}{c} \tag{7.8.1}$$

For a given medium and wavelength of light, the index of refraction is a function of density, $n = n(\rho)$. The refractive index of a gas is slightly greater than one and can be expressed as a function of density according to Snell's law,

$$n = 1 + \beta \frac{\rho}{\rho_s} \tag{7.8.2}$$

where β is a constant property of the gas and almost independent of wavelength, ρ is the local density, and ρ_s is the reference density at standard atmospheric conditions. The values of β for different gases at 1 atm and 273 K are tabulated in Table 7.3.

Table 7.3
Values of β for different gases at 1 atm and 273 K.

Gas	β
Air	0.000292
Nitrogen	0.000297
Oxygen	0.000271
Carbon dioxide	0.000451
Helium	0.000036
Water vapor	0.000254

equation (7.8.1) and equation (7.8.2) reveals that the speed of light in a medium increases as the density of the medium decreases. Also, changes in the density of a gas result in changes in its index of refraction, which in turn changes the direction of light rays passing through the gas. When these rays are projected onto a screen, the illumination intensity becomes sensitive to the direction of the light rays.

A slightly different form of equation (7.8.2) is sometimes used, called the Gladstone-Dale equation,

$$n = 1 + \kappa\rho \tag{7.8.3}$$

whereκ, the Gladstone-Dale constant, is determined from the measured values of n and ρ. The refractive index is variable in an inhomogeneous medium, such as a variable density flow, as shown by equation (7.8.2) or equation (7.8.3). A light ray that passes through one portion of the variable density flow is retarded differently than one that passes through another. As a result of this, two effects are observed:

1. A turning of wavefronts, that is, the refraction of light rays.
2. Different rays experience a relative phase shift.

The Shadowgraph and Schlieren techniques use the first effect, and the Interferometer technique is based on the second. We'll begin with the refractive effect. Consider a ray of light passing through two fluids of different densities, as shown in Figure 7.8.1(a). The light ray passing through fluid 1 will be refracted upon reaching fluid 2 owing to a change in density. Let i be the angle of incidence of the incoming ray to the normal at the point of incidence, and r be the angle of refraction to the same normal. Thus, from the Snell's law,

$$n_1 \sin i = n_2 \sin r \tag{7.8.4}$$

or

$$\frac{\sin i}{\sin r} = \frac{n_2}{n_1} \tag{7.8.5}$$

where n_1 and n_2 are the refractive indices of fluids 1 and 2, respectively. Notice that the Snell's law is based on Fermat's principle of least time, which is derived from the propagation of light as waves. equation (7.8.5) can be equivalently expressed with equation (7.8.1) as the speed of light in the two fluids,

$$\frac{\sin i}{\sin r} = \frac{\left(\frac{c_0}{c_2}\right)}{\left(\frac{c_0}{c_1}\right)} = \frac{c_1}{c_2} \tag{7.8.6}$$

Combining equation (7.8.3) and equation (7.8.6) yields

$$\frac{c_1}{c_2} = \frac{1+\kappa\rho_2}{1+\kappa\rho_1} \tag{7.8.7}$$

This is an important relationship, stating that light experiences a decrease in velocity as it travels from a lower-density medium, ρ_1, to a higher density medium, ρ_2. As a result, as shown in Figure 7.8.1, a light ray passing from a lower density medium to a higher density medium will bend toward the normal at the point of incidence.

Figure 7.8.1(b) shows the light traveling in a field in which the flow density varies slowly in the y−direction. The lines designated by W are the wavefronts, and R are the rays. Figure 7.8.1(c) shows the progress of a wavefront from position W_1 to W_2 in a small time interval,

$$dt = \frac{d\xi}{c} \tag{7.8.8}$$

If the density If the density increases in the y−direction, the speed of light eventually decreases along this direction; thus, $c_{R_1} > c_{R_2}$, where c_{R_1} and c_{R_2} are the light speeds along the rays R_1 and R_2, respectively. If $c_{R_1} - c_{R_2} \approx dc$, the turning of the light ray, in natural coordinates, is

$$d\phi = \frac{dt \times dc}{d\eta} \tag{7.8.9}$$

and its curvature is

$$\sigma = \frac{1}{R} = \frac{d\phi}{d\xi} = \frac{1}{c}\frac{dc}{d\eta} = \frac{1}{n}\frac{dn}{d\eta} \tag{7.8.10}$$

In Cartesian coordinates, this can be expressed as

$$\frac{1}{R} = \frac{\sin\alpha}{n}\frac{dn}{dy} \tag{7.8.11}$$

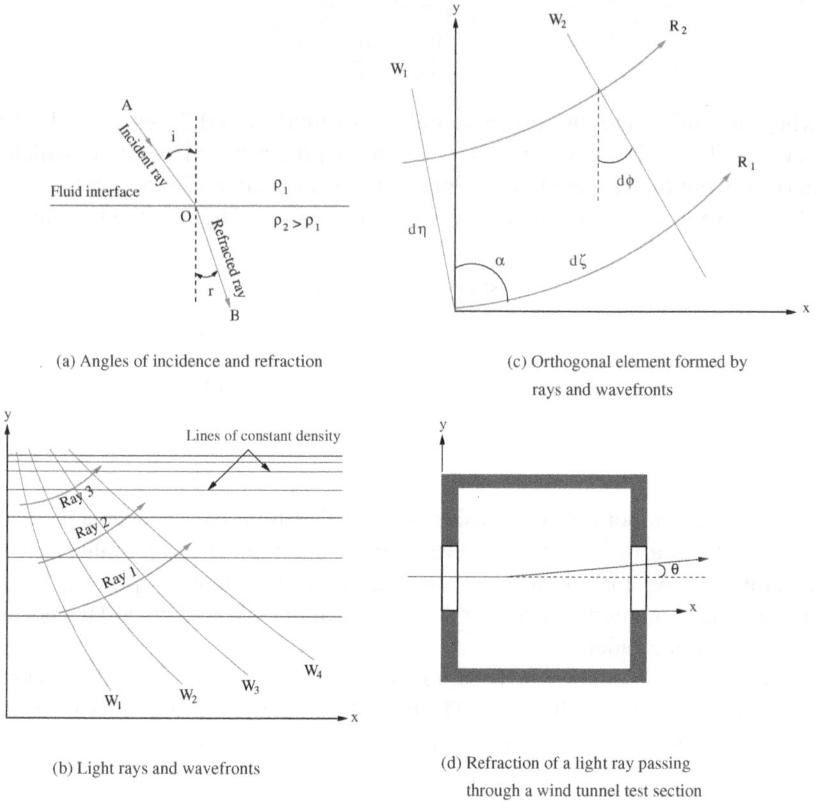

Figure 7.8.1 Refraction in a variable density field.

and in generalized three-dimensional form,

$$\frac{1}{R} = \frac{\sin\alpha}{n}|\nabla n| \tag{7.8.12}$$

where ∇n is the gradient vector in the W−field, and α is the angle between this vector and the ray. equation (7.8.12) indicates that the curvature of light rays (or the wavefronts) is in the direction of increasing flow density. Interestingly, this result is often referred to in optics as Fermat's principle, which states the path taken by a light ray between two points is the one that requires the least time.

For the variable density flow, like that in a wind tunnel, the light usually enters at right angles $(\alpha = 90^0)$ to a side wall, as shown in Figure 7.8.1(d). The curvature of the ray corresponds to density increasing in the y−direction.

Thus, the turning of the ray is

$$\theta = \int d\phi \tag{7.8.13}$$

where the integral is carried out over the length of the ray inside the test section, i.e., the test section width, L. For small ray turning angles, the density along the ray may be assumed constant. Hence,

$$\theta = \int_0^L \frac{1}{R} dx = \int_0^L \left(\frac{\sin 90^0}{n} \frac{dn}{dy} \right) dx = \int_0^L \left(\frac{1}{n} \frac{dn}{dy} \right) dx \tag{7.8.14}$$

equation (7.8.14) is the fundamental relation for the fields small density gradients, such as those occurring in isentropic or near-isentropic flows. For planar flows, where conditions are same in every x−plane, the turning angle, θ, is given by

$$\theta = \frac{L}{R} = \frac{L}{n} \left(\frac{dn}{dy} \right) \tag{7.8.15}$$

Differentiating Gladstone-Dale equation (equation (7.8.3)) and substituting the result in equation (7.8.15) yields

$$\theta = \frac{L\kappa}{\rho_s} \left(\frac{d\rho}{dy} \right) \tag{7.8.16}$$

This equation indicates that for planar flows the emerging ray deflection is proportional to the density gradient, $\frac{d\rho}{dy}$.

Summary

Optical flow visualization techniques are commonly used to visualize compressible flows. Shadowgraph, Schlieren, and Interferometer are the three widely used optical flow visualization techniques for visualizing shocks and expansion waves in supersonic flows. They are based on the refractive index variation, which Gladstone Dale's formula relates to the fluid density and consequently to the pressure and velocity of the flow. The three methods mentioned above are commonly used to make these variations visible. For a reference beam that has passed through a homogeneous medium with refractive index, n:

- Shadowgraph depicts the displacement of an incident ray as it passes through the high-speed flowing gas.
- Schlieren technique shows the incident ray's deflection angles.

- Interferometer shows the optical phase changes caused by the relative retardation of the disturbed rays.

These optical-visualization techniques have the advantage of being non-intrusive and avoiding the formation of unwanted shock or expansion waves in the supersonic flows. They also avoid issues associated with introducing foreign particles that, due to inertia effects, may not exactly follow the fluid motion at high speeds. These techniques, however, must provide information directly on the velocity field. Interferometer, Schlieren, and Shadowgraph optical patterns are sensitive to flow density, its first derivative, and its second derivative, respectively. The Interferometer is typically used for quantitative evaluation because it is based on the precise measurement of fringe patterns rather than the less precise measurement of change in photographic contrast, as in Schlieren and Shadowgraph techniques. However, because the Schlieren and Shadowgraph methods are useful and less expensive, they are frequently used to visualize flow patterns, particularly at supercritical Reynolds numbers. They clearly show the shock waves, and when combined with ultrashort duration recordings, they reveal the flow structure. Although these optical techniques are simple in theory, they are difficult to implement. Proper visualization necessitates high precision and optical quality of the setup components, including the test section windows.

7.8.1 SHADOWGRAPH SYSTEM

The principle of the Shadowgraph, as its name suggests, is to take an image of the shadow of the density change pattern in the wind tunnel. The optical arrangement of a typical Shadowgraph is illustrated in Figure 7.8.2(a). A point light source is located at A, and a lens at B. A visual or photographic screen D receives the light traveling through the test section at C.

The degree of darkness (resistance to light) at any point depends on the rate of change of density gradient with distance, i.e., $\frac{\partial^2 \rho}{\partial x^2}$. The reason for this is best understood by reference to Figure 7.8.3, which represents light rays through and around a cross-section of a shock wave (or any region of changing density) with flow densities ρ_1 and ρ_2 $(\rho_2 > \rho_1)$, upstream and downstream of the wave, respectively. The light rays passing through the region of ρ_2 will travel more slowly than those through the region of ρ_1 because of higher density (equation (7.8.7)). The density variation across the wave will be of the form shown. Hence, the pattern of the light wave front through the shock wave will be as indicated. The amount of the deflection of the light rays is a function of the change of density $\frac{\partial \rho}{\partial x}$. Now, where the density gradient is increasing, i.e., $\frac{\partial^2 \rho}{\partial x^2} > 0$, the rays will diverge, and less light will reach the screen (it will be darker in this region). Likewise, at the higher density edge of the shock wave,

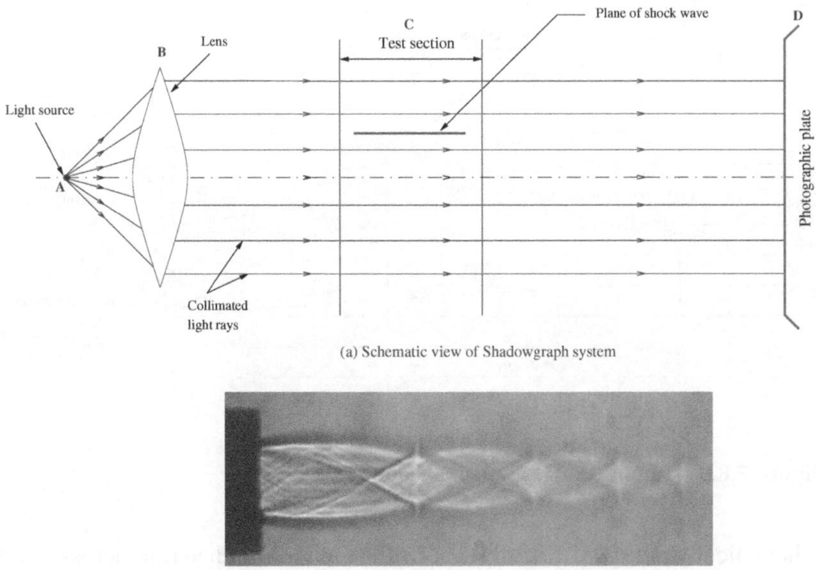

(a) Schematic view of Shadowgraph system

(b) Shadowgraph of Mach 1.73 jet at underexpanded conditions

Figure 7.8.2 Shadowgraph technique.

where the density gradient is decreasing, i.e., $\frac{\partial^2 \rho}{\partial x^2} < 0$, the rays will converge, and the region on the screen will be brighter. If the rate of change of density is constant between these two limits as indicated, the rays will all be deflected a constant amount, and the light intensity on the screen will be neither increased nor decreased. It is clear, then, that a Shadowgraph measures the rate of change of density gradient $\frac{\partial^2 \rho}{\partial x^2}$, and dark areas are regions of an increasing rate of change, and light areas are regions of decreasing rate of change. Therefore, the upstream side of a shock or expansion wave must represent a region of higher density (and pressure) than the downstream side (Figure 7.8.3).

7.8.2 SCHLIEREN SYSTEM

The Schlieren system is used for flow visualization and is based on the principle of refraction of light as a function of density gradient. It can be used to visualize boundary layers, combustion, shock waves, convection currents during heating and cooling, and airflow over models in wind tunnel testing. This visualization technique works on the basis that some deflected light is intercepted before it reaches the viewing screen or photographic plate, making this region of the test section appear darker. The Schlieren system is depicted

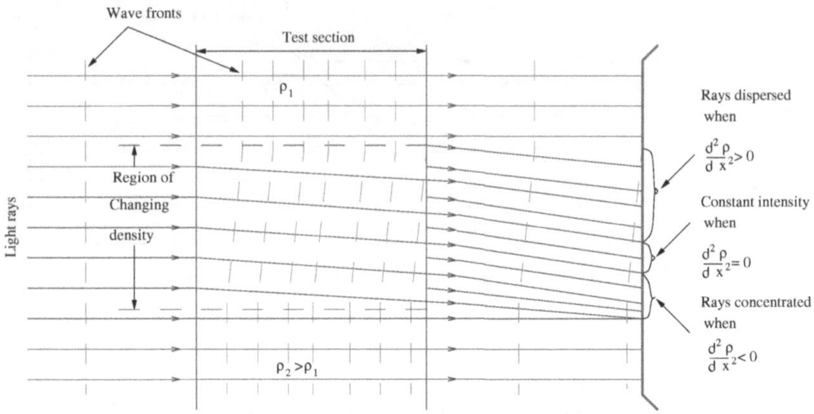

Figure 7.8.3 Turning of light rays in Shadowgraph system.

schematically in Figure 7.8.4(a), where it is used to simulate flow across a two-dimensional test section of a wind tunnel. A beam of parallel, monochromatic light from the source A is collimated by the first lens B and passes through the test section C to a second lens D. An image of the source is obtained by placing a viewing screen or photographic plate in the focal plane of D. Otherwise, light passes through this plane to a third lens, F, which directs it to a screen or plate located at the image plane of the test section. The screen will be uniformly illuminated when there is no flow in the test section. However, when the density of the flow varies, ray deviation occurs, and some rays are no longer concentrated at the focal point. If a knife-edge is introduced at the focal point of lens D so that some of the deviated rays that do not pass through the focal point are intercepted, the illumination at the final screen or photographic plate changes. Where the knife edge intercepts deviated rays, dark areas appear, and where such rays are not intercepted, bright areas appear.

If the source is not a point but has finite dimensions, an image of the source is generated at the focal point of the lens D; the knife edge can partially obstruct this image. In this manner, the total intensity of light in the sensitive apparatus and the intensity of uniform illumination on the screen can be gradually reduced without reducing the intensity of those rays that have deviated due to density variation but have not been intercepted by the knife edge. As a result, the contrast on the screen increases, and thus the system's sensitivity increases.

Since the rays that are eliminated from the screen pattern are those that deviate in the area occupied by the knife edge, changing the direction of the rays in the area in question in a direction perpendicular to the knife edge

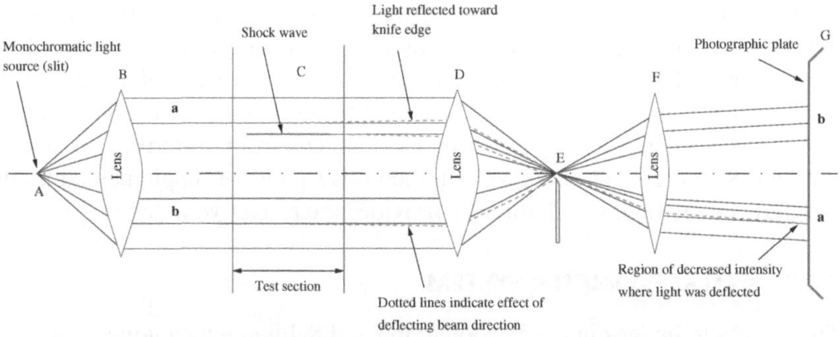

(a) Schematic view of Schlieren system (employing lenses)

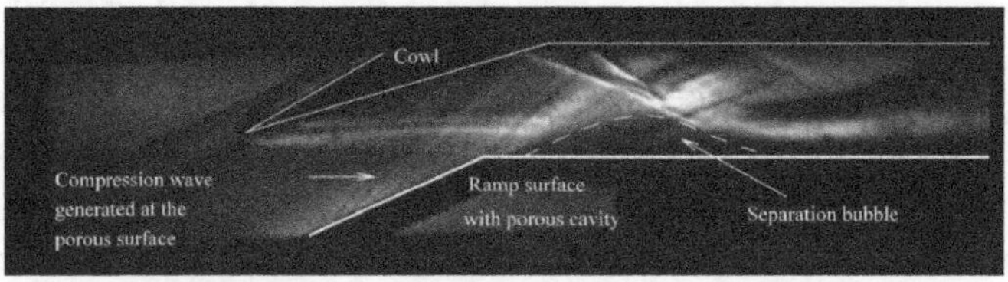

(b) Schlieren image of shock–wave/boundary–layer interactions in Mach 5.7 intake

Figure 7.8.4 Schlieren technique.

determines the increase or decrease in illumination on the corresponding area of the screen. In the Schlieren system, the variation in screen illumination intensity is proportional to the first derivative of the retardation of the light perpendicular to the knife edge. Also, because the deviations have opposite signs, the expansion and compression waves produce opposite variations of light if the density variation has opposite signs. The Schlieren system is typically used in supersonic flows to investigate the shape of shocks and the physics behind the phenomenon. The Schlieren image of shock-wave/boundary-layer interactions in a hypersonic intake at Mach 5.7 is shown in Figure 7.8.4(b).

7.8.2.1 Comparison of Shadowgraph and Schlieren Techniques

The Schlieren technique is dependent on the first derivative of the refractive index (flow density), whereas the Shadowgraph method is dependent on the second derivative. As a result, in phenomena where the refractive index varies slowly, the Schlieren method is preferred over the Shadowgraph method, all else being equal. On the other hand, the Shadowgraph method beautifully highlights the rapid changes in the index of refraction. This method also has the advantage of greater simplicity and a slightly broader range of potential applications. Therefore, the two methods complement each other and should be used whenever possible. Fortunately, in many cases, the same apparatus or optical parts can be used for both methods with minimal effort on the part of the experimenter. The refractive index can be obtained by integration in addition to the first and second derivatives. However, whenever possible, it is preferred to directly measure the density rather than calculate it from its derivatives. Therefore, the Schlieren and Shadowgraph methods should be supplemented by the interference method, which directly provides the refractive index.

7.8.3 INTERFEROMETER SYSTEM

The flow field images in the Shadowgraph and Schlieren techniques discussed in previous sections correspond to the second and first derivatives of the density, respectively. In theory, the field density can be calculated by integrating these derivatives; however, the accuracy of the results is poor. Fortunately, another optical technique, the Interferometer, is available, directly providing the flow density. It is based on the principle of light wave interference. Figure 7.8.5(a) shows a schematic diagram of a photographic density-indicating technique known as the Mach-Zehnder Interferometer, after its inventors.

The lens at B collimated (generates parallel) rays from a monochromatic (single color) light source at A. Special glass plates reflect part of the light and transmit the rest at C and G, while front mirrors (to cancel out the effects of glass thickness) reflect the light as shown at E and D. Suppose the density through path ABCEFGH is identical to that of the path ABCDGH and the mirrors are in perfect alignment (i.e., if both path lengths are identical), is identical to the density through path ABCDGH, and the mirrors are perfectly aligned (i.e., both path lengths are identical). In that case, the original light rays that were split at C are recollected at G and reflected on H as would they have remained as one continuous beam. The intensity of light across the photographic plate will be uniform at H, though slightly reduced due to the reflection of rays a′ and b′ and transmission of rays a and b at plate G.

Now, to understand the effects caused by imperfect plate alignment or changes in density in one of the paths, consider a light wave as an alternating wavelength λ, as shown in Figure 7.8.6. When another ray of light of the same wavelength and phase as the first is superimposed on the first, the intensity of the resulting image increases. If the second wave is out of phase by half a wavelength, the two offset each other and the intensity is zero, i.e., there is no light. As the phase of the two waves is shifted from 0 to $\lambda/2$, the intensity varies from bright to dark.

We can see that the light intensity on the plate is bright as long as the total distance traveled by the light rays ABCDGH from C to G through D is the same or differs by an even number of wavelengths from the distance from C to G through E. Similarly, if the distance difference is adjusted to an odd number of half-wavelengths (by moving mirror D toward C), all of the light is quenched.

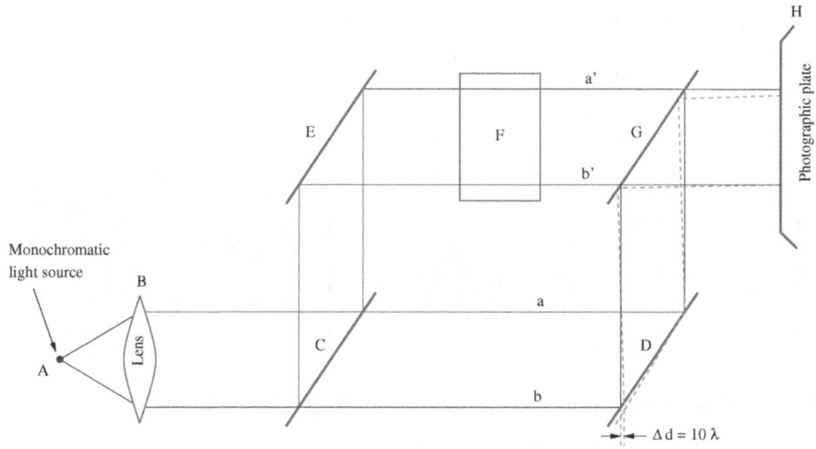

(a) Schematic view of Mach–Zehnder Interferometer

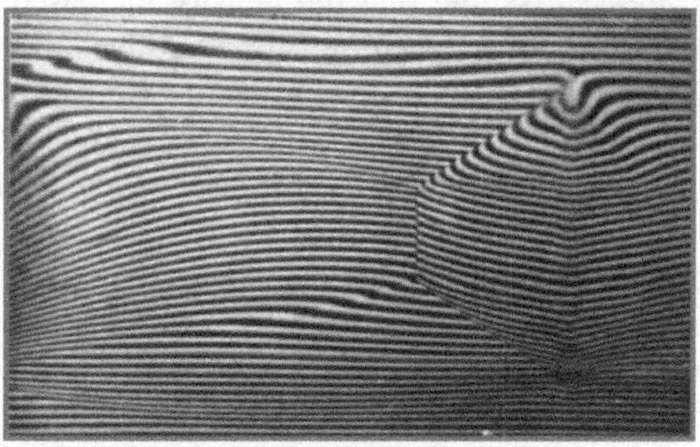

(b) Interferogram of Mach 1.62 jet

(Source: H. Shames, Mechanics of Fluids, McGraw–Hill, 1962)

Figure 7.8.5 Interferometer technique.

On the other hand, if the mirror spacing are kept constant, the density (which changes wavelength λ) is changed in one of the light paths, such at F, but not in the other. The length of the wave at G is again the same as that of the wave from mirror D; however, it was phase shifted due to the change in length as it passed through region F. If the density is uniform throughout the beam, the entire field will be darkened or brightened, depending on the magnitude of the shift caused by area F. Similarly, suppose the density varies in the region F in a direction normal to the beam. In this case, the waves between a′ and b′ will be shifted in phase by a variable amount at each point. Every ray shifted by an even number of half wavelengths produces a light spot on the photographic plate

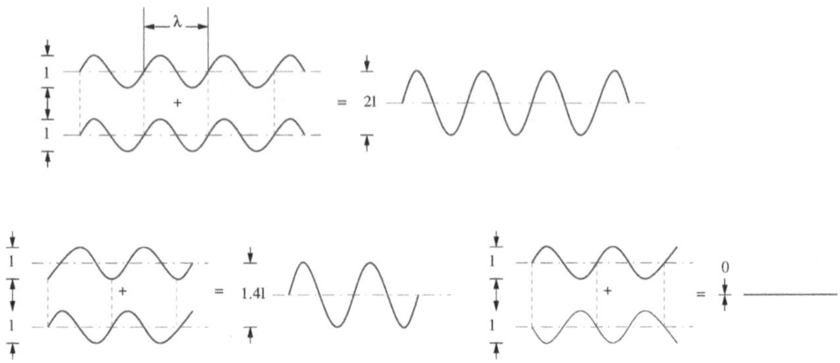

Figure 7.8.6 Reinforcement and cancellation effects of light waves.

H, while every ray shifted by an odd number of half wavelengths produces a dark spot (no light). Assume that the density increased at a constant rate across the beam. In this case, the phase shift could be either an even or an odd number of wavelengths, resulting in a series of evenly spaced fringes on the plate. Similarly, as the density gradient increases, so does the total number of phase shifts in a given distance, and the fringes become closer.

If the density change per phase shift for a given system is known, these fringes will simulate a density gradient of known magnitude. If region F is inserted into the path again, the resulting fringe shifts can be compared to the known variation, and the density can be measured. Figure 7.8.5(b) shows an interferogram of a two-dimensional Mach 1.62 jet.

7.9 SUMMARY

This chapter explored numerous ways to measure fluid properties in compressible flows concisely. The common techniques employed in wind tunnel measurements are covered. However, many of these methods apply equally to measurement problems in real flight situations.

Static pressure is measured on an aerodynamic surface, such as a wind tunnel wall or an airfoil, by drilling a small hole perpendicular to the wall and attaching a manometer or other pressure gauge. These orifices are also known as static pressure ports.

The total or stagnation pressure in a flow is defined as the pressure when the flow is brought to rest isentropically. A pitot probe, a simple thin tube with an opening at the nose tip, is oriented parallel to flow to measure stagnation pressure. The opposite end is attached to a pressure measurement equipment. The fluid within the tube remains at rest.

A pitot probe in supersonic flows does not detect the local stagnation pressure because a bow shock arises ahead of the probe nose. Thus, the probe measures the stagnation pressure downstream of the shock wave.

In compressible flows, the Mach number is one of the most important parameters that may be calculated from numerous equations in which it appears, given that other values, such as pressures, are measured. If the flow field at the measurement point is isentropic, the stagnation pressure will stay equal to the reservoir pressure, p_0. A static pressure measurement will then allow computing Mach number using the following relation (applicable for both subsonic and supersonic flows):

$$\frac{p_0}{p} = \left(1 + \frac{\gamma - 1}{2} M^2\right)^{\frac{\gamma}{\gamma - 1}}$$

The static pressure p at the wall is determined using static pressure ports and, for the flow field, using a static probe.

If the flow is supersonic, the pressure indicated by the pitot probe is p_{02}, the stagnation pressure downstream of the normal shock. In this case, since actual stagnation pressure p_{01} cannot be measured, the Mach number may be obtained using the following Rayleigh-pitot formula.

$$\frac{p}{p_{02}} = \frac{\left[\frac{2\gamma}{\gamma+1} M_1^2 - \frac{\gamma-1}{\gamma+1}\right]^{\frac{1}{\gamma-1}}}{\left[\frac{\gamma+1}{2} M_1^2\right]^{\frac{\gamma}{\gamma-1}}}$$

The velocity of a perfect gas can be calculated using the following equation:

$$v = M\sqrt{\gamma RT}$$

Furthermore, the isentropic relations can be used to calculate the Mach number using static and stagnation pressure measurements:

$$\frac{p_0}{p} = \left[1 + \frac{\gamma - 1}{2} M^2\right]^{\frac{\gamma}{\gamma - 1}}$$

and using static and stagnation temperature measurements:

$$\frac{T_0}{T} = 1 + \frac{\gamma - 1}{2} M^2$$

The inclination of shock waves in supersonic flow can also be used to compute the Mach number. The inclination of Mach waves, known as the Mach angle, is given by

$$\mu_M = \sin^{-1}\left(\frac{1}{M}\right)$$

A hot wire anemometer is a type of thermal transducer that measures flow velocity and direction. This instrument's sensing element is an electrically heated thin metal wire that is exposed to freestream and oriented directly toward the incoming flow. A Wheatstone bridge is connected to the wire, and an electric current is sent through it to elevate the wire temperature above the freestream temperature. When a heated wire is exposed to a flow, it cools as a result of convective heat transfer. Because the electrical resistance of most metals varies with temperature, the following relationship can be established between wire resistance, R, and airspeed, v.

$$Q = i^2R = \left(c_1 + c_2\sqrt{v}\right)(T_w - T_\infty)$$

This is also known as Kings law, where c_1 and c_2 are constants for a given wire and are estimated by calibrating the wire in a given flow. The theoretical values of c_1 and c_2 can be calculated using the heat transfer correlations for the flow past the heated wire. Hot wire anemometers are classified into two types: constant temperature anemometers (CTA) and constant current anemometers (CCA). Following Ohm's law, the voltage produced by these anemometers is thus the consequence of some form of circuit within the device attempting to maintain the constant of the specified variable (temperature or current).

Particle Image Velocimetry (PIV) is a non-intrusive flow visualization technology that allows viewing two or three-dimensional flow fields. It is a whole-flow-field technique for obtaining instantaneous velocity vectors over a flow cross-section. Because this technique is non-invasive, it can be used to investigate supersonic and hypersonic flows, and boundary layer fluid flows. A typical PIV setup consists of a high-speed camera, a high-power multi-phased laser, an optical arrangement to convert laser output to a light sheet to illuminate the region of interest, tracer particles, a synchronizer to control the synchronization between the laser and the camera, a fiber optic cable or liquid guide to connecting the laser to the lens setup, and a dedicated computer system (with PIV software installed) for saving the particle images and conduct image processing.

Laser Doppler Velocimetry (LDV) is a technique for determining a flow field's instantaneous velocity. Like PIV, this method is non-intrusive and can evaluate all three velocity components. The laser Doppler velocimeter projects a monochromatic laser beam onto an object and measures the reflected radiation. According to the Doppler effect, the change in wavelength of reflected radiation is a function of the target object's relative velocity. Thus, the object's velocity can be measured by detecting the change in wavelength of the reflected laser light.

The static temperature can be computed using the equation of state from the measured pressure and density; if the flow field is isentropic, one of these is sufficient. However, no direct method exists for determining static temperature in fluid flows. Because of the boundary layer on its surface, the temperature detected by a device, such as a thermometer, is higher than T. The static temperature rises from T at the boundary layer edge to T_r at the surface (also known as the recovery temperature). Because T_r varies across the surfaces of a measuring device, the thermometer will display a mean recovery temperature.

In theory, measuring stagnation or total temperature T_0 is straightforward. Because the flow inside a pitot probe comes to rest at equilibrium, the temperature inside the probe in both subsonic and supersonic flows should be the stagnation temperature. As a result, the temperature can be detected by inserting a thermometer into the probe. Because heat is conducted or radiated from the thermometer surface or probe wall, the equilibrium condition does not exist in reality. Consequently, the stagnation temperature decreases from T_0 to a lower value, T_r. A probe of this type is highly useful for determining the coefficient of thermal recovery (recovery factor),

$$r = \frac{T_r - T_\infty}{T_0 - T_\infty}$$

very close to 1.

In many applications, flow density must be determined in addition to pressure. All other variables can be easily determined after measuring pressure and density. For example, the speed of sound can be calculated by plugging the measured static pressure and density into the equation:

$$a = \sqrt{\frac{\gamma p}{\rho}} \tag{7.9.1}$$

The velocity is then calculated using the following equation:

$$v = aM \tag{7.9.2}$$

The following formulae are used to calculate the local values of static and stagnation temperature:

$$p = \rho RT$$

$$\frac{T_0}{T} = 1 + \frac{\gamma - 1}{2} M^2$$

In the case above, density measurements accompany pressure measurements, but density measurements alone are occasionally required. For measuring or visualizing the density alone in a variable density field, optical flow visualization techniques are commonly used. Shadowgraph, Schlieren, and Interferometer are the three widely used optical flow visualization techniques for visualizing shocks and expansion waves in supersonic flows. They are based on the refractive index variation, which Gladstone Dale's formula relates to the fluid density and consequently to the pressure and velocity of the flow. For a reference beam that has passed through a known refractive index homogeneous medium:

- Shadowgraph depicts the displacement of an incident ray as it passes through the high-speed flowing gas.
- Schlieren technique shows the incident ray's deflection angles.
- Interferometer shows the optical phase changes caused by the relative retardation of the disturbed rays.

EXERCISE PROBLEMS

Exercise 43. A supersonic wind tunnel test section has a dynamic pressure of 150 kPa. Find the Mach number and total pressure in the test section if a static pressure port on the test section wall provides a measurement of 80 kPa.

Exercise 44. If a wall-mounted static orifice shows a reading of 90 kPa, find the pressure reported by a pitot probe at Mach 2.5 in the test section of a wind tunnel.

Example 45. A blunt-nose pitot probe deployed in the test section of a supersonic wind tunnel reads 350 kPa. The static pressure is measured as 200 kPa using a wall-mounted static pressure port in the test section. Find the Mach number in the test section. Assume that the test section walls are appropriately distant from the probe nose.

Exercise 46. In a Mach 2 airstream, the temperature measured by a stagnation temperature probe is 230°C. If the probe has a recovery factor of 0.75, calculate the static and stagnation temperatures of the airstream.

8 Shock Wave and Boundary Layer Interaction in a Hypersonic Intake

8.1 INTRODUCTION

In addition to developing advanced combat aircraft in modern warfare, research into passenger aircraft flying at supersonic and hypersonic speeds is gaining importance due to its obvious advantages. Recently, fighter jets, missiles, and reentry vehicles flying at supersonic and hypersonic Mach numbers have been the area of particular research attention in the aerospace industry. In order to have stable and safe flight operations, the vehicle should be designed to withstand high-pressure and -thermal loads at high Mach numbers. Also, the effective design of a flight improves its aerodynamic performance. Although there are several advantages associated with high-speed aircraft engines, they are susceptible to significant disadvantages that should be carefully considered. The vehicle moving at speeds above the speed of sound often results in the most undesirable consequence, the generation of a non-isentropic shock. A shock wave is always produced whenever a supersonic or hypersonic flow encounters a turning in its path like that due to the geometrical corner of a body. It also forms when the flow experiences a pressure imbalance along its path, like that between the nozzle exit and ambient pressure. Because a boundary layer forms on the solid surface, its interaction with a shock wave in a supersonic or hypersonic flow is unavoidable. This type of interaction is commonly referred to as the shock wave/boundary layer interaction (SBLI).

8.1.1 SHOCK-WAVE/BOUNDARY-LAYER INTERACTION

For a compressible flow, the physical nature is described by the speed of sound; hence, it is given utmost importance. Essentially, the propagation of a weak perturbation occurs through molecular collisions at the speed of sound in a fluid. When a solid body is introduced into a fluid, the momentum of the fluid molecules surrounding the body changes, this change in momentum propagates as a disturbance that informs the body's existence to the upstream flow. So for a subsonic flow, the flow is aware of the body well in advance and

DOI: 10.1201/9781003139447-8

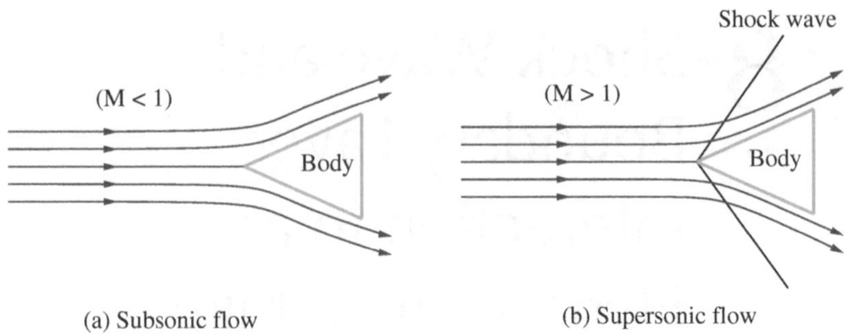

Figure 8.1.1 Subsonic and supersonic flows past a wedge.

adjusts accordingly into a smooth flow past the body (Figure 8.1.1). But in the case of supersonic flow, this disturbance gets piled up ahead of the body as it cannot propagate upstream. Thus, a thin compression front, defined as a shock wave, is generated, and since the upstream flow would receive no information about the body's presence, the flow would abruptly change direction across the shock wave. This results in high gradients in the flow properties such as pressure, density, temperature, etc. This region will also have viscous dissipation because of sudden velocity and temperature changes.

The adverse pressure gradient generated across the shock-impinging region in an SBLI thickens the boundary layer. Sometimes, it could result in boundary layer separation if the gradient is high enough. The upstream flow assumes the boundary layer thickening as a geometric inclination, and the flow starts to deflect toward the freestream. This results in the generation of compression waves that, in turn, merge into a shock wave, and the same can be noticed once the boundary layer reattaches. The shock, which is incident on the boundary layer, reflects from the sonic line as expansion waves. As a result, a series of compression and expansion waves are generated because of the SBLI, causing a huge total pressure loss, drag increment, unsteady oscillations of shock waves, and so on that could ultimately induce engine failure. Turbulence intensity may also be increased because of the SBLI; thus, heavy viscous dissipation can occur.

The SBLI pattern for a two-dimensional domain is depicted by Figure 8.1.2(a) for a relatively weak shock. The reduction in local flow speed within the boundary layer eventually bends the shock toward itself ultimately decreasing the shock strength rapidly till sonic line. Thus, the downstream and upstream directions of the subsonic boundary layer experience a smooth pressure rise induced by the shock. Figure 8.1.2(b) illustrates this in lengthwise direction. This results in the thickening of boundary layer's subsonic portion.

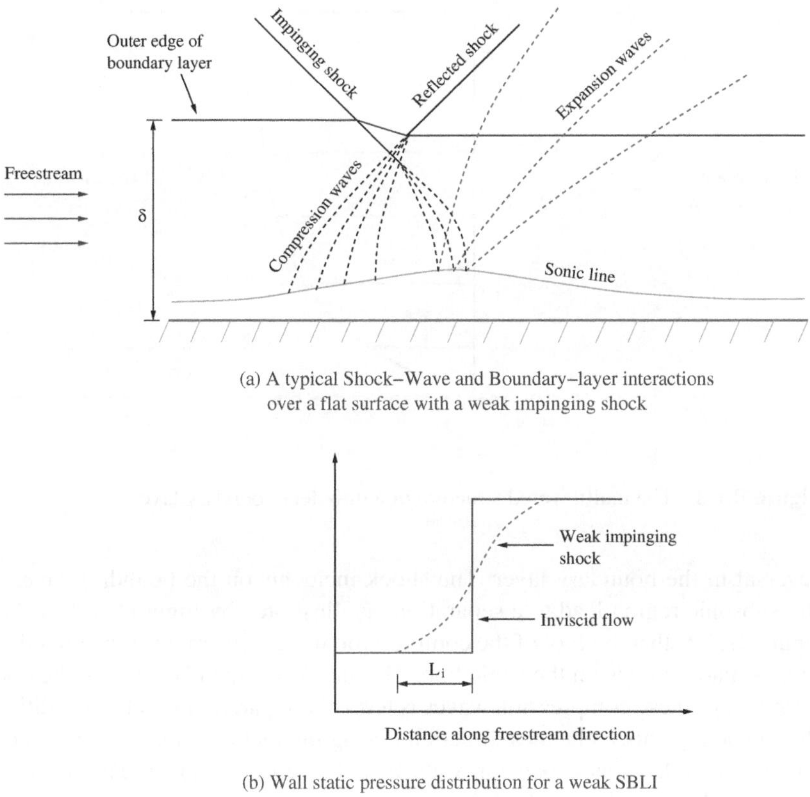

Figure 8.1.2 SBLI due to a weak shock.

While the flow above the sonic line being supersonic gets deflected and produce compression waves. This series of compression waves merge together into a single reflected shock outside the boundary layer as the physical phenomenon behind this merging resembles the conditions of an inviscid flow. This is known as weak viscous-inviscid interaction. Note that the incoming flow conditions strongly affect the upstream interaction length, denoted by L_i in Figure 8.1.2(b).

Figure 8.1.3 shows that a turbulent boundary layer consists of three different layers: rotational-viscous layer (inner layer), rotational-inviscid layer (middle layer), and irrotational-inviscid layer (outer layer). The mutual interactions of these layers and the incoming flow behavior cause a significant change in the upstream interaction length.

Considering a flow situation in which the unfavorable pressure gradient induced by the shock is intense enough, the subsonic region experiences a flow

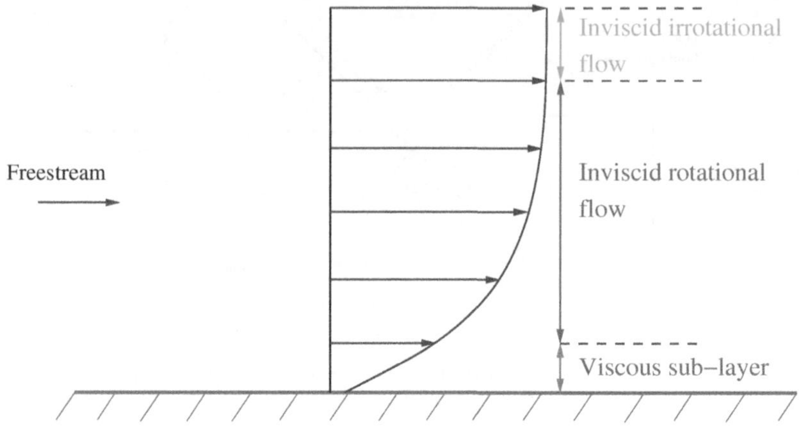

Figure 8.1.3 The multilayered structure of a turbulent boundary layer.

reversal in the boundary layer. The shock incidents on the boundary layer at the subsonic region lead to a separation, as illustrated by Figure 8.1.4(a). The figure depicts that the foot of the compression waves generated at the boundary layer separation lies on the sonic line. The shock obtained because of the convergence of these compression waves is termed a separation shock. In addition, the impacting shock changes direction during interaction with the detachment shock and reflects as expansion waves by the separated layer. The detached boundary layer is thus deflected and accumulated after some distance downstream, resulting from interchanging fluids across the detached shear layer. This again generates a train of compression waves which eventually converge, forming a reattachment shock; thus, the boundary layer reattaches. This interaction process is defined as a strong viscous-inviscid interaction since strong viscous effects dominate it. The pure inviscid flow behavior is incomparable with this phenomenon. Figure 8.1.4(b) shows a strong SBLI-induced static pressure increment over the wall. Two stages of pressure increment separated by a plateau can be observed here, unlike the previous one. The two jumps indicate both separation and reattachment processes, respectively. Similar to the weak interaction process, the properties of upstream flow and viscous-inviscid interactions govern the pressure rise at the separation region.

SBLIs may lead to many unwanted effects which can cause the entire engine failure, and thus it is given sufficient attention in high-speed aerodynamics studies. Stagnation pressure loss, boundary layer separation, flow discontinuity, heavy thermal loads, and so on are some major consequences of an SBLI. A lot of noise is generated when the SBLIs occur on the helicopter

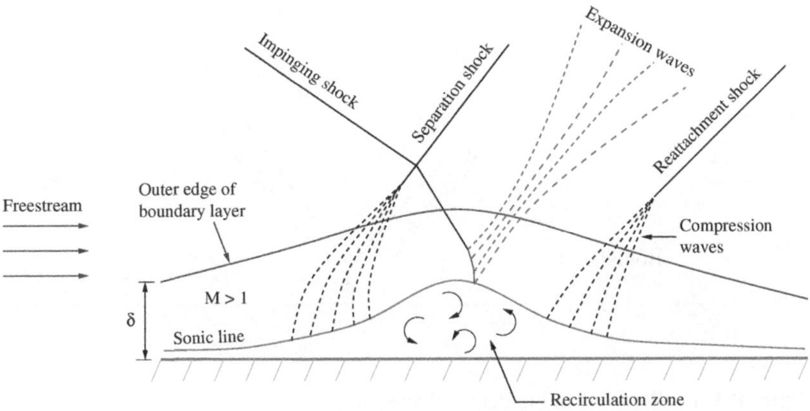

(a) A typical Shock–Wave and Boundary–layer interactions over a flat surface with a strong impinging shock

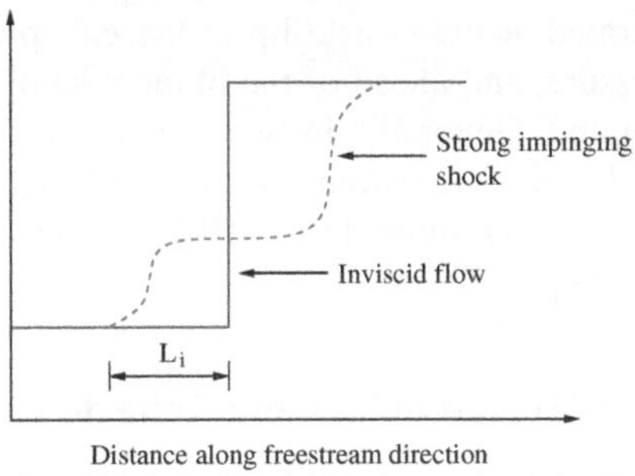

(b) Wall static pressure distribution for a strong SBLI

Figure 8.1.4 SBLI due to a strong shock.

blades and reduces overall efficiency in the case of turbomachine blades. It can also increase aerodynamic heating because of high-viscous dissipation as it increases the near-wall temperature. This becomes even worse in the case of hypersonic flows, as it consists of viscous and vortex interaction and higher shock layer temperatures. Therefore, suppression of shock wave/boundary layer Interaction (SBLI) is crucial.

8.1.2 TYPES OF SBLIS

Recently, many situations can be specified where SBLIs are frequently observed. Examples include: over a transonic airfoil, normal shock is generated

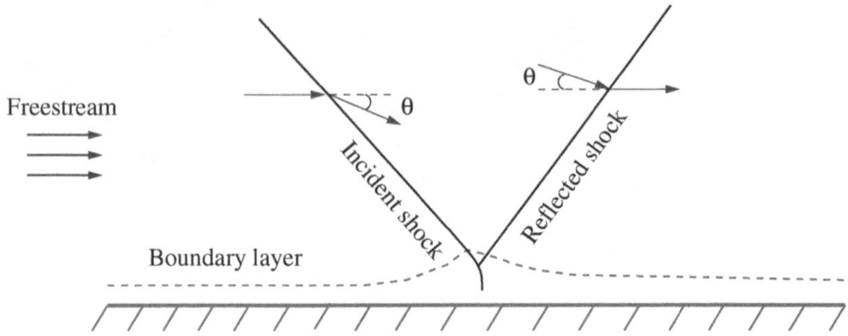

Figure 8.1.5 SBLI due to incident-reflected shocks.

at the small pocket of the supersonic flow region, combinations of oblique and normal shocks are observed in high-speed intakes, and oblique shock waves are generated at the nozzle lip at the exit pressure, which is lower than the back pressure, and ahead of the blunt nose of a hypersonic body, bow shocks are generated. Generally, in two-dimensional flow situations, SBLI phenomena are classified into four categories: SBLI due to a shock reflected by an event, SBLI in a ramped-flow, SBLI due to a normal shock, and SBLI due to a pressure jump.

8.1.2.1 SBLI Due to Incident-Reflected Shocks

The impingement of an oblique shock over the boundary layer deflects the incoming flow at a given angle, as seen in Figure 8.1.5. As an outcome, the reflected shock is produced in order to keep the flow parallel to the wall surface. The boundary layer thickens or, in the worst-case scenario, detaches as the pressure above the shock wave abruptly increases. This phenomenon is frequently found in the intake's isolator region.

8.1.2.2 SBLI in a Ramped-Flow

In the compression ramp, the abrupt changes in the wall inclination generate the oblique shock wave over which flow passes. This oblique shock induces unfavorable pressure over the boundary layer, causing it to bulge or could also result in separation if the shock is sufficiently strong (Figure 8.1.6). This leads to more complex shock patterns and discontinuities in the flow structure. In this situation, the complex shock patterns due to reattachment and separation shocks and the intense vortex activity inside the separation bubble dominate the interaction.

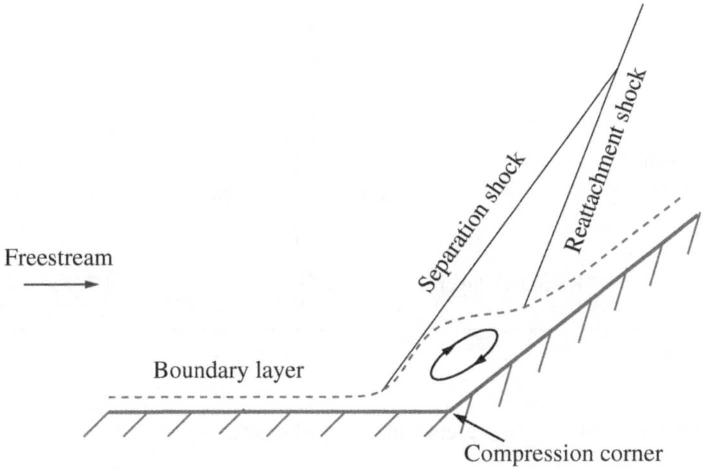

Figure 8.1.6 SBLI over a single ramp surface.

8.1.2.3 SBLI Due to a Normal Shock

Within a duct or over a transonic wing, the normal shock appears to end the locally established supersonic flow regime (Figure 8.1.7). In such a situation, the downstream disturbance can interact with the normal shock and cause large-scale flow instability.

8.1.2.4 SBLI Due to Pressure Jumps

The inequality between the nozzle exit and the ambient pressures causes the formation of oblique shocks at the tip of a supersonic nozzle at overexpanded conditions. The interaction of oblique shock with the jet boundary results in SBLIs, as shown in Figure 8.1.8.

8.1.3 CONSEQUENCES OF SBLIS

The undesirable consequences of SBLIs are numerous and very severe, which may even lead to the failure of the entire structure. Therefore, quantifying these harmful phenomena is essential. These consequences include flow field instability, dynamic pressure loss, boundary layer separation, and huge thermal stress. Normally the losses are divided into two categories, total pressure loss, and viscous loss, that result in wave drag and viscous drag, respectively. The SBLI extends the shock-induced adverse pressure gradient, and the propagation of pressure disturbance also influences the upstream flow field. Thus, the

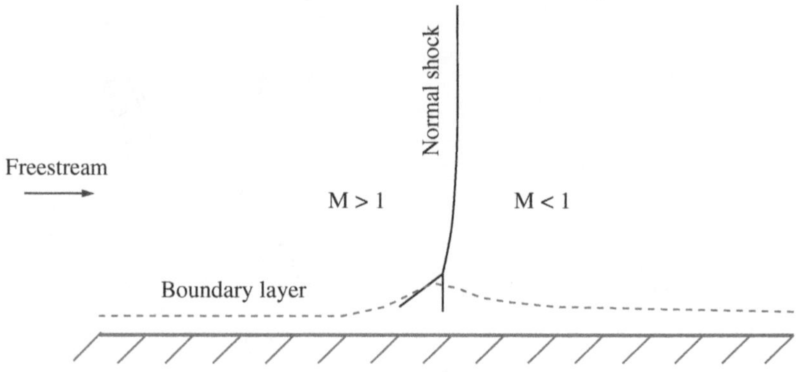

Figure 8.1.7 SBLI in the presence of a normal shock.

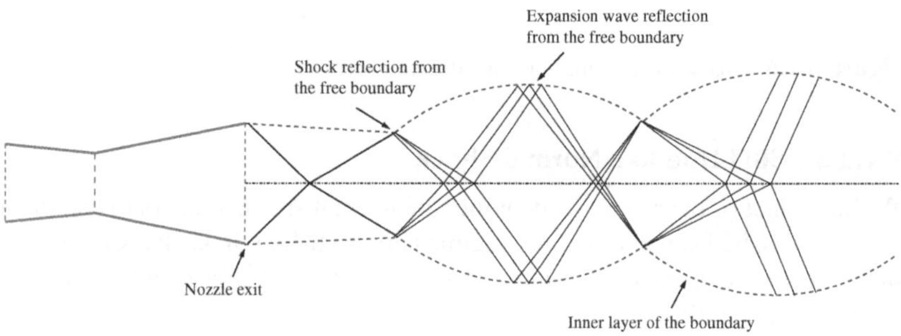

Figure 8.1.8 SBLIs at the exit of an overexpanded supersonic nozzle.

finite area of the boundary layer experiences this unfavorable pressure gradient. Therefore, the boundary layer starts bulging or may even detach, resulting in a separation bubble. In an inlet isolator, the thick boundary layer or separated region changes the inlet geometry, reducing the effective mass flow rate, and can lead to the intake unstart. For these reasons, a decrease in propulsive efficiency is usually observed due to uncontrolled interactions in the intake. In addition, these interactions result in flow field discontinuity and fluctuating pressure loads, resulting in wing buffeting or intake buzz. In a transonic wing, the locally generated shock wave interacts with the boundary layer causing wave drag and instability of the flow field. The SBLIs occurring on the helicopter rotor blades cause a lot of noise. The SBLIs occurring over turbo machine blades also reduce the overall efficiency. In addition, it increases aerodynamic heating as the near-wall temperature of the body increases because of the high-viscous dissipation. These disadvantages become more prominent in the case of hypersonic flows because of the eddy viscous interactions and increased shock layer temperature.

8.2 NEED AND OBJECTIVE OF SBLI CONTROL

Because SBLIs are connected with many penalties, and it is critical to limit the consequences, control is required to reduce the shock strength or suppress the flow separation. As a result, overall pressure recovery is improved, and viscous losses are reduced. An efficient control mechanism reduces thermal load and instability in the flow field. The control strategies are primarily introduced before or at the place of the encounter, as the nature of the incoming flow influences the severity of the interaction. Depending on the individual requirements of an aircraft component, various control systems can be used. Nonetheless, the controls should be deployed in an optimum manner to improve an aircraft's overall efficiency.

8.2.1 CLASSIFICATION OF SBLI CONTROLS

SBLI controls are classified as shock or boundary layer controls based on their specific use. However, depending on the amount of energy required, they are also categorized as active or passive controls.

8.2.1.1 Shock Control

Shock control techniques essentially deal with reducing shock strength, thereby resulting in total pressure recovery. The main principle of this mechanism is to smear or split a strong shock into several weak shocks. For example, a single normal shock can be smeared onto a shock system through passive control, reducing shock strength. The well-established shock control techniques employ a porous surface over a cavity or the surface bump at the point of shock impact. Figure 8.2.1(a) illustrates a typical shock control mechanism where a cavity is covered with a porous surface. The basic mechanism of surface perforation with the cavity is to recirculate the high-pressure fluid downstream of the shock to relatively low-pressure fluid upstream via the cavity. Fluid is injected into the boundary layer upstream of the shock to increase the thickness at the interaction region. This causes strong shocks to transform into weaker shocks; thus, interaction strength is reduced in turn wave drag.

8.2.1.2 Boundary Layer Control

The other type of SBLI manipulation technique is boundary layer control. Because the resistance to separation is affected by the energy level of the incoming boundary layer, in this control technique boundary layer is excited.

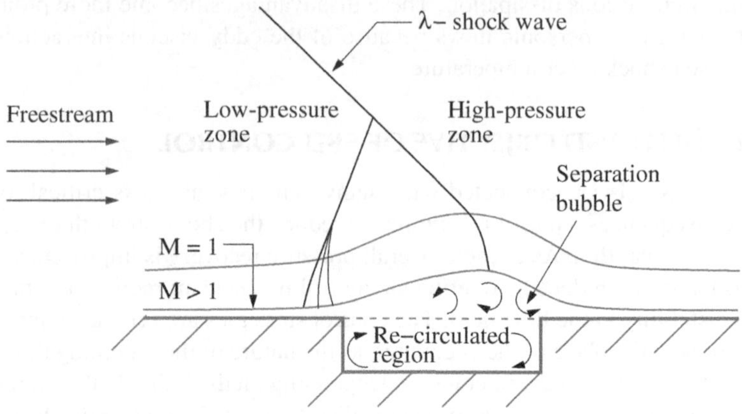

(a) Porous surface deployed over a cavity

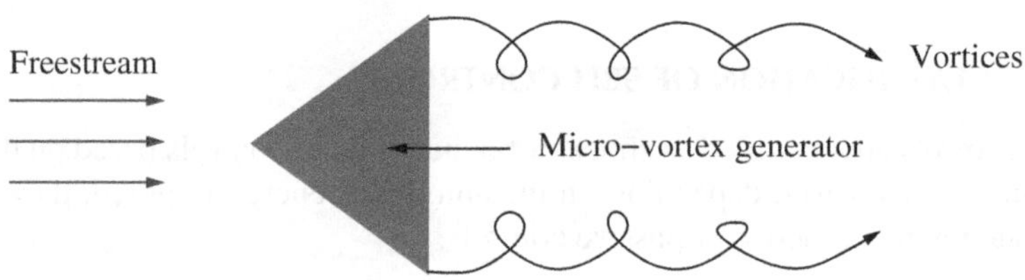

(b) Streamwise vortices generated from the vortex generator

Figure 8.2.1 Effects of a shock and a boundary layer control deployed in an intake.

The conventional boundary layer control method involves the introduction of vortex generators into the flow field. Generally, the vortex generators induce vortices in the flow direction by creating a wake behind. These vortices enable momentum transfer between fluids in the freestream and the near-wall region (Figure 8.2.1(b)). Therefore, the high-momentum fluid gets mixed with boundary layer fluid at low momentum, reenergizing the boundary layer and delaying the separation.

8.2.1.3 Active Control

Additional energy is required to work in the active control technique. Active controls include injection of fluid into the boundary layer, suction of the boundary layer, plasma actuator, etc. In this technique, extra power is necessary, which may cause a loss in the overall engine efficiency. Also, the engine configuration becomes complex and bulky due to the employment of additional components.

8.2.1.4 Passive Control

With passive control techniques, no additional power is required as the mechanism draws power from the mainstream. Because of this, the mechanism is simple, compact, and inexpensive. A cavity covered with a porous surface, a vortex generator, and a surface hump are some frequently used passive control mechanisms.

8.3 SOME SPECIFIC STUDIES ON SBLIS

Few predominant studies on SBLIs using shock or boundary layer controls deployed in a hypersonic intake are discussed in the following articles.

8.3.1 MOTIVATION AND SCOPE OF INVESTIGATIONS

Advances in supersonic and even hypersonic flight have attracted the attention of several researchers over the past several decades. However, at supersonic and hypersonic speeds, the undesirable consequences of generating a non-isentropic shock often result in an SBLI. The negative consequences of an SBLI are mainly boundary layer thickening or detachment, total pressure loss, massive flow distortion, instability in the flow field, aerodynamic heating, etc. Sometimes they could be exacerbated, especially for hypersonic flow, because of the viscosity/turbulence interactions and significantly hotter shock layers. Therefore, a thorough understanding of SBLIs must have a fail-safe aircraft engine design for the smooth and safe operation of high-speed military or transport aircraft.

An aircraft intake is a component of immense importance. Providing required mass flow with minimal total pressure as well as viscosity losses while maintaining a stable shock structure is the primary purpose of an intake. Because for a supersonic or hypersonic aircraft, intake power is directly proportional to propulsion efficiency, and it is the subject of extensive research attention. Although the intentionally generated shock waves by intakes are highly efficient in compressing the flows, they come up at the expense of SBLIs. The SBLI effects are large at hypersonic speeds due to the thick viscous layer developed on intake surfaces and the predominance of aerothermal effects. Therefore, special attention is always paid to high-speed intakes, especially hypersonic intakes, to minimize the impact of SBLIs.

Literature reveals that the investigations on SBLI characteristics are normally conducted in simplified models such as a flat surface subjected to externally generated shock or the compression corner, where the shocks are induced due to the abrupt change in wall inclination. However, this situation may differ in real operations since the intake geometry is complex and dominated by multiple shocks and expansion waves. In a single ramp intake, a strong oblique shock is generated at the entrance resulting in a higher drag. Thus, the mixed compression intake double ramp is considered to study the SBLI characteristics inside the intake in the following investigations. The study investigates the efficacy of the control mechanisms deployed over the ramp surface to reduce SBLIs. Though the computational techniques recently employed to study the hypersonic flow characteristics over simple geometries, they are less effective in accurately capturing the flow physics in complex geometry like double ramp mixed-compression intake. Specifically, in the hypersonic domain, the computational resources could be more robust for visualizing the SBLI's unsteady nature. Hence, the present investigations aim to explore the unsteady nature of SBLIs in a hypersonic intake by experimental means. These investigations employed a shallow cavity covered with a porous surface and micro vortex generators (MVGs) as shock and boundary layer controls, respectively, to examine and compare the uncontrolled and controlled interactions inside a hypersonic intake. At first, the study is carried out by varying the surface perforation limit over the control area to investigate their control efficacy. After that, the effectiveness of MVGs in manipulating the SBLIs in the hypersonic intake by varying MVGs height, placed separately at two different locations, is investigated experimentally. In both controlled and uncontrolled intakes, the wall static pressures on the ramp surface are measured. Schlieren technique is used to visualize the prevailing waves.

8.3.2 EXPERIMENTAL METHODOLOGY

In the following investigations, the hypersonic shock tunnel facility of the Indian Institute of Science, Bangalore, is utilized for conducting experiments, and Figure 8.3.1 shows the photographic view of the facility.

The principle behind a shock tube operation is compressing a working fluid by generating a moving shock when the fluid from the high-pressure section is instantaneously released into a low-pressure section. A metal diaphragm is

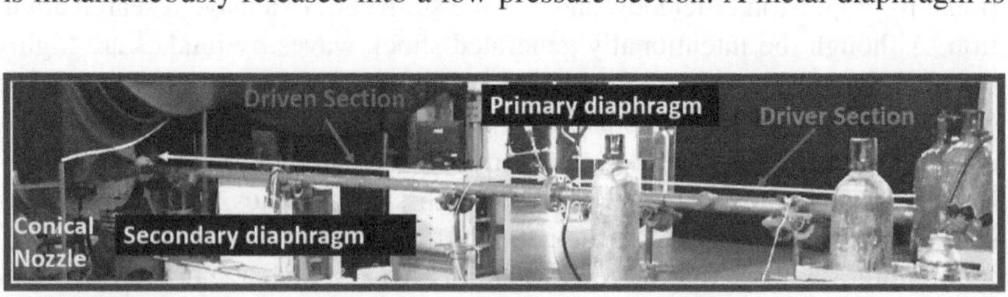

Figure 8.3.1 Photographic view of a hypersonic shock tunnel.

utilized to separate the high-pressure (driver) and the low-pressure (driven) regions. A non-reacting gas like helium is utilized as driver gas at sufficiently high pressure at which the diaphragm ruptures. Besides, the driven section consists of low-pressure test gas, which is less than atmospheric pressure. One end remains closed while the other is attached to a conical convergent-divergent nozzle. Generally, the nozzle's divergent section is longer than the convergent section to keep the divergence angle small. This essentially helps reduce or mitigate flow from separating from the walls of the divergent section. Depending upon a high or low Mach number range, the nozzle can be operated with the throat insert (reflected mode) or without the throat insert (straight-through mode). A secondary diaphragm is utilized to separate the shock tube's driven section from the nozzle at the start. It is made of tracing paper; hence it is known as the paper diaphragm. Separation of test gas from other sections (operating at vacuum) like nozzle, dump tank assembly, and test section is the sole purpose of the secondary diaphragm. The secondary diaphragm ruptures when the driven section pressure exceeds atmospheric pressure slightly (soon after the primary diaphragm ruptures). The nozzle, dump tank assembly, and test section are maintained at a high-vacuum pressure to achieve a high difference in pressure of the nozzle inlet and exit sections. Generally, the dump tank pressure is maintained at 10^{-8} bar (considered a vacuum). Table 8.1 gives dimensional information about the shock tunnel.

Table 8.1
Typical dimensions of the hypersonic shock tunnel facility.

Length of the test section	450 mm
Cross-sectional area of the test section	300 mm × 300 mm
Diameter of the shock tube	50 mm
Length of the driven section	2 m
Length of the driven section	5.12 m

The primary diaphragm controls the pressure of the helium gas when discharged into the driver section. A certain pressure ruptures the primary diaphragm leaving the driver gas at high pressure into the driven section, which is nearly vacuum. The instantaneous expansion of driver gas into the vacuum generates shock and expansion waves. The shock wave propagates till the downstream end of the driven section and gets reflected. So both primary and reflected shock waves increases the temperature and pressure of the test gas (air is used in the current study). With the help of a convergent-divergent nozzle, this high-pressure and high-temperature test gas is expanded to achieve the required Mach number (essentially hypersonic – Figure 8.3.2). The high

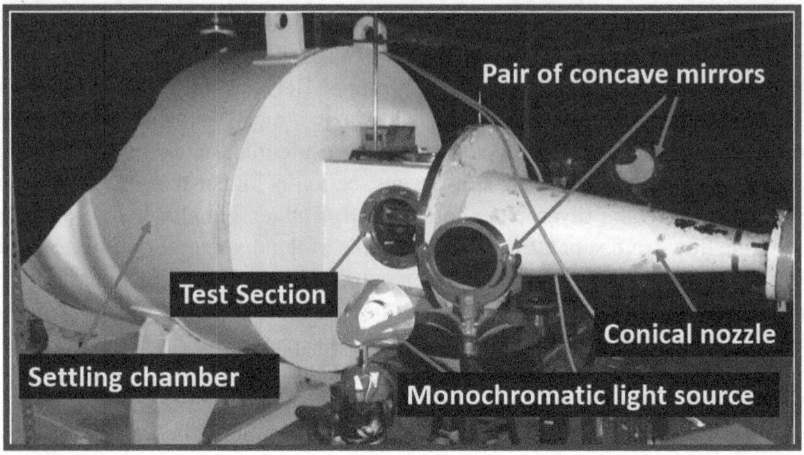

Figure 8.3.2 Photographic view of conical nozzle, test section, and dump tank assembly.

temperature of driver gas leads to the stimulation of much higher enthalpy in a shock tunnel than that produced in a wind tunnel. The enthalpy can be altered by varying the driver gas pressure.

Note that the diaphragm's material and thickness are vital in controlling the driver's gas pressure. Indeed, the state of the driver's gas is determined by the pressure at which the diaphragm bursts. Therefore, an impurity in the diaphragm material may cause a large deviation in the driver's gas pressure from its designed value. V−grooves are provided over the diaphragm to resolve this issue, as depicted in Figure 8.3.3(a). These grooves over the diaphragm confirm its rupture at a controlled rate (Figure 8.3.3(b)) and thus prevent the splintering of the diaphragm material in the shock tube.

The calibrated test section Mach number of the tunnel is 5.7 for the experiment. Therefore, the intake configuration is called the Mach 5.7 intake in this study. In the test section, the stagnation pressure is 92 kPa at Mach 5.7 downstream of the bow shock. Using the Rayleigh-Pitot formula (equation (8.3.1)), the test section's static pressure is observed to be 2.179 kPa. At Mach 5.7, 1.1 $^{MJ}/_{\text{kg}}$ of stagnation enthalpy was stimulated in the shock tube with an uncertainty ±3.5.

$$\frac{p_{02}}{p_\infty} = \left[\frac{(\gamma+1)^2 M_\infty^2}{4\gamma M_\infty^2 - 2(\gamma-1)}\right]^{\frac{\gamma}{\gamma-1}} \left[\frac{1-\gamma+2\gamma M_\infty^2}{\gamma+1}\right] \tag{8.3.1}$$

The experiments are also performed at Mach 7.9, specifically for the uncontrolled and controlled intake (with a thin porous surface placed over a flat

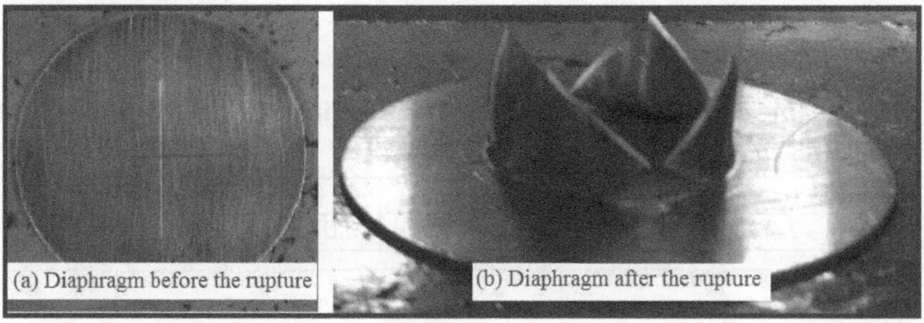

Figure 8.3.3 Photographic views of primary diaphragm before and after the rupture.

cavity). A Ludwig tube is used to generate the Mach 7.9 stream. In Ludwig mode, the tube is subjected to a higher pressure and thus acts as a reservoir. The high-pressure tube is separated using a diaphragm from the low-pressure nozzle, test section, and dump tank. A weak shock wave translates into the test section through the nozzle when the diaphragm ruptures. Simultaneously, the unsteady expansion waves propagate in the opposite direction (away from the test section and nozzle) and reflect off the downstream wall of the tube. A constant flow of the test gas through the test section and nozzle is established. This process is steady only for a certain period until the reflection of expansion waves from the end wall arrives test section. The static pressure corresponding to Mach 7.9 is 0.1 kPa.

Furthermore, the Reynolds number corresponding to the Mach numbers 5.7 and 7.9 are significantly high (7.04 $^{million}/_{m}$ and 11.0 $^{million}/_{m}$, respectively), which exceeds the limiting Reynolds number 500 [12]. Thus, viscous influences over the pressure measurements were negligible for the present study. Notice that, instead of estimating the Reynolds number at every location (since it varies with length scale) in the test section, it is a standard in gas dynamics to define this per unit length. This avoids its dependency on the test section model size or shape. Tables 8.2 and 8.3 contain freestream conditions measured with uncertainties for Mach 5.7 and 7.9, respectively.

Table 8.2

Freestream conditions and uncertainties in the measurement at Mach 5.7.

M_∞ (±1%)	p_∞ (±4%)	T_∞ (±4%)	ρ_∞ (±5.5%)	Re_∞ (±5.5%)
5.7	2.179 kPa	147.6 K	0.0515 $^{kg}/_{m^3}$	7.04 $^{million}/_{m}$

Table 8.3
Freestream conditions and uncertainties in the measurement at Mach 7.9.

M_∞ (±1%)	p_∞ (±4%)	T_∞ (±3%)	ρ_∞ (±4%)	Re_∞ (±5%)
7.9	0.1 kPa	22 K	0.01 kg/m^3	11 $million/m$

8.3.2.1 Instrumentation

In the current investigation, a pressure sensor is used to measure the test section's total pressure. The piezoelectric crystal inside the sensor produces an equivalent voltage signal while experiencing a change in pressure. The voltage signal is amplified by the built-in amplifiers and recorded by a data acquisition system. The typical rise time of the sensor is around 5 μs. The total pressure was calculated by taking the time average of pitot pressure values. In a supersonic flow, the pitot tube measures the total pressure behind the bow shock formed ahead of its nose (Figure 8.3.4). The mid-portion of the bow shock is usually considered as a normal shock across which we can use the Rayleigh-pitot formula to measure the stagnation pressure ratio (equation (8.3.1)). At six different locations and the intake core, the wall static pressures are measured using Kulite sensors. These Kulite sensors have a sensitivity of 11.1 mV/psi. Time-averaged static pressures at each location are taken into consideration.

Because of the very-short testing time, a high-frequency data acquisition system, as shown in Figure 8.3.5, is used to collect the data in the shock tunnel.

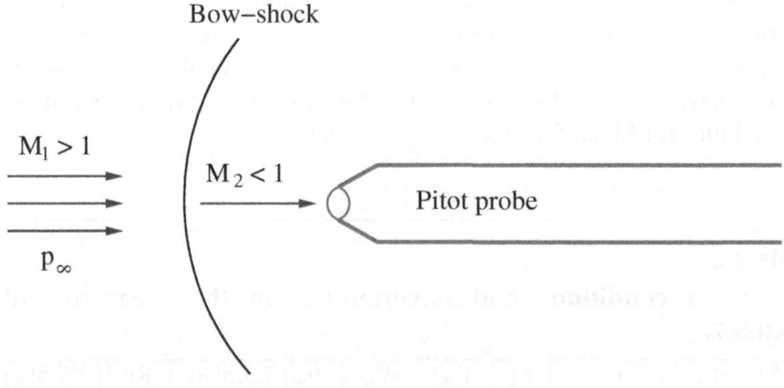

Figure 8.3.4 Stagnation pressure measurement by a pitot probe.

Figure 8.3.5 Data acquisition system.

Besides, estimating the actual test time in the unsteady flow field is a cumbersome task. Fortunately, this can be ascertained by analyzing the wall pressure value. A typical wall pressure signal during the experiments is schematically described in Figure 8.3.6. Note that the entire duration is divided into two times; rise time – time taken for the pressure to reach the peak from zero value and test time – time taken for the pressure to become nearly steady. It is observed that a rise time of 600–700 μs and a testing time of 800–900 μs is obtained for this study. An uncertainty of about ±0.5% is measured for wall pressures.

Schlieren technique is used to visualize the shock and expansion waves in both the intake models. The setup consists of two concave mirrors, a point light source, a high-speed camera, and a knife edge. At first, using the first concave mirror, the light from the point source is made parallel. The parallel light rays cross the test section and fall on the second concave mirror. The second mirror is used to refocus the light rays at the knife edge. The knife edge blocks half of the light intensity just before their passage to the camera. It is well known that when a light ray passes through mediums of different densities, they bend toward the higher density. Shock and expansion waves change the density in the shock tunnel, causing the light rays to get deflected while passing through the test section. The knife edge blocks the reflected light rays or completely misses them. As a result, the regions of the density gradient may appear darker or brighter in the image.

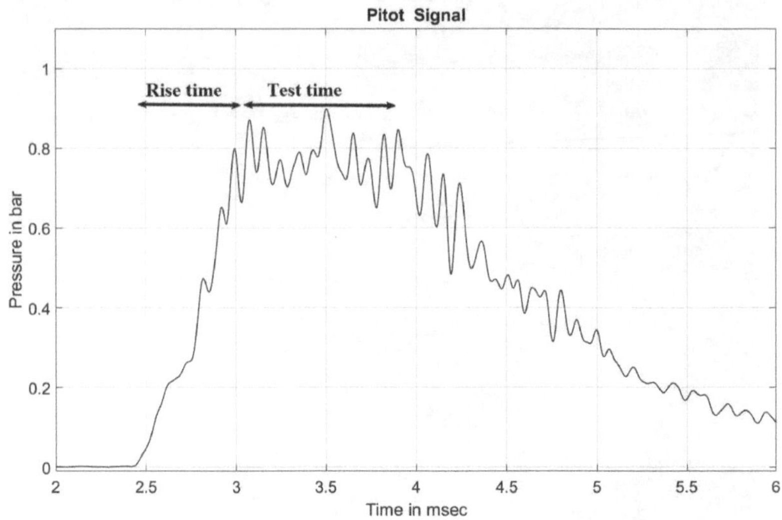

Figure 8.3.6 Typical wall pressure signal characteristics.

A high-speed Phantom V-310 camera, capable of 10,000 FPS, is used to record the Schlieren images (resolution 640 × 480 pixels) with time. Figure 8.3.8 shows the recirculation zone in an uncontrolled hypersonic intake. In order to quantify the separation bubble, the distance between separation (downstream location) and reattachment points is considered separation length, and the height of the shock impingement point from the surface is termed separation height.

8.3.3 DOUBLE RAMP MIXED COMPRESSION INTAKE

A dual ramp mixed-compression intake constant across a third dimension (2D design) is designed for the experimental investigation (Figure 8.3.9(a)). Rather than an isentropic compression ramp, the double ramp intake is used because the ramp surface with isentropic compression would have more blockage in the tunnel's test section when mounted. The mixed compression intake is designed following the oblique shock theory to achieve the shock-on-lip condition [14]. The total pressure drop can be minimized following a criterion proposed by the Oswatitsch (1947), which states that if $(n-1)$ oblique shocks combined with a normal shock results in minimum pressure loss when their strengths are similar [15]. This is identical to the following:

$$M_\infty \sin\beta_1 = M_1 \sin\beta_2 = M_2 \sin\beta_3 = \ldots = M_{n-1} \sin\beta_n \tag{8.3.2}$$

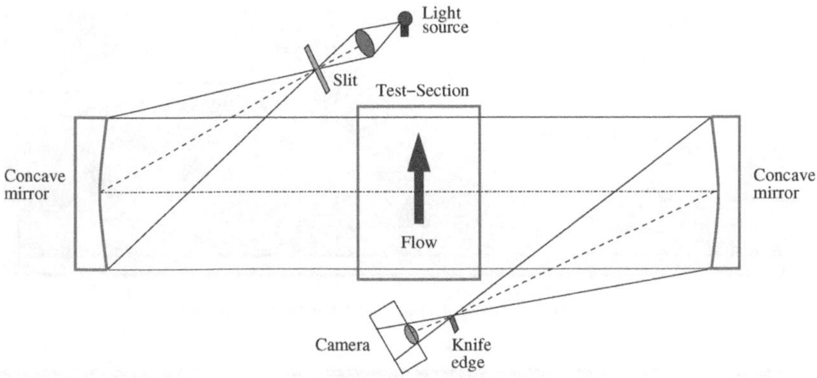

Figure 8.3.7 Schematic layout of the Schlieren system.

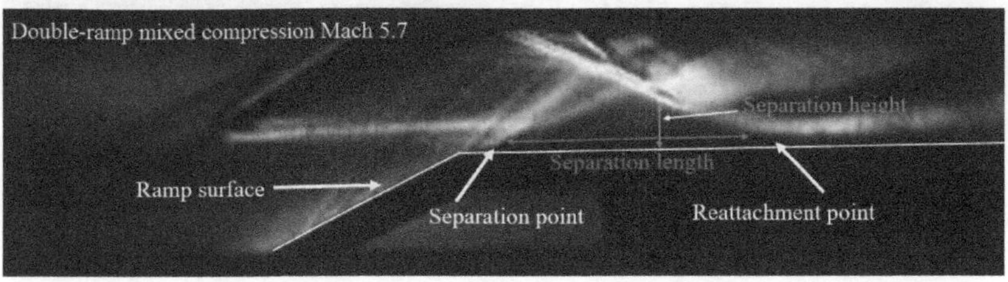

Figure 8.3.8 Schlieren image of a separation bubble in a hypersonic intake.

The flow characteristics through an intake strongly depend upon its contraction ratio, defined as

$$R_c = \frac{A_c}{A_t} \tag{8.3.3}$$

where A_c is the flow is capture area and A_t is the throat area of the isolator. Accordingly, a series of shock waves are generated in all the models externally and internally. The pressure losses are minimal (however, there will be no normal terminal shock in this case). The shock waves generated over the double ramp have nearly equal strengths for maximum pressure recovery [16]. Also, a ratio of 0.38 is maintained for the Mach numbers between the burner and intake entrances to keep the temperature rise in the combustion chamber optimum ([1], [17]).

Kantrowitz's condition imposes a further constraint on the contraction ratio; it states that a value of R_c higher than R_ℓ results in the compression of a higher

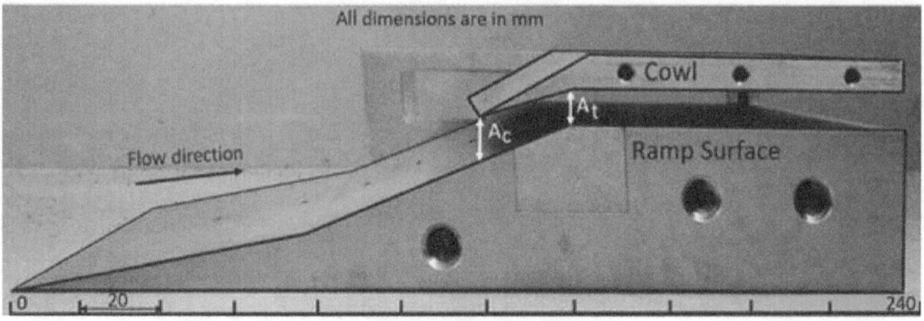

(a) Double–ramp mixed compression intake

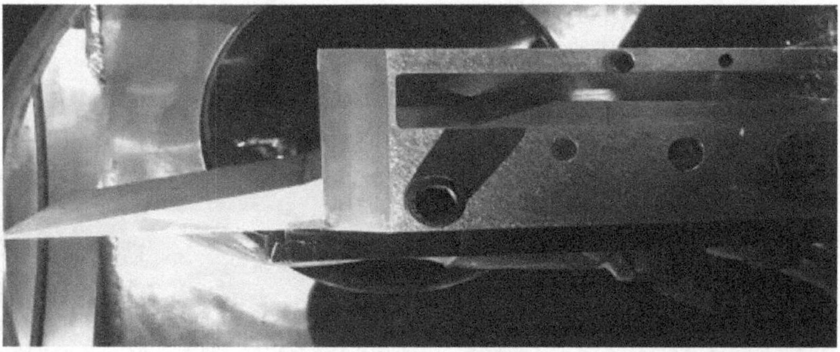

(b) Intake model mounted inside the test–section

Figure 8.3.9 Double ramp mixed compression intake mounted in the shock tunnel.

degree that could lead to intake unstart and is expressed as follows ([18], [19]):

$$R_\ell = \left(\frac{A_c}{A_t}\right)_{\text{limiting}} = \left[\frac{\gamma-1}{\gamma+1} + \frac{2}{(\gamma+1)M_\infty^2}\right]^{-\frac{1}{2}} \left[\frac{2\gamma}{\gamma+1} - \frac{\gamma-1}{(\gamma+1)M_\infty^2}\right]^{-\frac{1}{\gamma-1}} \tag{8.3.4}$$

Therefore, the decrement in the throat area, either by geometric constraints or by an SBLI, reduces the mass flow rate and could lead to intake unstart. In the present investigation, all configurations of the intake model are designed with a contraction ratio of 1.25. The dual ramp has a width of 100 mm and a length of 234.20 mm, and it has two inclinations of 10° and 20°. The angle made by the cowl lip with freestream flow is 15°. The width-to-height ratio at the intake entrance section is kept at 10. Figure 8.3.9(b) illustrates the intake model mounted in the test section. A metal side plate is used to hold the cowl and core together.

Estimating the blockage ratio of the intake model, deployed in a supersonic or hypersonic test section, is essential to prevent the unstart. The blockage ratio is calculated using equation (8.3.5) and is usually kept below 10% for a stable supersonic/hypersonic tunnel operation. In the present experiments, the blockage is kept below 8% for the intake model and metal side plates.

$$\text{Blockage ratio } (\%) = \frac{\text{Projected area of the model}}{\text{Area of the test section}} \times 100 \qquad (8.3.5)$$

8.3.4 SBLI CONTROL TECHNIQUES

The research conducted with dual ramped mixed-compression hypersonic intake shows that the generation of a large recirculation bubble and multiple shock waves significantly disrupts the flow structure at both downstream and upstream locations of the ramp shoulder. Therefore, the present study aims to control interactions, especially near the ramp shoulder. With this in mind, a porous surface-covered cavity (with different surface perforation levels) is implemented at the interaction point, specifically where the cowl lip shock meets the ramp surface. In addition, the micro vortex generators will be placed at different heights in front of and at the point of interaction to control the SBLIs near the ramp shoulder.

8.3.4.1 A Shallow Cavity Covered with Thin Porous Surface

Because the cowl-generated shocks induce an abrupt rise in the adverse pressure gradient near the ramp convex corner, this passive control technique's prime motto is to reduce the cowl shock's strength to improve pressure recovery. A shallow cavity with an aspect ratio of 1.46 covered by a thin porous surface is fixed beneath the impingement location of the first cowl shock (generated from the lip), as shown in Figure 8.3.10(a). The pores are uniformly distributed with a variation in diameters, such that the porosity of the cavity's upper surface is modified as 4.5%, 7.5%, 17%, 21.6%, and 25%. Two different Mach numbers are used in the experiment to confirm the optimum porosity value in controlling SBLI and to observe the significance of porosity in different conditions of freestream flow. For all five porosity cases, the experiments are carried out at Mach 5.7 and obtained values of optimum porosity limit as 17% and 25%. For Mach 7.9, experiments are conducted at these two values to check the reliability. The formula for surface porosity calculation may be written as follows:

$$\text{Porosity } (\%) = \frac{\text{Total surface area of pores}}{\text{Total control area of ramp surface}} \times 100 \qquad (8.3.6)$$

Six wall-mounted pressure ports ($P_1 - P_6$) are used across the wall to estimate the static wall pressure along with the inlet core (Figure 8.3.10(a)).

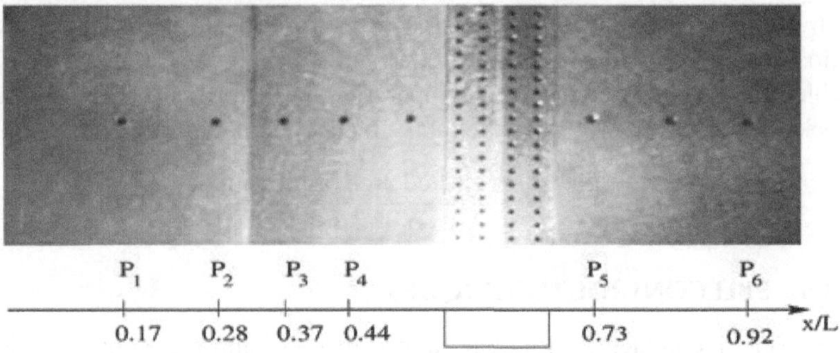

(a) Top view of dual–ramp with surface perforation

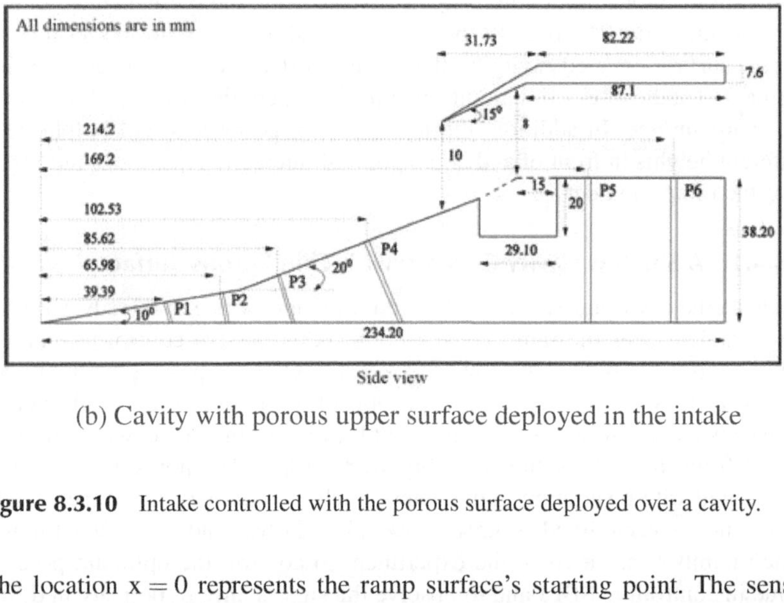

(b) Cavity with porous upper surface deployed in the intake

Figure 8.3.10 Intake controlled with the porous surface deployed over a cavity.

The location x = 0 represents the ramp surface's starting point. The sensor P4 measures the static pressure upstream of the impacting point, while sensor P5 measures the pressure at the reattachment point. Figure 8.3.10(b) provides the dimensions of the porous surface covering a cavity-controlled intake. The configurations of pores over the control surface for different levels of surface perforation are shown in Figure 8.3.11.

8.3.4.2 An Array of Micro Vortex Generators

The investigations are performed with micro vortex generators (MVGs) of 0.5 mm, 0.7 mm, and 1 mm heights, deployed separately upstream and at the

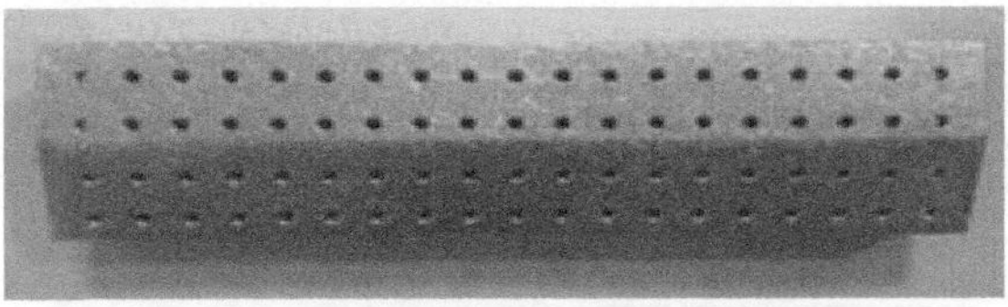

(a) 4.5% Surface perforation

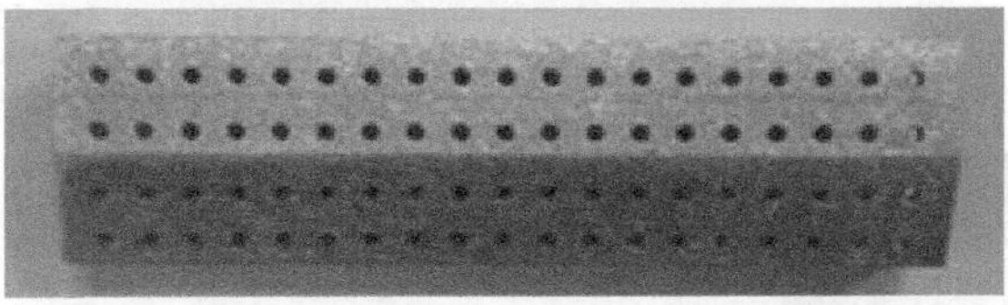

(b) 7.5% Surface perforation

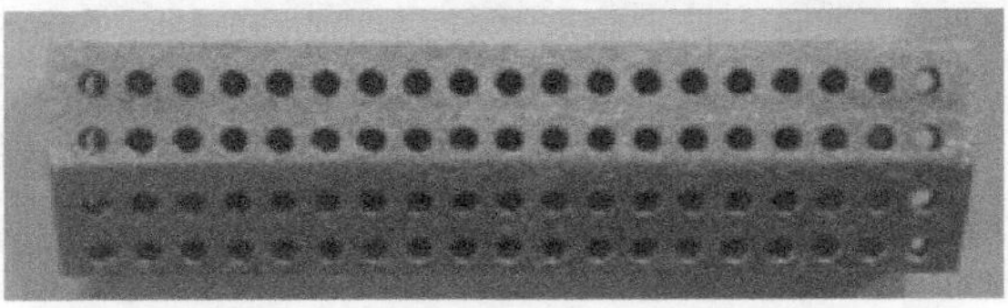

(c) 17% Surface perforation

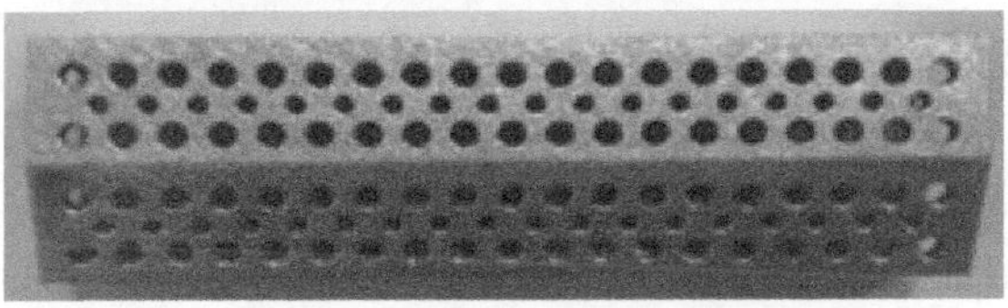

(d) 21.6% Surface perforation

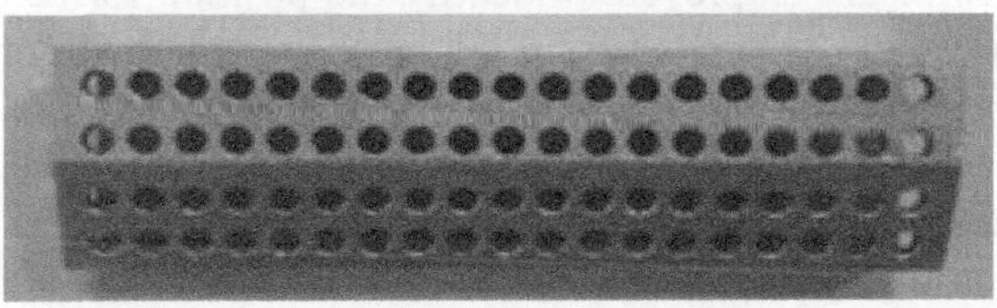

(e) 25% Surface perforation

Figure 8.3.11 Cavity covered with different surface porosities.

interaction regions. Figure 8.3.12(a) shows the schematic diagram of a conventional MVG, and its typical dimensions are tabulated in Table 8.4. The front and plan views of the MVG assembly above the ramp surface are shown in Figure 8.3.12(b) and Figure 8.3.12(c), respectively. The undisturbed

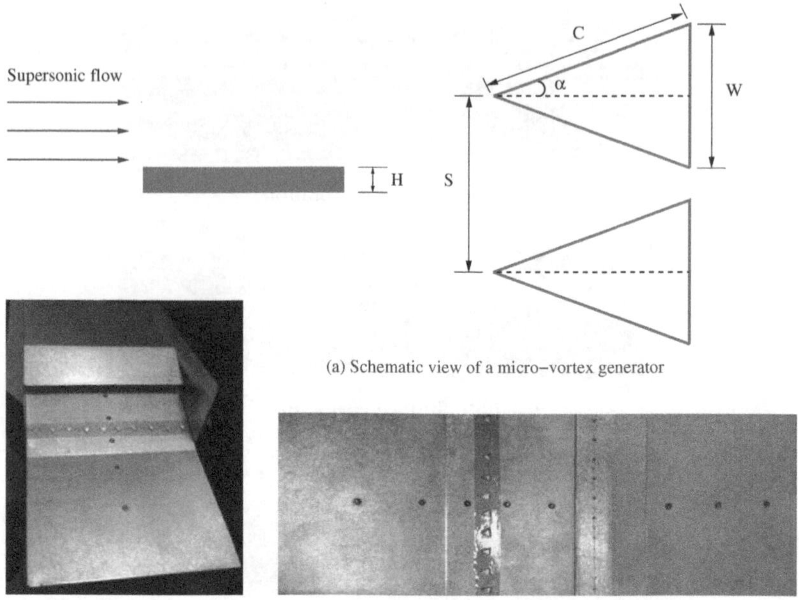

(a) Schematic view of a micro–vortex generator

(b) Array of MVGs (front view)

(c) Array of MVGs (top view)

Figure 8.3.12 An array of MVGs deployed above the intake ramp.

boundary layer thickness (δ) in the plain intake is about 3 mm measured from the Schlieren image. The height of the vortex generator (H) in terms of boundary layer thickness varies at 0.17δ, 0.23δ, and 0.33δ. In any case, ten MVGs arrayed spanwise are employed to identify the position for better control efficiency. The shocks generated from the cowl lip and cowl corner strike the surface near the shoulder and $x/L = 0.59$, respectively. Therefore, negative gradient pressure is imparted to the boundary layer over the intake core. Since there is good evidence in the literature that the MVGs placed in front of or at the interaction regions are advantageous for regulating the interactions, they are used at the sites; $x/L = 0.41$ and $x/L = 0.59$ [20].

8.3.5 SBLI CONTROL USING A SHALLOW CAVITY COVERED WITH THIN POROUS SURFACE

This study examines the effectiveness of a shallow cavity covered with a thin porous surface to control SBLIs at Mach 5.7 and 7.9 for a mixed-compression intake to reduce shock resistance and improve pressure recovery. The configurations of the intake models are already discussed in section 8.3.4.1. At Mach 5.7, the surface porosity is altered by modifying the pore diameter to 4.5%,

Table 8.4
Typical dimensions of the conventional MVGs.

Half angle (α)	24°
Side length (C)	7.2H
Spanwise pitch (S)	7.5H
Base width (W)	6H

7.5%, 17%, 21.6%, and 25%. The investigations at Mach 7.9 is conducted, specifically for the controlled intake with 17% and 25% surface porosities, to confirm the effectiveness of porous surfaces in controlling the interactions at high Mach numbers. The control efficiency of the porous surface is examined quantitatively and qualitatively. Static pressures above the ramp surface are measured through wall-mounted ports, and Schlieren images are captured at various time instants during the flow evolution.

8.3.5.1 Wall Static Pressure Measurement

The current study considers six axial locations along the centerline for pressure measurements (Figure 8.3.10(a)). Here, the passive cavity effects far upstream are investigated using the 1st port (x/L =0.17). The upstream effects of the cavity are studied using 2nd and 3rd ports (x/L = 0.28 and 0.37, respectively). The 4th and 5th ports (x/L = 0.44 and 0.73, respectively) will explain the near field region effects (upstream and downstream) of the cavity. However, far fields influences of the cavity are studied with the help of the final port at x/L =0.92.

The static pressures measured at these locations are non-dimensionalized by dividing them with the freestream pressure (p_∞). The axial lengths (x) are non-dimensionalized with model length (L). These non-dimensional static pressures are plotted against non-dimensional axial locations for all the intakes operating at Mach number 5.7 (Figure 8.3.13). It is observed that the pressure at the first two ports is not significantly affected because of the porous cavity and showed a pressure rise indicating oblique shock at the ramp leading edge. An increment in pressure rise is observed in all the intakes in the next two ports owing to the oblique shock-induced compression at the second ramp. Also, only a minute variation between uncontrolled and controlled intake pressure values is observed. At 0.73L, maximum static wall pressures are achieved for all intake configurations indicating the presence of a reattachment shock wave. This static pressure is also termed reattachment peak pressure.

However, a controlled intake has lower static wall pressures than an uncontrolled intake. Also, this decrement is increased with porosity, having a

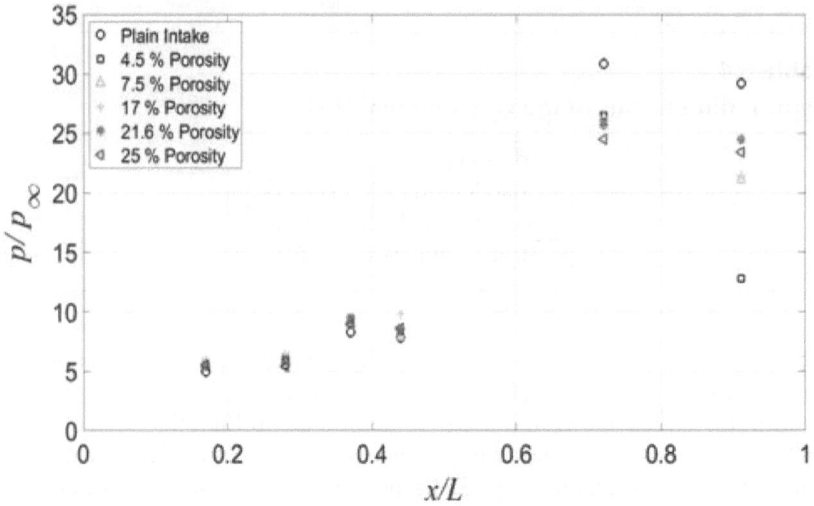

Figure 8.3.13 Wall Static pressure variation at Mach 5.7.

maximum drop in pressure at 25% porosity. For controlled intake, the percentage variation in wall pressure is calculated using equation (8.3.7) at $x/L = 0.73$ (reattachment location). The calculated values are tabulated for both Mach 5.7 and 7.9 in Tables 8.5 and 8.6, respectively.

$$\% \frac{\Delta p_w}{p_w} = \left[\frac{(p_w)_{uncontrolled} - (p_w)_{controlled}}{(p_w)_{uncontrolled}} \right] \times 100 \qquad (8.3.7)$$

Table 8.5
Percentage drop in the wall static pressure due to cavity at $x/L = 0.73$ (Mach 5.7).

Surface porosity	Percentage variation in p_w
4.5%	14.10%
7.5%	14.70%
17%	16.30%
21.6%	16.64%
25%	20.53%

A maximum pressure drop of about 20.53% near the reattachment point, corresponding to 0.73L, is observed when a 25% porous surface is used across

Table 8.6
Percentage drop in the wall static pressure due to cavity at $x/L = 0.73$ **(Mach 7.9).**

Surface porosity	Percentage variation in p_w
17%	14.76%
25%	20.20%

the cavity (Table 8.5). A decrease in pressure essentially confirms the decrease in shock strength due to the influence of the cavity. Besides, a high-pressure drop is observed for the controlled intakes at 0.92L than plain intake, confirming the cavity's influence downstream. The static pressure drops at this location for all the intakes are faster than that of the reattachment location. Also, with all the controlled intakes, a small pressure increase is observed ahead of the cavity. The upstream communication is improved as the porous top surface results in the expansion of the interaction zone.

For Mach 7.9, the dimensionless wall static pressure distribution for plain and controlled intakes is depicted in Figure 8.3.14. Compared to Mach 5.7 intake, higher wall pressures are observed at Mach 7.9 as the shock strength is correspondingly increased. The pressure variation among the first four ports is

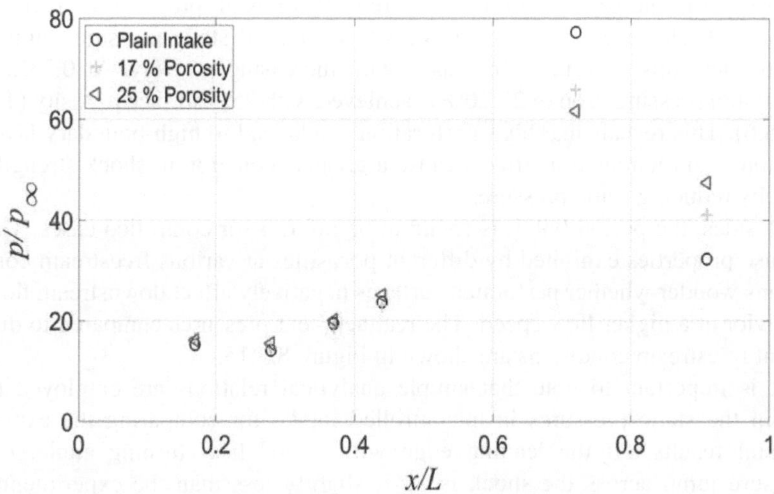

Figure 8.3.14 Wall Static pressure variation at Mach 7.9.

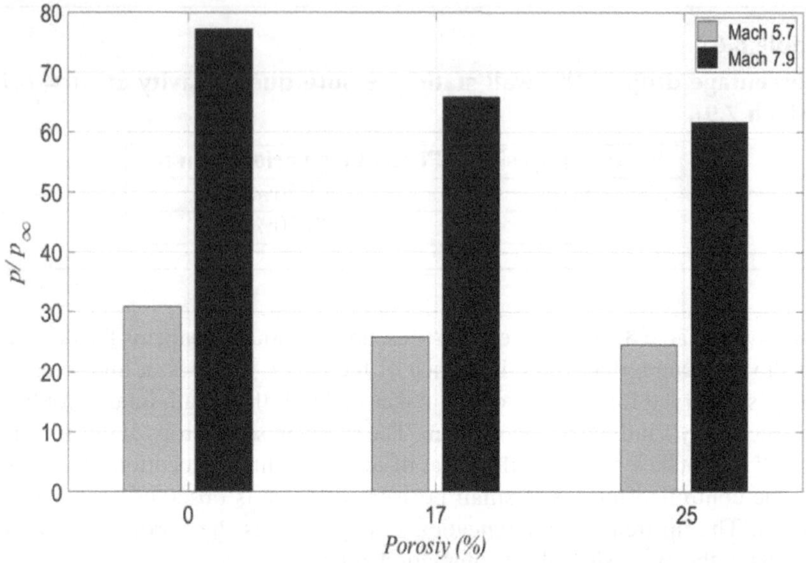

Figure 8.3.15 Comparison of peak non-dimensional static pressures at varied surface porosities.

the same as in the previous case. So it is clear that the influence of the cavity over the upstream pressure distribution is minimal, even at Mach 7.9.

For plain intake, a significant increment in wall pressure is seen at $x/L = 0.73$. However, introducing porosity decreased static pressures significantly, and this decrement increased with increasing porosity. At 0.73L, a maximum pressure drop of 20.20% is achieved with 25% surface porosity (Table 8.6). This reveals that high-perforation levels lead to high-boundary layer suction, and recirculation effects cause a greater reduction in shock strength, thereby reducing static pressure.

Besides, the port at 0.92L is found to be inferior for controlled cases. The intense properties exhibited by different porosities at various freestream conditions wonder whether perforated surfaces negatively affect downstream flow behavior at a higher flow speed. The reattachment pressures compared to different freestream conditions are shown in Figure 8.3.15.

It is important to note that simple analytical relations are employed to obtain the static pressures in uncontrolled intake for comparing the experimental results. At the leading edge with a 10° flow turning angle, the pressure jump across the shock is 3.47, slightly less than the experimental value. This is a consequence of the influence of upstream since experimental results automatically incorporate the viscous effects. The variation in pressure at the second port is insignificant as any additional shock wave does not encounter this port. The first and second ports are located upstream of the ramp concave corner shock (edge of first and second ramp). Since the separation of flow occurs at the ramp corner due to SBLI, a significant upstream influence of the interaction is inevitable. This may be why the static pressures measured at the first two ports are higher. However, experiments indicate an eight times pressure jump at the third port indicating ramp-induced compression, while the inviscid relations provide 9.4. The same trend is observed for the next port except for a slight pressure drop, which can be considered experimental uncertainty. Unlike the first and second ports, the experimentally obtained pressures at third and fourth ports are smaller than the ones calculated from inviscid relations. Also, no significant shock is observed at these ports near downstream locations. Thus, measured static pressures at third and fourth ports are lesser than those obtained from inviscid relations, as there is no significant upstream influence at these locations.

At the ramp shoulder, a cowl lip shock interacts with the expansion waves and reduces its strength. Therefore, the first cowl shock and expansion waves are neglected in analytical calculations, and the second cowl shock is given primary significance. With a cowl angle of 15°, the inviscid flow relations compute the Mach number as 3.62. This gives the pressure spike at the fifth port as 31.75, which is almost in accordance with the experimental values. The wall static pressures at ports; 1, 3, and 5 are found to be 5.1 kPa, 17.74 kPa, and 74.11 kPa, respectively, at Mach 7.9.

In order to examine control efficacy dependency on Mach number, a ratio of maximum non-dimensional static pressure, P_R at Mach 7.9 and 5.7, is calculated using equation (8.3.8) and tabulated in Table 8.7.

$$P_R = \frac{\left[\left(\frac{p}{p_\infty}\right)_{max}\right] \text{ at Mach 5.7}}{\left[\left(\frac{p}{p_\infty}\right)_{max}\right] \text{ at Mach 7.9}} \tag{8.3.8}$$

Table 8.7
Variation of P_R with surface porosities.

Surface porosity	P_R
0	2.50
17%	2.55
25%	2.51

It is observed that when the intake Mach number is increased from 5.7 to 7.9, P_R for plain intake is 2.5, clearly indicating the increase in shock strength.

With different surface perforation levels, the higher P_R values confirm the surface perforation's superiority at Mach 5.7 compared to Mach 7.9. The phenomenon behind this is that the porous surface covered over a shallow cavity produces the recirculation effect, decreasing the wall pressures. However, simultaneously because of the pores on the upper surface, many weak compression waves are generated, incrementing the static pressures. This pressure rise would be even more at a high Mach number. Therefore, the effect of recirculation combined with compression waves in reducing pressure is inferior at Mach 7.9 than at 5.7.

8.3.5.2 Schlieren Flow Visualization

The Schlieren technique visualizes the time-resolved flow evolution and wave structures in plain and controlled intakes.

Steady Flow Field Morphology

Figure 8.3.16(a) shows the Schlieren, while Figure 8.3.16(b) shows the schematic views of a steady flow through a mixed-compression intake at Mach 5.7. Notice that the expansion waves are generated both at the ramp shoulder and the location at which incident shock undergoes reflection from the

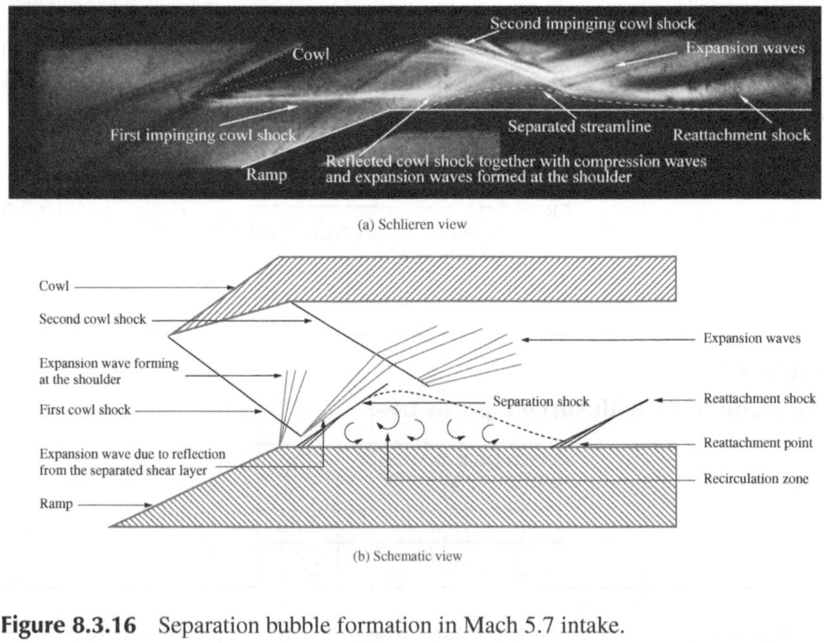

Figure 8.3.16 Separation bubble formation in Mach 5.7 intake.

detached boundary layer. Two cowl shocks are incident on the surface of the ramp at two locations. The first cowl shock incident at the ramp shoulder interacts with the expansion waves, reducing the shock strength.

On the other hand, the second cowl shock, being at a distance from the shoulder, possess a higher strength as it does not encounter the expansion waves. Hence, it is termed the primary shock and becomes a key factor in separation bubble formation. This cowl shock imparts a negative pressure gradient onto the boundary layer, resulting in a separation bubble on the ramp surface. Since the pressure signal is conveyed upstream through the boundary layer's subsonic region, the pressure spike that occurs in shock downstream is likewise experienced upstream of the place of impingement. The boundary layer slows down and gradually enlarges upstream of the impinging location. The boundary layer thickening produces a sequence of compression waves that converge into a single shock. In addition, the reflection of cowl shock from the shear layer into an expansion wave pushes the shear layer near the surface for reattachment. This process ends with a series of compression waves that converge into a reattachment shock.

The interaction of the boundary layer with shock waves leads to flow separation in viscous flows. This interaction results in a larger separation bubble with a suitably high-shock strength. Thus, with separated flows, pressure increases gradually in two steps (in conjunction with separation and reattachment shocks) rather than abruptly, as happens with inviscid flows. The first and second jumps are caused by the separation and reattachment shocks, respectively. The maximum pressure is seen in viscous flows downstream of the reattachment point. This conclusion is supported by the pressure diagrams shown in Figures 8.3.13 and 8.3.14.

Flow Field Development

Figure 8.3.17 illustrates the flow field evolution at various instants of time at Mach 5.7. Here, the reference time at taken as zero milliseconds, at which flow starts. We can see that the cowl shock starts to form only after 0.8 milliseconds from the start, and till then, there exists only a shock-free flow field. Interestingly, the second cowl shock developed earlier than the first cowl shock incidents first at the ramp shoulder. The reason could be the development of powerful expansion waves, which initially reduce the first cowl shock intensity. The second cowl shock is anticipated to have a greater impact on the expansion of the recirculation bubble (otherwise known as the separation bubble) because the first shock is feeble. The phenomenon of the separation bubble could be described in the following manner.

Initially, the cowl shock is reflected from the ramp because of the flow field's initial inviscid behavior. The boundary layer later develops and reacts with the cowl shock. Partially, the cowl shock is reflected as expansion waves

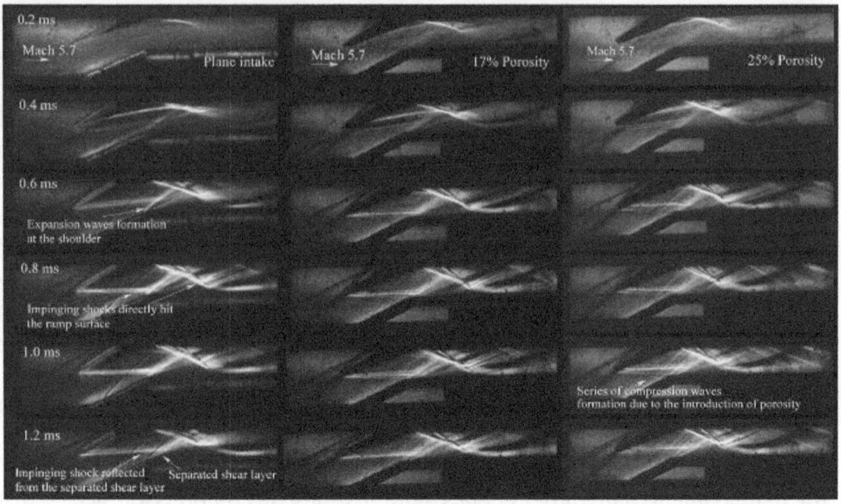

Figure 8.3.17 Flow development in plain and controlled intakes at different surface porosities.

due to unlike-reflection. After 1.0 ms, the boundary layer begins to decelerate because of an unfavorable gradient in pressure brought on by impinging shock waves. In the worst case, the separated boundary layer attaches to the surface because of the balance in pressure between the inviscid and viscous zones of SBLI's downstream position, creating a recirculation zone or separation bubble. There are two distinct regions in the boundary layer over the surface of the ramp, near the ramp – subsonic layer and near the inviscid region–supersonic layer, distinguished by a sonic line [21]. The upstream location receives the interaction information through the subsonic layer, thereby increasing the thickness of the boundary layer. In simple terms, the boundary layer momentum decreases because of upstream pressure signal transmission before the interaction region. This causes the boundary layer to separate early, which lengthens the separation. At first, a small pocket-size separation bubble is formed locally at the impingement location. Later, the bubble grows and extends to a certain portion of the interaction region. Because of the growing bubble size, the reattachment point moves further downstream during the run. Besides, controlled intakes possess several less intense compression waves because of the fluid injection upstream of shock using a porous cavity. As a result, the oncoming Mach number decreases.

Variation in the Separation Bubble Size

Figures 8.3.18 and 8.3.19 are Schlieren views of separation bubble formation in plain and controlled intakes at Mach 5.7 and 7.9, respectively.

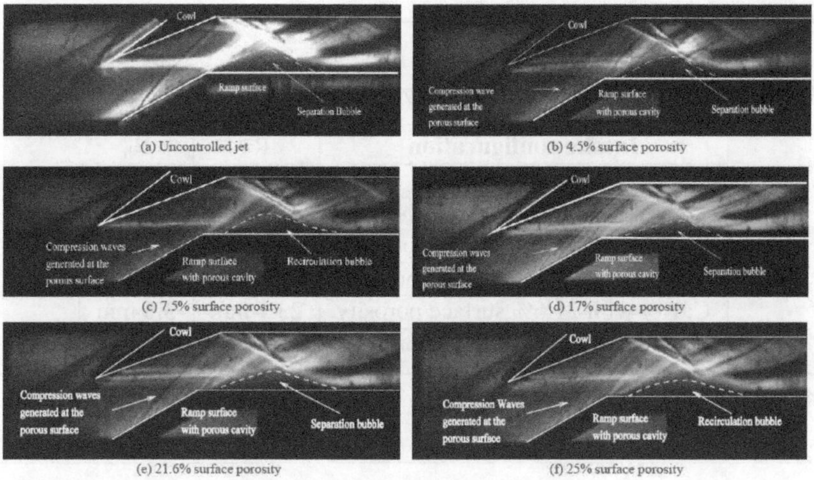

Figure 8.3.18 Schlieren pictures of plain and controlled intakes at Mach 5.7.

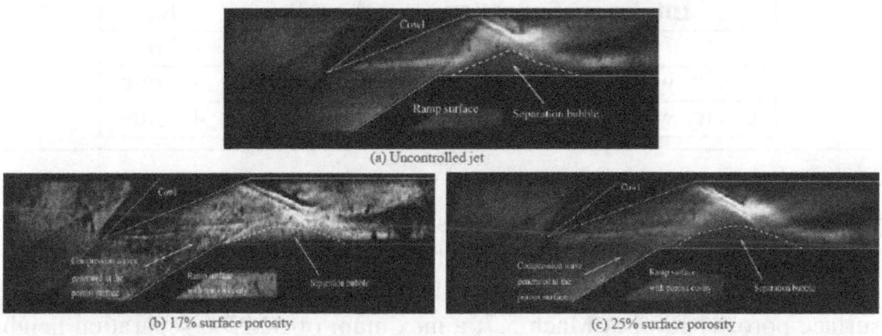

Figure 8.3.19 Schlieren pictures of plain and controlled intakes at Mach 7.9.

Figures reveal that at the cowl lip, the cowl shock generated is shifted marginally from its location due to shock unsteadiness.

The fluid is decelerated because of a series of oblique shocks, including a leading-edge shock of the ramp, ramp corner shock, cowl lip shock, and second cowl shock. Subsequently, before reaching the compressor, the flow experiences a shock train in the isolator region. Thus, it leads to a continuous unsteady variation of separation bubble size, which makes the quantification difficult. Therefore, standard deviations are used to estimate the recirculation zone's average length and height for uncontrolled and controlled intakes. The dimensions of the separation bubble are mentioned in Tables 8.8 and 8.9 for Mach 5.7 and 7.9, respectively. Bar plots (Figures 8.3.20 and 8.3.21) are used

Table 8.8
Separation bubble size at Mach 5.7.

Intake configuration	R_ℓ	R_h
Plain intake	26.9 mm	4 mm
Cavity with 4.5% surface porosity	23.2 mm	3.7 mm
Cavity with 7.5% surface porosity	23.2 mm	3.4 mm
Cavity with 17% surface porosity	22.4 mm	3.1 mm
Cavity with 21.6% surface porosity	23.1 mm	3.6 mm
Cavity with 25% surface porosity	23.3 mm	3.7 mm

Table 8.9
Separation bubble size at Mach 7.9.

Intake configuration	R_ℓ	R_h
Plain intake	19.6 mm	4.2 mm
Cavity with 17% surface porosity	18.1 mm	4.1 mm
Cavity with 25% surface porosity	21 mm	4.3 mm

to plot these data, where the red lines indicate uncertainties associated with separation bubble length and height estimation.

It is observed that the recirculation bubble shrank with a 17% increment in surface porosity level at Mach 5.7; a maximum of 22.25% separation height reduction and 16.73% separation length reduction is observed. With the same porosity level, 7.65% and 2.38% reductions in separation length and height, respectively, are noticed for Mach 7.9. The boundary layer profile degrades as the porosity increases (above 17%), and bubble size grows again. Essentially, the shock strength is decreased by dividing it into weak λ−shock with the help of surface perforation, ultimately leading to a diminished separation bubble. This, indeed, is favorable for controlling SBLI. However, the upstream boundary layer gets thickened when the fluid is injected through the porous cavity, promoting interactions. Optimizing these two effects is the main purpose of a control technique. With a porosity level of up to 17%, the fluid injected is not significant enough to cause a boundary layer as it is suppressed by smearing. However, advancing the porosity level further will, in turn, increase fluid injection resulting in a significantly thick boundary layer. Since the SBLI depends on both adverse effects of boundary layer degradation as well as shock

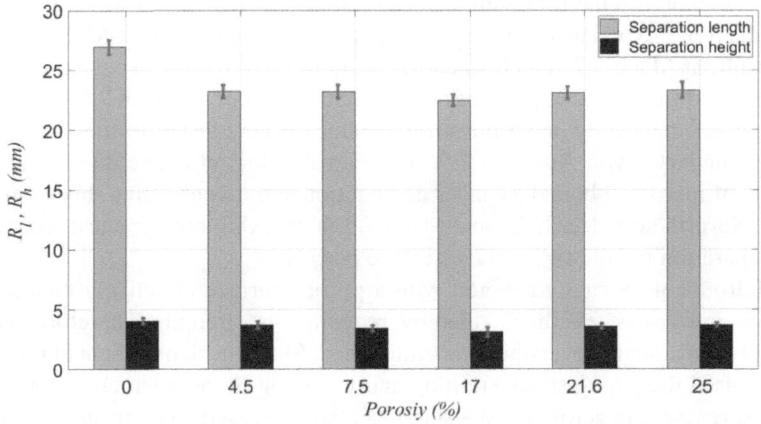

Figure 8.3.20 Variation in separation bubble size at different surface porosities at Mach 5.7.

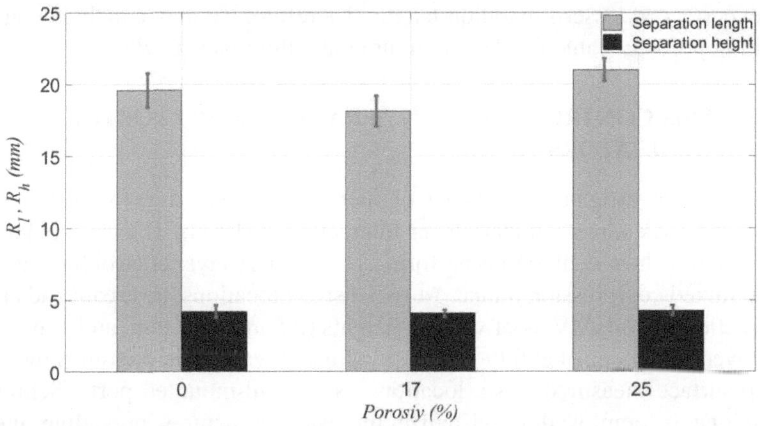

Figure 8.3.21 Variation in separation bubble size at different surface porosities at Mach 7.9.

splitting phenomena, the control technique should not be implemented to an extent where the associated disadvantages compensate for the advantages of one phenomenon. At Mach 5.7, the boundary layer degradation and its negative effects are dominated by shock-splitting phenomena, even with porosity beyond 17% (which is 25%) for controlled intake. While it is the opposite for Mach 7.9 flow, the degraded boundary layer predominantly thickens the boundary layer for a surface porosity level of 25%. This intriguing phenomenon has the following explanation. The compression waves that flow at Mach 5.7 experiences are significantly weaker than those at Mach 7.9. As a result, at Mach 5.7, shock smearing is more effective for the same porosity level. Additionally, due to higher downstream pressure at Mach 7.9, recirculation inside the cavity under the stronger shock occurs faster, further degrading the boundary layer. At Mach 7.9, the control is less effective due to the impacts of increased boundary layer degradation and less effective shock smearing. Nevertheless, Mach 5.7 and Mach 7.9 flows exhibit the greatest reduction in separation bubble size at 17% surface porosity.

Introducing a cavity covered with a porous surface effectively breaks the shock into weak $\lambda-$shock, thereby reducing its strength. Therefore, large shock-associated irreversibility is diminished. Since the deployment of the cavity caused the strong shock splitting and widening of the interaction zone, the pressure rise rate across the $\lambda-$shock is also decreased. As a result, a significant decrement in characteristic impedance is accomplished.

The intake mass flow is directly linked to the separation bubble size – as the bubble increases, effective mass flow decreases. This investigation repeatedly observed that the controlled intake with a 17% porous surface-covered cavity experienced lesser separation length. Therefore, we may conclude that the limiting porosity value for the maximum mass flow rate is 17%.

8.3.6 SBLI CONTROL USING AN ARRAY OF MICRO VORTEX GENERATORS

This article investigates the efficacy of micro vortex generators (MVGs) in suppressing shock wave/boundary layer Interactions (SBLIs). The main objective is to examine how well MVGs perform as a boundary layer control for a double ramp mixed compression intake. At two distinct locations, upstream and at the interaction region, MVGs of varying heights (0.5 mm, 0.7 mm, and 1 mm) are deployed. The control ability is quantified using wall static pressures over the ramp surface measured at six locations using wall-mounted ports. Schlieren technique is employed to visualize the wave structures prevailing inside intakes.

8.3.6.1 Wall Static Pressure Measurement

The static pressures recorded at various wall ports are plotted versus the intake-core length. The measured static pressures (p) are made non-dimensional with the freestream static pressure (p_∞), demonstrating how those plots can quantify the control effectiveness of traditional MVGs, likewise the previous investigation. The intake length (L) renders the axial distance (x) non-dimensional. The fluctuation of non-dimensional pressure (p/p_∞) is displayed against the variation of non-dimensional axial distance (x/L). As shown in Figure 8.3.22, the

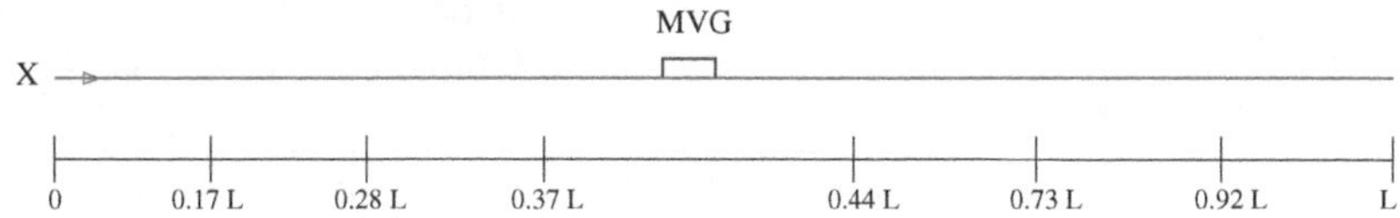

Figure 8.3.22 Static pressure ports (MVGs installed at x = 0.41L).

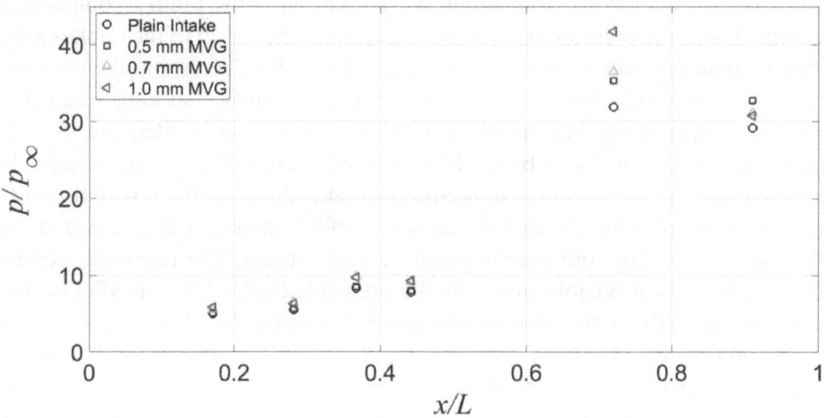

Figure 8.3.23 Wall static pressure variation (MVGs installed at x = 0.41L).

static pressures are monitored at six separate pressure ports installed at 0.17L, 0.28L, 0.37L, 0.44L, 0.73L, and 0.92L over the surface. The MVGs are placed in the interaction region upstream, as shown in the Figure. In order to look at the static pressure spike over the oblique shocks produced at the leading edge of the ramp, the initial two ports are positioned at 0.17L and 0.28L. Additionally, two more were set at 0.37L and 0.44L to explore the impact of MVGs in the vicinity of the location upstream of SBLIs. Two more were positioned at 0.73L and 0.92L to study the flow dynamics downstream of the separation bubbles.

For the MVG-controlled and uncontrolled intakes, the wall static pressure variations are shown in Figure 8.3.23. Both controlled and plain intakes encounter a comparable pressure spike at the initial pressure port ($x/L = 0.17$) because of an oblique shock at the ramp's leading edge. However, a little increment in static pressures for the regulated intakes is observed at the second and third positions (x = 0.28L and x = 0.37L, respectively). The pressure increase for all intakes is greater than at close-by upstream ports at

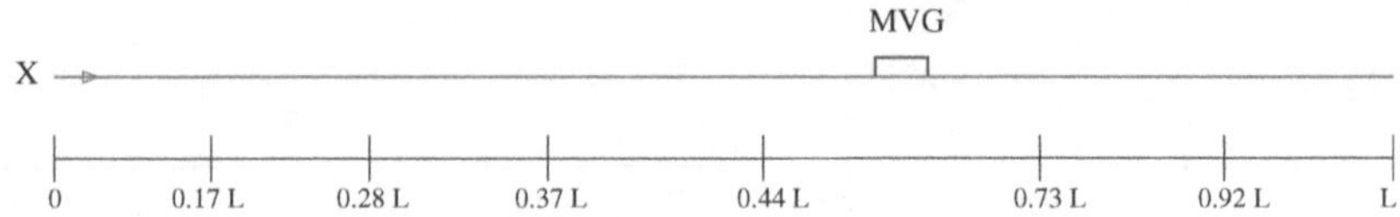

Figure 8.3.24 Locations of static pressure ports (MVGs installed at x = 0.59L).

$^x/_L = 0.37$. This results from shock wave creation at the ramp's compression corner. The static pressure rise is essentially the same for all of the intakes at the downstream pressure port of MVGs ($^x/_L = 0.44$). By decreasing the upstream influence of SBLI, MVGs installed in the plain intake shorten interaction length. However, the benefit of interaction suppression is adequate to make up for the drag brought on by the MVGs profile. Thus, the overall pressure increase is unchanged from the unregulated intake. At $^x/_L = 0.73$, which is close to the region of reattachment in the uncontrolled intake, it is discovered that the increase in static pressure is greater for all intakes. The maximal pressure rise is particularly visible in the intake controlled with 1.0 mm MVGs. This may be due to the generation of additional shocks on the MVGs geometry (as the height of the MVG inside the boundary is beyond the sonic line), which eventually results in a high reattachment pressure for all the controlled intakes. In controlled intakes, the static pressure is higher than the uncontrolled intake at the last pressure port, which is positioned at 0.92L.

MVGs are mounted at the interaction region in the second scenario (Figure 8.3.24). Figure 8.3.25 depicts the estimation of static pressures over the wall at different pressure ports in the preceding situation. The pressures at the first four ports are the same as in the preceding case. At $^x/_L = 0.73$, pressure decreased for all MVG-controlled intakes relative to plain intake. Table 8.10 shows the results of equation (8.3.7) for wall static pressure variation in percentage at the reattachment region ($^x/_L = 0.73$).

Static pressure decline is estimated for 0.5 mm and 0.7 mm height MVGs. A maximum of 13.57% is observed for 1.0 mm height MVGs in wall static pressure decrement relative to uncontrolled intake. From the results, we can infer that the shedding of counter-rotating vortices improved the performance of micro devices by promoting high-momentum fluid entrainment toward the adjacent wall region. Additionally, distortion of impinging and reattachment shocks because of the presence of vortices at the interaction region, in turn, reduced the shock strengths. Moreover, deploying MVGs inside the recirculation zone offers less wave drag. This may be the reason for the lesser reattachment pressure. Also, enhanced pressure recovery is noticed at $^x/_L = 0.92$ for controlled intakes.

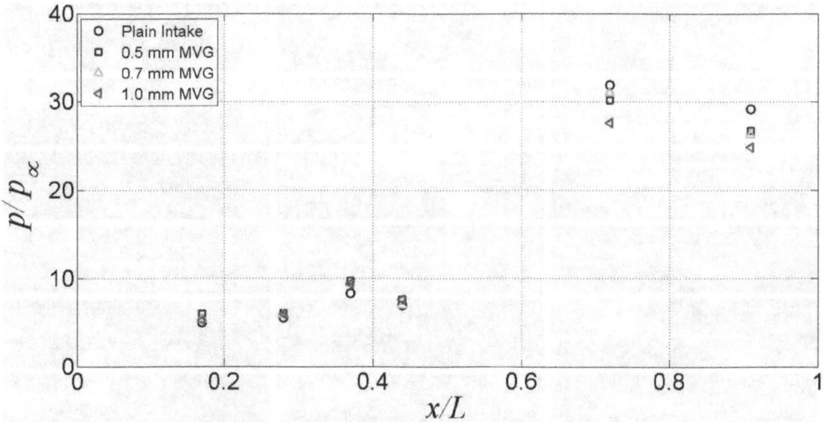

Figure 8.3.25 Wall static pressure variation (MVGs installed at x = 0.59L).

Table 8.10

Percentage variations in the wall static pressure at x = 0.73L.

Height of MVGs	Percentage variation in p_w
0.5 mm	5.42%
0.7 mm	2.57%
1.0 mm	13.57%

8.3.6.2 Schlieren Flow Visualization

The Schlieren technique visualizes the time-resolved flow evolution and wave structures in plain and controlled intakes.

Flow Field Development

Figures 8.3.26, 8.3.27, and 8.3.28 demonstrate the evolution of the flow field at Mach 5.7 in uncontrolled and controlled intakes for all three heights of MVGs (0.5 mm, 0.7 mm, and 1.0 mm). The flow starts at 2.2 milliseconds. So, considering this as a reference time and an interval of 200 microseconds, frames of flow visualizations have been captured. The flow field evolution inside intake considered in this study is between 2.4 and 3.8 milliseconds. This flow field evolution period is divided into three zones; flow rising time, steady test time, and flow termination process. All these three zones are exhibited

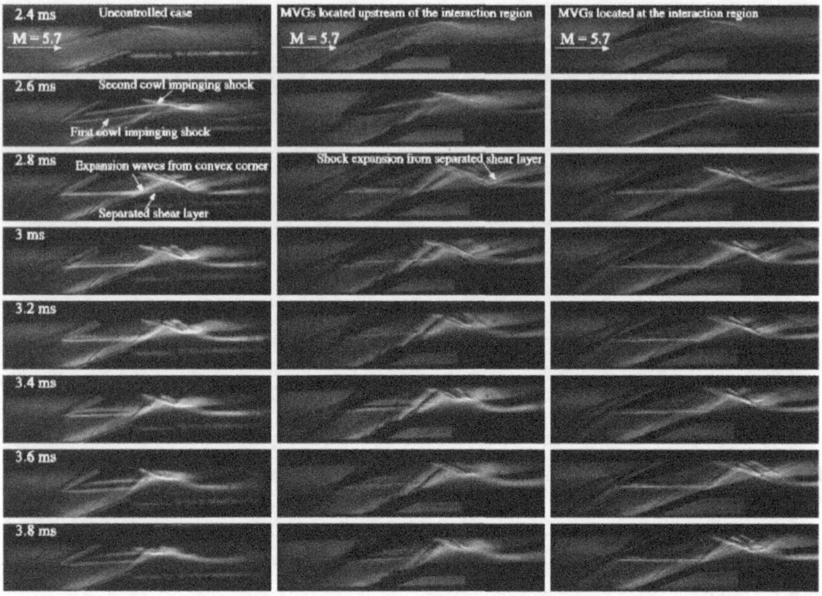

Figure 8.3.26 Flow field development in plain and controlled intakes (0.5 mm MVGs).

with the help of time-resolved Schlieren images for the hypersonic intake. At 2.4 milliseconds, impingement of second cowl shock initiation can be seen. Following this, at 2.6 milliseconds, the boundary layer development is observed. We can witness the flow separation as soon as the cowl shocks are formed (at 2.8 milliseconds). At the same time, the second cowl shock reflection from the separated shear layer resulted in the formation of expansion waves.

Interestingly, the separation bubble develops between 2.8 and 3.4 milliseconds, nearly the same as observed in the porous cavity-controlled intakes (where the reference time is considered zero milliseconds). As soon as the test time begins, it is noticed that the reattachment and separation point starts moving to and fro. After some time, the reattachment point moves downstream and stays at a certain location. Nevertheless, the separation point is also stabilized at the ramp shoulder once the flow is fully developed for all the investigations.

The Schlieren images show that, as far as the flow evolution is concerned, no significant deviations in the flow structures between the uncontrolled and the controlled intakes (MVGs are installed at the interaction region) are observed. This may be due to the deployment of the MVGs at the recirculation zone. In addition, for 0.5 mm MVGs installed before the interaction region, the shocks induced by the MVGs, are of lower strength. Therefore, the flow field evolution

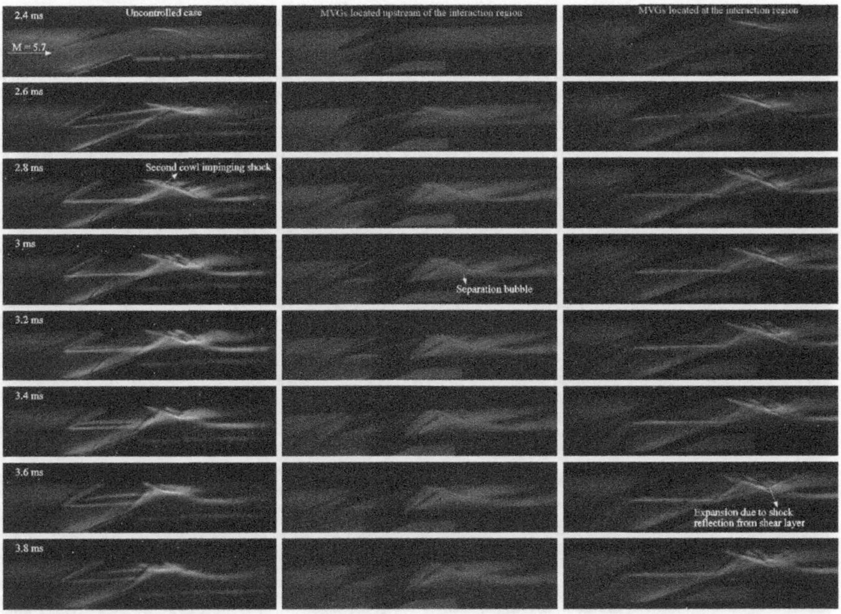

Figure 8.3.27 Flow field development in plain and controlled intakes (0.7 mm MVGs).

and shock structure, and the shock/shock interaction process in uncontrolled intake and intakes controlled with MVGs of height 0.7 mm and 1.0 mm, are compared in Figure 8.3.29. In the figure, the evolution of the shock waves from the MVGs (MVGs are placed upstream of the interaction region) describes the shock/shock interaction between the cowl and MVGs generated shocks. The shock at the leading edge of MVGs is seen to interact with the cowl lip shock whereas the shock at the trailing edge of MVGs interacts with the expansion waves and hence experiences a reduction in its strength. This reveals that the shock at the leading edge of MVGs is mainly responsible for the static pressure rise.

Variation in the Separation Bubble Size

Figure 8.3.30(a) shows the Schlieren image for the steady flow field in the plain intake. It is observed that the first cowl shock generated at the cowl lip incidents over the ramp shoulder and interacts with the expansion waves, eventually reducing the shock strength. A second cowl shock, generated at the cowl corner inside the isolator region, has a greater influence in forming a separation bubble. It can be observed that after interacting with the second cowl shock, the boundary layer over the intake core thickened.

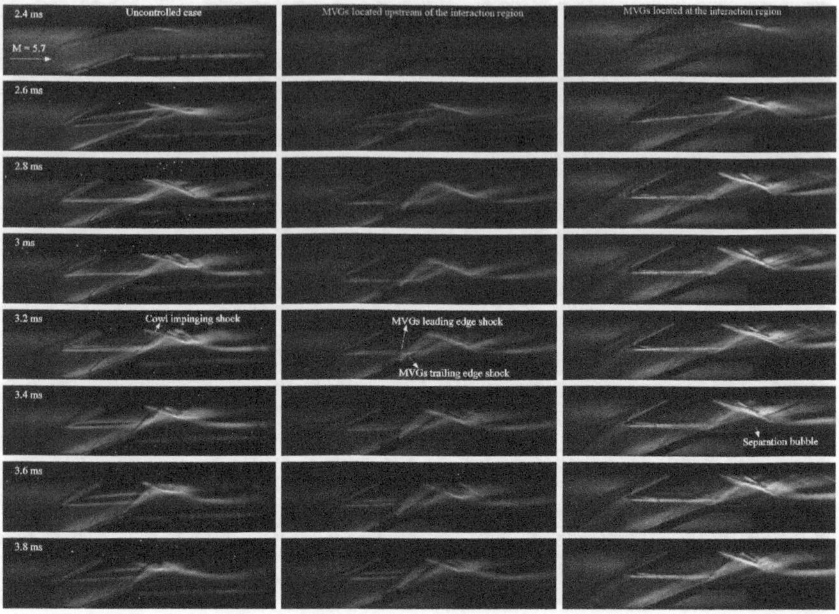

Figure 8.3.28 Flow field development in plain and controlled intakes (1.0 mm MVGs).

The negative pressure gradient imparted by the second cowl shock detaches and reattaches the boundary layer leading to the separation bubble formation (Figure 8.3.30(a)). However, introducing 0.5 mm and 0.7 mm height MVGs before the interaction region substantially decreased the boundary layer's thickness, as depicted in Figure 8.3.30(b)–(c). A considerable reduction in separation length is also observed in these two intakes. Figure 8.3.30(d) shows the Schlieren image of 1 mm height MVGs controlled intake. In contrast to the previously controlled intakes, a growing separation bubble is observed as the geometric blockage offered in this case is higher than the previous ones. A higher geometric blockage provokes a higher momentum deficit; the advantages of MVGs cannot compensate for it. Hence, experiencing higher drag, 1 mm high MVGs causes a larger separation bubble. On the other hand, the performance of MVGs of height 0.7 mm is superior in shedding smaller and more efficient counter-rotating vortices into the flow. So, the entrained fluid with higher momentum is mixed with lower momentum boundary layer fluid adjacent to the wall. The efficacy of MVGs of height 0.7 mm in SBLI controlled is explained as follows. In each vortex generator's wake, the counter-rotating vortices are created (Figure 8.3.31(a)). They move upward at first due to upwash induced by these vortices from the MVG surface, and later flow inertia

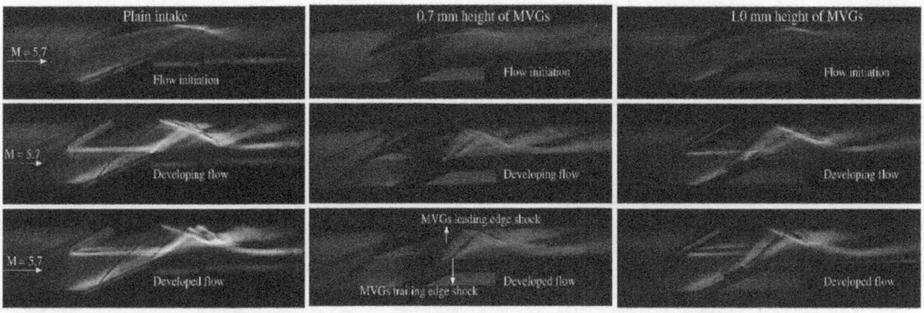

Figure 8.3.29 Comparison of shock structures in uncontrolled and controlled intakes (MVGs installed before the interaction zone).

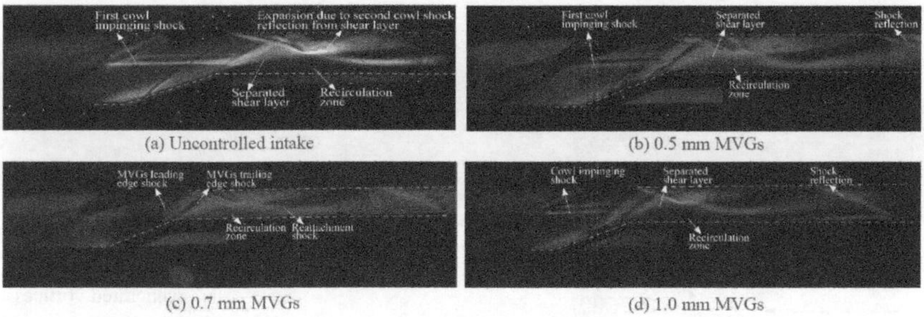

Figure 8.3.30 Schlieren pictures of plain and controlled intakes (MVGs installed at x = 0.41L).

makes them streamwise. These vortices play a major role in transferring momentum from the freestream to fluid adjacent to the wall. The fluid with low momentum near the wall is moved upward from the boundary layer, while the fluid with higher momentum beyond the boundary layer is entrained. The extent of the separation zone is reduced as the fluid within the boundary layer is energized by the entrained higher momentum fluid. Figure 8.3.31(b) shows the schematic illustration of streamwise shedding vortices by micro vortex generators (MVGs) array.

For controlled intakes when MVGs are placed at the interaction region, Figure 8.3.32 shows the images of flow visualization. A triangle shape separation bubble is noticed for the three controlled intakes. Also, an expansion fan is generated at the apex of the separated shear layer. A thinner boundary layer downstream of the separation bubble is observed here. Essentially, 0.5 mm, 0.7 mm, and 1 mm height MVGs are responsible for separation length reduction compared to plain intake (Figure 8.3.32(a)–(c)). It is suggested that deploying shorter MVGs at the interaction region rapidly mixes the near wall

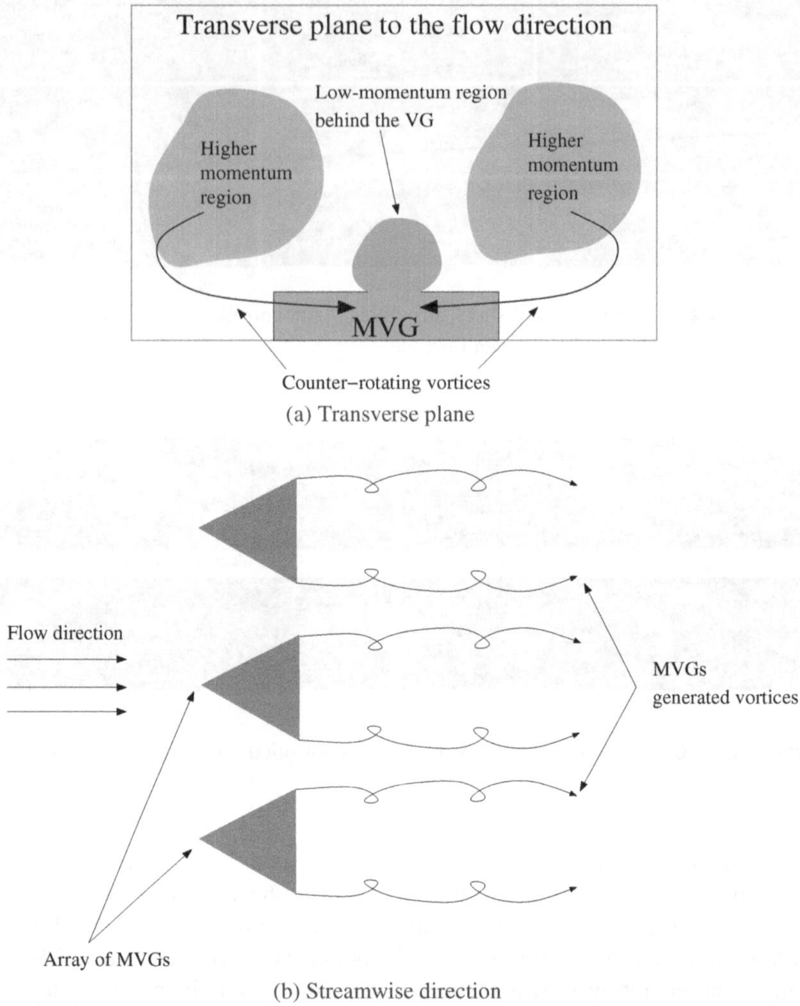

Figure 8.3.31 Schematic diagrams of vortex shedding from MVGs.

region fluid with high-momentum fluid, reducing separation length. However, the deployment of MVGs of 1 mm increased the size of the separation bubble (Figure 8.3.32(c)), which infers that large-size MVGs cause low-momentum flow. Besides, additional drag is the compensation of high-strength vortices. However, compared to upstream MVGs, deploying MVGs at interaction

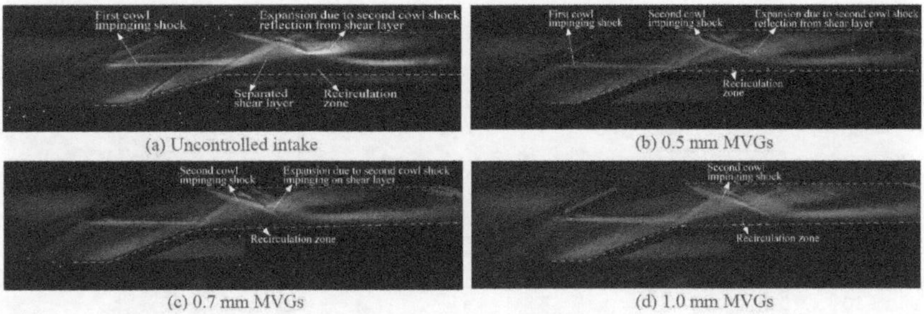

Figure 8.3.32 Schlieren pictures of plain and controlled intakes (MVGs installed at x = 0.59L).

resulted in higher separation length (except 1 mm high MVGs where similar sizes of recirculation bubble are observed).

Table 8.11
Separation bubble size (MVGs installed before the interaction region).

Intake configuration	R_ℓ	R_h
Plain intake	28.8 mm	4.3 mm
MVGs of height 0.5 mm	22 mm	3.7 mm
MVGs of height 0.7 mm	19.8 mm	3.6 mm
MVGs of height 1.0 mm	24.3 mm	4 mm

The quantification of MVGs efficacy is done by calculating the separation lengths and heights for all the intakes and tabulated in Table 8.11. In the Schlieren view, marking out the reattachment and separation points of the recirculation zone gives us the separation length. It is important to note that the estimation of separation bubble size precisely is very hard because of its transient nature. Hence, considering many observations, the average estimation of separation bubble size is exhibited with measured standard deviations, providing a qualitative description. Figure 8.3.33 shows the separation length and height fluctuation for all the controlled intakes deploying MVGs upstream of the interaction region (the red line indicates uncertainty). The separation length for the uncontrolled intake is 28.8 mm, but the separation lengths for the controlled intakes with MVGs of heights 0.5 mm, 0.7 mm, and 1 mm are measured as 22 mm, 19.8 mm, and 24.3 mm, respectively. For the MVGs heights of 0.5 mm, 0.7 mm, and 1 mm, respectively, the separation lengths are reduced by around 23.61%, 31.25%, and 15.63%, respectively.

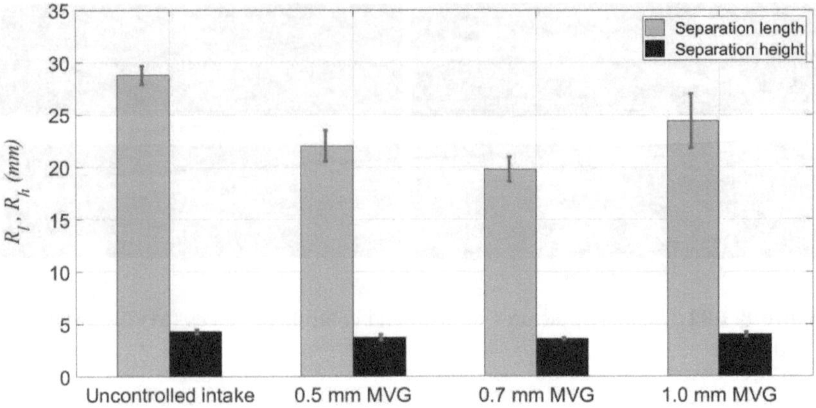

Figure 8.3.33 Variation in separation bubble size (MVGs installed at x = 0.41L).

Uncontrolled intake is less effective than controlled intake in all three of the examined MVG designs; however, 0.7 mm MVGs have demonstrated the greatest capacity to shorten the separation length. The separation height is measured to be 4.3 mm for controlled intake, whereas the heights of separation are measured to be 3.7 mm, 3.6 mm, and 4 mm for controlled intakes with MVGs of heights 0.5 mm, 0.7 mm, and 1 mm, respectively. The greatest reduction in separation height was seen with MVGs that were 0.7 mm in height. The average length of the separation bubble in the Mach 5.7 controlled case can be as low as 26.9 mm, as observed in section 8.3.5. This may be because of the sensitivity of the recirculation zone size to the small variations in test conditions.

Table 8.12 and Figure 8.3.34 show the separation lengths for the simple intake and the one controlled by various MVGs used in the interaction region (the red line indicates uncertainty). The lengths of the separation bubbles are 25.4 mm and 24.1 mm, respectively, for the controlled intakes with 0.5 mm and 1 mm height MVGs. The separation length was 23.3 mm for the controlled intake with 0.7 mm height MVGs. Therefore, for 0.5 mm, 0.7 mm, and

Table 8.12
Separation bubble size (MVGs installed at the interaction region).

Intake configuration	R_ℓ	R_h
Plain intake	28.8 mm	4.3 mm
MVGs of height 0.5 mm	25.4 mm	3.8 mm
MVGs of height 0.7 mm	23.3 mm	3.7 mm
MVGs of height 1.0 mm	24.1 mm	3.9 mm

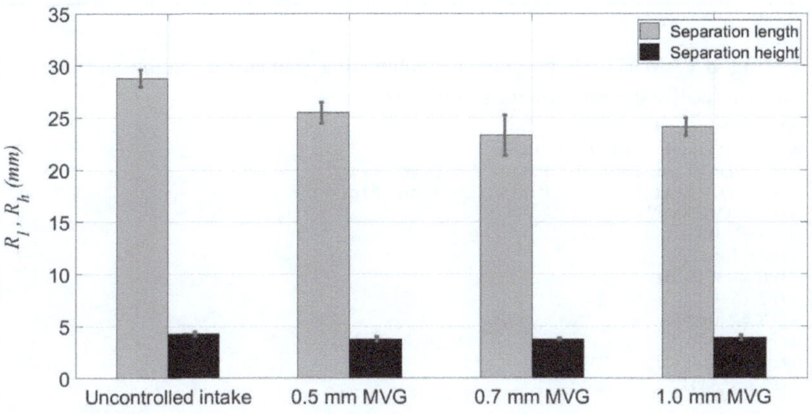

Figure 8.3.34 Variation in separation bubble size (MVGs installed at x = 0.59L).

1 mm height MVGs, the separation length reductions in percentage are roughly 11.81%, 19.10%, and 16.32%, respectively. The separation heights are 3.8 mm, 3.7 mm, and 3.9 mm, respectively, for the intake regulated by 0.5 mm, 0.7 mm, and 1 mm height MVGs. The second cowl shock and the reattachment shock are distorted and eventually lose some strength when the MVGs are positioned in the interaction region because of the vortices the MVGs shed into the environment. Reduced reattachment pressure is the result of this occurrence.

The MVGs within the interaction zone offers a significantly lower-wave drag. However, compared to MVGs placed upstream of this region, the reduction in separation length caused by the deployment of MVGs at the interaction zone is lesser. Since the double ramp mixed compression intake is considered in this experiment, numerous internal and external oblique shock waves are generated. As a result, as the oncoming wave travels through these oblique shocks, its Mach number gradually drops. Therefore, the intensity of shock created at the cowl lip is diminished by the incoming flow that slows down just before the cowl lip.

The expansion waves produced at the shoulder of the ramp also lessen the intensity of this shock. Finally, the local heating at the interaction zone is reduced due to the decreased intensity of shock inside the intake. However, the placement of MVGs over the surface of the ramp sheds counter-rotating vortices that increase flow field disturbances. The MVGs extra disruptions effectively accelerate the rate of heat transfer. Additionally, it would be advantageous if the separation length is shortened, as doing so may result in a slower heat transmission rate. As a result, placing MVGs at their ideal height of 0.7 mm could increase local heat flux and reduce local heat load.

Listing 8.1 MATLAB program for calculating standard deviation and uncertainty in wall static pressure measurements.

```
clc; clear all;
N = 5; %number of test runs
str1='D:“Text Data“1. Plain“T = 8 mm“Text“';
str2='D:“Text Data“1. Plain“T = 8 mm“Plot“';
str3='Ori˙XvsV';
str4=' 001';
str5='.txt';
str6='˙WError';
str7=' 002';
str8=' 003';
str9=' 004';
str10=' 005';
str11='XvsP';

textname=strcat(str1,str11,str6,str5);

filename1=strcat(str1,str3,str4,str5);
filename2=strcat(str1,str3,str7,str5);
filename3=strcat(str1,str3,str8,str5);
filename4=strcat(str1,str3,str9,str5);
filename5=strcat(str1,str3,str10,str5);

M1=load(filename1);
length(M1);
x=M1(:,1);
v1=M1(:,2);
M2=load(filename2);
length(M2);
v2=M2(:,2);
M3=load(filename3);
length(M3);
v3=M3(:,2);
M4=load(filename4);
length(M4);
v4=M4(:,2);
M5=load(filename5);
length(M5);
v5=M5(:,2);

%average
v = (v1+v2+v3+v4+v5)/N

%deviation from the average
d1=(v1-v)./v
d2=(v2-v)./v
d3=(v3-v)./v
d4=(v4-v)./v
d5=(v5-v)./v

%average deviation
ad = (abs(d1)+abs(d2)+abs(d3)+abs(d4)+abs(d5))/N

%standard deviation
sd = sqrt(((d1.*d1)+(d2.*d2)+(d3.*d3)...
+(d4.*d4)+(d5.*d5))/(N-1))

%standard error or uncertainity in voltage
```

```
se'v = sd/sqrt(N)

%Conversion of voltage to pressure
p=0.6v

%standard error or uncertainity in pressure
dp'dv=0.6
se'P=(abs(dP'dV)).* se'v

C=[x; v; sd; se; p; sep]
fileID = fopen(textname,'w');
fprintf(fileID, '%4s %6s %6s %6s %6s %6s"r"n','x','v',...
'sd','se','p','sep');
fprintf(fileID,'%1.4f %1.4f %1.4f %1.4f %1.4f %1.4f"r"n',C);
fclose(fileID);
```

8.4 SUMMARY

In this chapter, a generic 2D planner dual ramp design of mixed compression inlet is studied experimentally using a hypersonic shock Tunnel. To control the interactions of Shock wave and boundary layer, a cavity covered using a porous surface is deployed (shock control), and the intake-mounted micro vortex generators (boundary layer control) are studied at hypersonic Mach numbers. The investigation is performed by deploying a porous surface over a 1.46 aspect ratio cavity to control the SBLIs occurring in the hypersonic intake, mounted in a shock tunnel with Mach 5.7. Five different surface porosities (4.5%, 7.5%, 17%, 21.6%, and 25%) are investigated at Mach 5.7. The surface porosity variation is attained by altering pore diameters. It is observed that an increase in porosity effectively decreased the size of the separation bubble, with the highest suppression of 17% porosity. Surprisingly, the bubble grew again for higher porosity cases (21.6% and 25%). We performed the investigations by increasing the Mach number to 7.9 to validate this optimum porosity limit in suppressing SBLIs efficiently, particularly for 17% and 25% surface porosity. In addition, utilizing MVGs of varied heights (0.5 mm, 0.7 mm, and 1.0 mm) positioned at two distinct positions (upstream and at the interaction zone), the MVG's efficiency in modifying the shock-wave/boundary-layer interactions in the hypersonic intake has been tested experimentally. In order to quantify their control ability, the wall static pressures over the ramp surface are measured at six distinct pressure ports on a double ramp for all the intakes. Visualization of wave structures and formation of a boundary layer is performed using the Schlieren technique. The summary of specific results of the current investigation is as follows:

- The intake controlled with a shallow cavity covered with a porous surface is observed to have a greater ability to reduce the shock strength and suppress the recirculation zone than uncontrolled intake. By splitting the interaction zone using λ−shock, pressure jump increases more uniformly across the shock, leading to a near isentropic compression.

- At Mach 5.7, a maximum drop of 20.53% is observed in reattachment static pressure for controlled intake with a 25% porous surface cavity. At the same time, a pressure drop of about 20.2% is seen in controlled intake with the same porosity at Mach 7.9. Additionally, it has been found that separation length decreases when the Mach number increases, but shock strength increases. The less significant impacts of the pressure signal delivered upstream are the obvious cause.
- The recirculation bubble size reduces with an increment in surface porosity up to 17%, according to the Schlieren views taken at Mach 5.7. In the intake controlled with 17% surface porosity at Mach 5.7, a maximum reduction of around 22.25% in the separation height and approximately 16.73% in the separation length is noted. The decreases in separation length and height for intake regulated with 17% surface porosity are 7.65% and 2.38%, respectively, at Mach 7.9. Beyond a certain amount of porosity (17%), the boundary layer deteriorates, and the size of the bubble increases once more. The interaction between the thickening of the upstream boundary layer and the division of the shock structure into the λ−shock causes this.
- Even in the greatest porosity (25% surface porosity) instance, at Mach 5.7, the beneficial shock splitting effects significantly dominate the detrimental boundary layer deterioration effects compared to plain intake. For the 25% porosity instance, however, at Mach 7.9, the negative impacts of the deteriorated boundary layer in terms of enlarging the size of the separation bubble predominate. However, at 17% surface porosity, both Mach 5.7 and Mach 7.9 flows exhibit the greatest decrement in the separation bubble size.
- Time-resolved Schlieren technique shows that the boundary layer develops only beyond a specific time-lapse, and hence shock structure in the flow is initially dominated by inviscid flow characteristics.
- The losses due to the boundary layer separation are effectively reduced by micro vortex generators (MVGs). Deploying 0.7 mm MVGs before the region of interaction decreased the separation length to a maximum of 31.25%. However, when static pressure rises across the incident shock is considered, they fail. The static pressure rise is higher for all the MVGs when stationed before the interaction zone. The generation of additional shock on MVG geometry could be the reason (since inside boundary, MVGs are taller than a sonic line), which eventually results in higher reattachment pressure for all

the control intakes. Alternatively, deploying MVGs at the interaction zone showed a 13.57% reduction in static pressure near the reattachment point with 1.0 mm MVGs. MVGs showed improved performance by shedding counter-rotating vortices, encouraging the high-velocity fluid entrainment toward the nearby wall. Also, the wave drag offered by MVGs inside the interaction zone is relatively lesser. Additionally, the reattachment and incident shock waves may be distorted and have weaker forces due to the vortices created at the interaction region, which will lower the static wall pressure close to the reattachment point.

- Unlike 1.0 mm MVGs, deploying MVGs before the interaction zone effectively shortens the separation length than deploying at the interaction zone. The boundary layer momentum is enhanced by the vortices the MVGs emit into the boundary layer. This works especially well with shorter MVGs because they efficiently shorten the separation length by shedding smaller, more efficient vortices into boundary layer flow. According to the Schlieren pictures, 0.7 mm high MVGs enhance mixing and minimize separation losses.
- The MVGs of 1 mm height show a growing trend in the separation bubble size for both deployment locations, which indicates that a large MVG causes a low-momentum flow. Besides, longer MVGs shed stronger vortices at the expense of additional drag, increasing the separation bubble size.
- A pair of counter-rotating vortices, which first rise from the surface of each vortex generator due to upwash caused by the vortex and then become streamwise due to the inertia of the flow, form the wake behind each vortex generator. In essence, these vortices do transmit fluid momentum to the near wall region fluid from the freestream. The boundary layer achieves greater momentum as the suction produced by these vortices draws within the outside fluid. The separation zone shrinks in size due to the reenergizing of the approaching boundary layer.

9 Mixing Characteristics of Supersonic Multiple Jets

9.1 INTRODUCTION

Many researchers have been studying the multijets phenomenon in the realm of supersonic Mach numbers for a long time. This is used in many engineering problems of practical significance, such as the flow field of a spacecraft booster motor. There is a wealth of literature available for both subsonic and supersonic jet propagation, studying the mixing phenomena of multijets and revealing the decay, spread, and entrainment properties. However, the majority of the researchers concentrated on multijets arrangements, which are solely of academic relevance. Aside from the jet aspects, they have yet to concentrate on the base pressure distribution. The base pressure is important in the case of rockets and missiles because the high-temperature plumes stimulating the nozzle will be attracted toward the base region when the base pressure is at a low sub atmospheric level. This is concerning because the hot base may harm the electronic equipment placed within the rocket shell. As a result, care must be made to keep the base pressure at a reasonable level in order to prevent base heating. The merging location and entrainment characteristics determine the base flow. The multijets greatly influence the base pressure since they decays faster and engulf mass from the surrounding environment. If there is a lot of entrainment right after the nozzle exit, the pressure at the base will be quite low. The hot stream, approaching the nozzle, will be redirected toward the base. The associated thermal radiations caused by the large temperature differential between the hot jet and the environment will induce severe heating of the base. As a result, a full understanding of the merging, decay, and influence properties of multijets is required. Because the realistic configurations will have significant technical significance for launch vehicle applications. With these considerations in mind, this chapter investigates a configuration comparable to that of the Indian Space Research Organization's Geostationary Satellite Launch Vehicle (GSLV) Mark II satellite launch vehicle [11].

9.2 EXPERIMENTAL METHODOLOGY

The experiments in this study are conducted in the supersonic open jet facility located in the Supersonic Jet Research Laboratory at the Indian Institute of Technology Kanpur. The facility consists of an air supply system (compressor

DOI: 10.1201/9781003139447-9

and storage tanks) and an open jet setup, which consists of a settling chamber with provision to mount the jet nozzle at its end plate. The settling chamber is fed with compressed dry air at high pressure (up to 30 bar) through a pressure regulating valve, which controls the settling chamber pressure at any desired level before expansion through the jet nozzles.

The pressure was measured using a 16-channel pressure transducer with a range of 0 to 300 psi. To connect the transducer to a computer, graphical user interface (GUI) software was used. The menu-driven programme collects data and displays pressure readings from all 16 channels on the computer monitor in a window-style display. The programme can select the pressure units from a list and perform a re-zero/full calibration. The transducer also allows you to select the number of samples to average using dip-switch settings. The transducer's accuracy (after re-zero calibration) is specified as 0.15% of full scale. The pitot pressure was calculated using the mean pressure data averaged over 250 samples per second. Because the jet is fundamentally unsteady, this is accomplished. The frequency and amplitude of the unsteadiness, however, were not examined because the goal of this study is to investigate the mixing of the core jet in the presence and absence of strapon jets. Pressure measurement repeatability was found to be $\pm 3\%$.

The pitot probe utilized in the investigation had an outside diameter of 0.6 mm and an inner diameter of 0.4 mm. Thus, the ratios of the core nozzle and strapon nozzle exit areas to the probe exit area are 1600 and 469.5, respectively. These numbers are significantly over the cutoff of 64 for minor probe blockage [12]. The pitot probe was mounted on a rigid three-dimensional traverse with a linear translation resolution of 0.1 mm. In all measurements, the sensing probe stem as held parallel to the jet axis, with the sensing hole toward the flow. The pitot pressure measured was accurate to within 2%. Furthermore, the Reynolds numbers for nozzle pressure ratios[1] (NPRs) 4 and 7 based on the outer probe diameter of 0.6 mm are 30000 and 60000, respectively. Viscous effects will not produce mistakes in pitot pressure measurements because these values are significantly higher than the maximum Reynolds number of 500 [22].

9.2.1 MULTIJETS CONFIGURATION

Figure 9.2.1 shows the schematic and photographic views of the Mach 1.6 and 2.5 convergent-divergent nozzles utilized in the multijets. The Mach 1.6 (core or central) and Mach 2.5 (strapon) nozzles had exit diameters of 24 mm and 13 mm, respectively. The strapon nozzles (two nozzles on either side of the central nozzle) were installed on a base plate with the core nozzle axis canted 6° outward.

[1]Nozzle pressure ratio (NPR) is defined as the ratio of settling chamber pressure (P_{01}) and the back pressure (P_b) i.e., NPR $= \frac{P_{01}}{P_b}$.

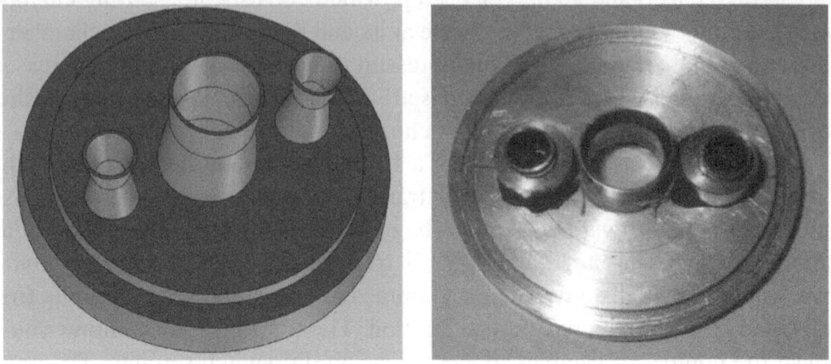

(a) Schematic view (b) Photographic view

Figure 9.2.1 Schematic and photographic views of the multijets.

9.3 QUANTITATIVE AND QUALITATIVE ANALYSES

Aerodynamic and aeroacoustic characteristics are two crucial aspects that must be properly understood for the efficient use of jets in general and multijets in particular. Multijets are generally employed in all new vehicles introduced since the early 1990s. Only underexpanded jets were deemed efficient thrust generators by launch vehicle researchers, and hence they were used in launch vehicle motors. However, this way of thinking has evolved and become overly broad. For launch vehicle operations, highly enlarged jets are used. In India's GSLV, for example, both the core and all four strapon motors are operated at severely overexpanded conditions. It is also usual practice to use distinct supersonic Mach numbers for the core and strapon of the launch vehicle. Even though the acoustical characteristics of multijets have been well examined, the aerodynamic properties must also be considered. Thus, the current study intends to acquire insight into the aerodynamic properties of a core and two strapon jets with a 6^{0} outward canting, similar to the upgraded GSLV design. The focus is on jet propagation in the combined mode, emphasizing the merging and confluence of independent jets under overexpanded conditions. Additionally, mass entrainment is quantified. The exit Mach numbers of the core and strapon nozzles are 1.6 and 2.5, respectively. Measurements were taken at NPR 4, 5, and 7 to determine the level of overexpansion with varying degrees of unfavorable pressure gradient. Measurements of velocity profiles in the transverse plane (i.e., normal to all three jet axes) and pressure measurements at specific grid points in the jet field were performed to estimate the mass flow rate at various axial positions.

9.3.1 CENTERLINE PRESSURE DECAY

Centerline pressure decay is a reliable way to quantify jet mixing. In the case of subsonic jets, the centerline velocity decay is used to calculate the potential core length, characteristic decay, and fully developed flow zone. However, due to wave dominance, determining the Mach number or velocity of a supersonic jet using measured pressure data is problematic. As a result, it is common practice to plot the non-dimensional centerline pitot pressure variation to quantify jet characteristics such as core length, characteristic decay, and fully developed zone [22]. The potential core of a subsonic jet is defined as the axial extent from the nozzle exit to which the axial velocity remains constant. Alternatively, the core of a supersonic jet can be defined as the axial distance from the nozzle exit to the point where the characteristic decay begins. The flow field turns subsonic after the supersonic core.

The measured pitot pressure distribution (P_{02}) along the core jet centerline is non-dimensionalized by dividing it by the settling chamber pressure (P_{01}). The corresponding axial distance (x) is non-dimensionalized by the core nozzle exit diameter (D_e). Figures 9.3.1, 9.3.2, and 9.3.3 depict the variation of centerline pitot pressure with axial distance for NPRs 4, 5, and 7. The individual decay of core and strapon jets are also included for comparison. It is worth mentioning that the Mach 1.6 core nozzle at NPRs 4, 5, and 7 operates under overexpanded, nearly correctly expanded, and underexpanded conditions.

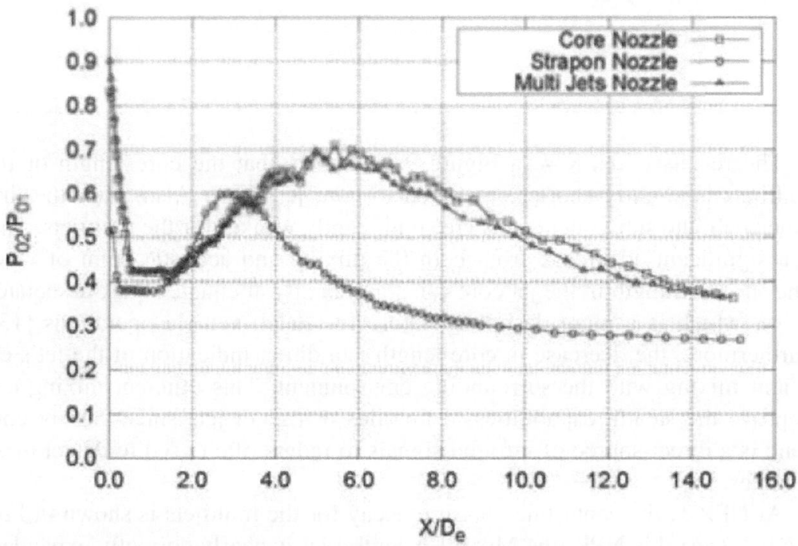

Figure 9.3.1 Centerline pressure decay of multijets at NPR 4.

In contrast, at these NPRs, the Mach 2.5 strapon nozzles always operate in an overexpanded state.

Listing 9.1 Fortran program for calculation of stangnation pressure ratios.

```
C         For calculating P'02/P'01 for all the nozzles

          DIMENSION X(100), P'02 (100), PTR (100)

          REAL NX
          OPEN (1, FILE = 'file1 ')
          OPEN (2, FILE = 'file2 ')

C         WRITE(*,*) 'NUMBER OF ROWS FOR NX'
          READ (1,*)NX

          DO I = 1, NX
          READ (1, *)X (I) , P'02 (I)
          ENDDO

          P'atm = 'Atmospheric pressure '
          T'atm = 'Atmospheric temperature '
          P'01  = 'Stagnation chamber pressure '
          DIA   = 'Core nozzle exit diameter '

          DO I = 1, NX
          PTR (I)= (P'02(I)+P'atm)/(P'01+P'atm)
          WRITE(2, 1000)X(I)/DIA,PTR(I)
          ENDDO

1000 FORMAT (2F12.4)

          STOP
          END
```

The results of NPR 4 in Figure 9.3.1 shows that the core length of the multijets is slightly shorter compared to core jet alone. Also, after the first shock cell, the subsequent cells are significantly weaker for the multijets. This is a significant advantage from both the mixing and acoustic point of view. The shock strength in the jet core can significantly attenuate shock-associated noise, which is a more desirable feature in launch vehicle operations [13]. Furthermore, the decrease in core length is a direct indication of the jet's efficient mixing with the surrounding environment. This efficient mixing can improve the stealth capabilities of missiles or fighter jets since the hot core zone is a direct source of infrared signals to radars often used to detect these vehicles.

At NPR 5, the centerline pressure decay for the multijets is shown in Figure 9.3.2. At this NPR, the Mach 1.6 central jet is nearly correctly expanded, but the strapons are overexpanded. However, the behavior of the combined jet is almost identical to those at NPR 4. This may be because even though the

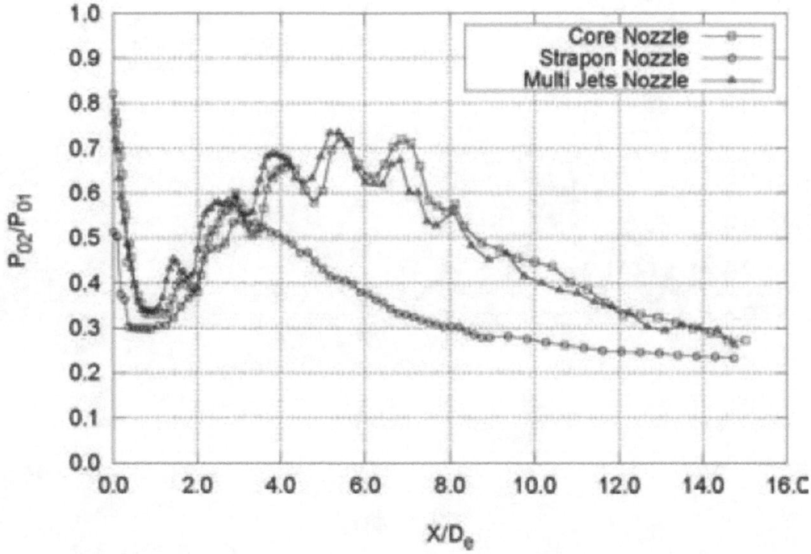

Figure 9.3.2 Centerline pressure decay of multijets at NPR 5.

central nozzle experience almost zero pressure gradient and hence a shorter core length, the strapon nozzles with adverse pressure gradient could efficiently promote mixing in supersonic core and characteristics decay zones.

When the NPR is increased to 7, the decay of the combined jet is drastically different from NPR 4 and 5. At this NPR, the unfavorable pressure gradient for the core jet is reduced significantly; hence, the shock waves in the supersonic core are not forced to increase the pressure to the extent of lower NPRs. Because of this, the shocks in the cells up to $\frac{x}{D_e} = 7$ are of significant strength. However, compared to the central jet, the multijets experience a reduction in supersonic core length at about two times D_e; this can be considered a significant advantage of the multijets. The physical reasoning for this behavior is that the strapon offers to shield the core jet propagation, which restricts the latter jet from regaining its energy in the subsequent cells, and the supersonic core becomes shorter.

From the above discussion, it is evident the presence of neighboring strapon jets augments the mixing of the central jet, and likewise, the presence of the core jet help enhance the mixing of the strapon jets. Thus, it may be concluded that the multijets is a superior configuration for aerodynamic mixing compared to core and strapon jets alone.

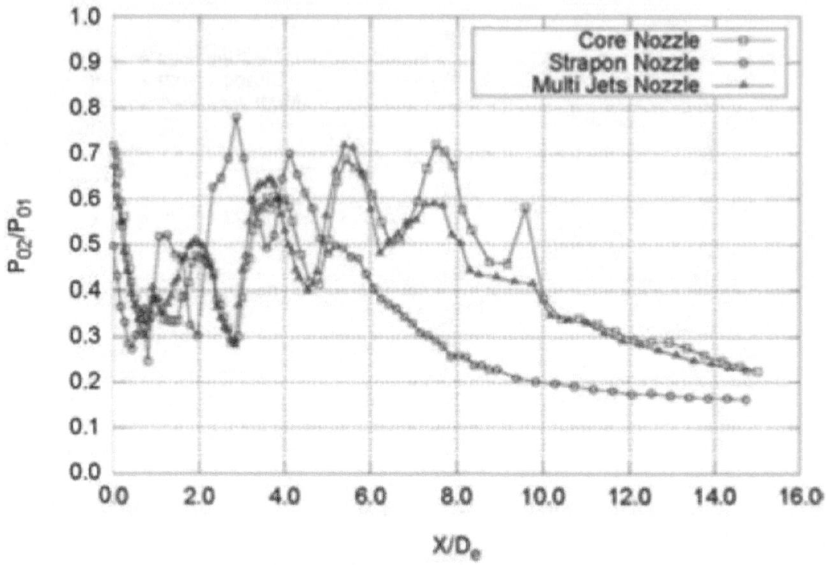

Figure 9.3.3 Centerline pressure decay of multijets at NPR 7.

9.3.1.1 Uncertainty in Stagnation Pressure Ratio Calculation

This section outlines a method for estimating the uncertainty for the non-dimensionalized stagnation pressure ratio $\frac{P_{02}}{P_{01}}$, which is obtained from the measured pitot pressure (P_{02}) and settling chamber pressure (P_{01}). The stagnation pressure ratio is denoted as

$$\frac{P_{02}}{P_{01}} = P_R = \frac{P_{02}}{NPR \times P_a} \tag{9.3.1}$$

where P_a is the atmospheric pressure. Differentiating equation (9.3.1) yields

$$dP_R = \frac{\partial P_R}{\partial P_{02}} dP_{02} + \frac{\partial P_R}{\partial P_{01}} dP_{01} + \frac{\partial P_R}{\partial P_a} dP_a \tag{9.3.2}$$

$$= \frac{1}{P_{01}} dP_{02} + \left(\frac{-P_{02}}{P_{01}^2} \right) dP_{01} + \left(\frac{-P_{02}}{NPR \times P_a} \right) dP_a \tag{9.3.3}$$

equation (1.6.6) can be used to calculate the expression for uncertainty in $\frac{P_{02}}{P_{01}}$:

$$u_{P_R} = \pm \left[\left(\frac{P_{02}}{P_R} \frac{\partial P_R}{\partial P_{02}} u_{P_{02}} \right)^2 + \left(\frac{P_{01}}{P_R} \frac{\partial P_R}{\partial P_{01}} u_{P_{01}} \right)^2 + \left(\frac{P_a}{P_R} \frac{\partial P_R}{\partial P_a} u_{P_a} \right)^2 \right]^{1/2} \tag{9.3.4}$$

The uncertainty in the measured pitot pressure is

$$u_{P_{02}} = \pm\frac{\text{Expected error in } P_{02}}{\text{Measured value of } P_{02}} \tag{9.3.5}$$
$$= \pm 0.15$$

and the uncertainty in the settling chamber pressures is

$$u_{P_{01}} = \pm\frac{\text{Expected error in } P_{01}}{\text{Measured value of } P_{01}} \tag{9.3.6}$$
$$= \pm 0.15$$

Similarly, the uncertainty in the atmospheric pressure (Barometric height of the mercury column) is

$$u_{P_a} = \pm\frac{\text{Expected error in } P_a}{\text{Measured value of } P_a} \tag{9.3.7}$$
$$= \pm\frac{0.5 \text{ in}}{29 \text{ in}}$$
$$= \pm 0.0172 \tag{9.3.8}$$

Also,

$$\frac{P_{02}}{P_R}\frac{\partial P_R}{\partial P_{02}} = \frac{P_{02}}{P_R}\frac{1}{P_{01}} = 1 \tag{9.3.9}$$

$$\frac{P_{01}}{P_R}\frac{\partial P_R}{\partial P_{01}} = \frac{P_{01}}{P_R}\left(\frac{-P_{02}}{P_{01}^2}\right) = -1 \tag{9.3.10}$$

$$\frac{P_a}{P_R}\frac{\partial P_R}{\partial P_a} = \frac{P_a}{P_R}\left(\frac{-P_{02}}{\text{NPR} \times P_a^2}\right) = -1 \tag{9.3.11}$$

From equation (9.3.4), we get

$$u_{P_R} = \pm\left[u_{P_{02}}^2 + \left(-u_{P_{01}}\right)^2 + \left(-u_{P_a}\right)^2\right]^{1/2} \tag{9.3.12}$$
$$= \pm\left[(0.15\%)^2 + (-0.15\%)^2 + (-0.0172\%)^2\right]^{1/2}$$
$$= \pm 0.213$$

Thus, the uncertainty in calculating the stagnation pressure ratio is ±21.3%.

9.3.2 VELOCITY PROFILES

It is worth noting that in a wave-dominated complex field, such as the jet field studied here, the pitot probe measures the total pressure behind a normal shock

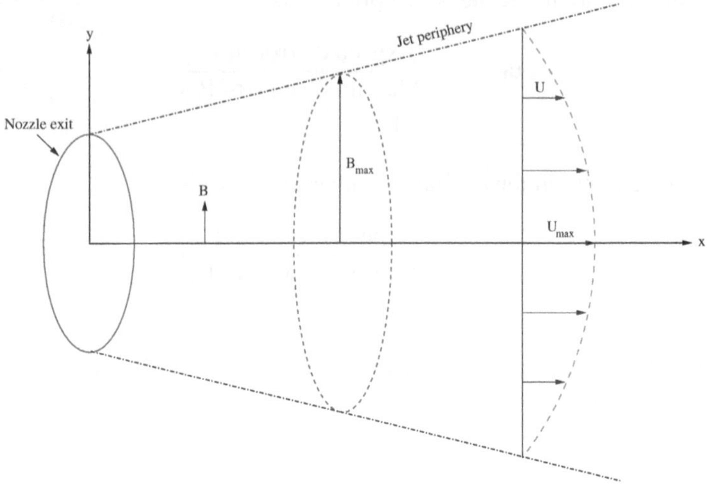

Figure 9.3.4 Schematic diagram showing B, B_{max}, U and U_{max}.

at its nose (the actual shock at the nose is detached bow shock, but the portion of the bow shock can be approximated as a normal shock). Although the computed value of Mach number and velocity with the measured pitot pressure is not exact, it is sufficient for examining the jet structure. The Rayleigh-Pitot formula can be used to determine the Mach number along the jet centerline, which is given by,

$$\frac{p}{p_{02}} = \frac{\left[\frac{2\gamma}{\gamma+1}M_1^2 - \frac{\gamma-1}{\gamma+1}\right]^{\frac{1}{\gamma-1}}}{\left[\frac{\gamma+1}{2}M_1^2\right]^{\frac{\gamma}{\gamma-1}}} \tag{7.4.1}$$

After that, the Mach number is converted to velocity using,

$$U = M_1\sqrt{\gamma R T_1} \tag{2.7.7}$$

The calculated local velocity U in the jet field was made non-dimensional with the local centerline velocity U_{max}. The results are presented in $\frac{B}{B_{max}}$ versus $\frac{U}{U_{max}}$, where B is the local off-width, and B_{max} is the total off-width as shown in Figure 9.3.4.

Listing 9.2 Fortran program for calculation of velocity profile.

```
C        For calculating Mach Number for all the nozzles.

         PARAMETER (N=100)
         DIMENSION X (N), PTR(N), P'02 (N), T1 (N), V1 (N)
         DIMENSION P'1 (N), RHO (N)
         REAL NX, M, MACH (N), P, RHOTOTAL
         OPEN (1, FILE = 'p10d4')
         OPEN (2, FILE = 'V10d4')
         OPEN (3, FILE = 'v10d4')

C        WRITE(*,*) 'NUMBER OF ROWS FOR NX'
         READ (1,*)NX

         DO I = 1, NX
         READ (1, *)X (I) , P'02 (I)
         ENDDO

         P'atm = 'Atmospheric pressure'
         T'atm = 'Atmospheric temperature'
         P'01  = 'Stagnation chamber pressure'
         DIA   = 'Core nozzle exit diameter'
         G     = 1.4
         R     = 'Gas constant'

         DO I = 1, NX
          IF (I . LE.O) THEN
         WRITE(*,*) I
           M  =  1.0
           PTR (I)= (P'02(I)+P'atm)/(P'01+P'atm)
           VA = G+1
           VB = G-1
             DO ITER = 1, 10000
             VC = 1+ (2*G*(M*M-1))/VA
             VD =(VA*M*M)/((VB*M*M)+2)
             VE =VC**(-1.0/VB)
             VF =VD**(G/VB)
             SU =(VE*VF) - PTR (I)
               IF(SU.LE.0.00001) THEN
                MACH(I) = M
                 T1 (I) = T'atm/(1.0+VB/2*MACH(I)**2)
                 V1 (1) = MACH(I)*SQRT(G*R*T1(I))
                 VG = (2.0*G*MACH(I)**2)/VA
                 VH = VB/VA
                 VI = (VA*MACH(I)**2)/2.0
                 VK = (VG-VH)**(1.0/VB)
                 VL = VI**(G/VB)
                 P      = ((P'02(I)+P'atm)*VK)/VL
                 P1(I) = (P*101325)/29.92
                 RHO(I)= P1(I)/(R*T1(I))
                 GO TO 100
               ELSE
                 M = M + 0.001
                 ENDIF
               ENDDO

           ELSE
             A = G-1
             B = G+1
             C = 2.0/A
```

```
              D = A/2.0
              MACH(I)   = SQRT ((((P'02(I)+P'atm)/P'atm)**B
         1    1.0)*C)
              T1 (I) = T'atm/(1.0 + D * MACH(I) + MACH(I))
              V1 (I) = MACH(I) * SQRT(G*R*T1(I))
              P      = (P'02(I)+P'atm)/(1.0+D*MACH(I)**2)**(G/A)
              P1 (I) = (P*101325)/29.92
C             RHOTOTAL = (P'atm*101325)/(29.92*R*T'atm)
C             RHO (I)  = RHOTOTAL/(1.0+D*MACH(I)**2)**(G/A)
              RHO (I)  = (P'atm*101325)/(29.92*R*T1(I))
              GO TO 100
              ENDIF

100           WRITE(2,100)X(I), X(I)/DIA,MACH(I),P1(I)
         1    T1(I), V1(I), RHO(I)
              WRITE(3,2000) V1(I), RHO(I)
              ENDDO

1000          FORMAT (7F12.4)
2000          FORMAT (7F12.4)

STOP
END
```

The velocity profiles at axial locations, $\frac{x}{D_e} = 2.0$, 3.0, 5.0, and 9.0 at NPR 4 are shown in Figure 9.3.5. At $\frac{x}{D_e} = 2.0$, it is seen that the jets retain individual characteristics even though they show a tendency to merge, exhibiting a nominal value of slightly more than 10% of the centerline velocity at the merging point. At $\frac{x}{D_e} = 3.0$, the jets are merged, and the velocities have gone up to as high as 27% of the merging location. At the same time, the maximum velocity ratio of the strapon jets, which was 0.6 at $\frac{x}{D_e} = 2.0$ has come down to 0.5 at $\frac{x}{D_e} = 3.0$. As the jet travels downstream, the strapon jet slowly loses its identity showing a tendency to mix with the core jet. This exhibits a progressive reduction in the velocity ratio of the strapon jet continuously with an increase in the velocities of the merged jet. At $\frac{x}{D_e} = 5.0$, the strapon jets lose their identity, and the combined jet tends toward the nature of a single jet. This process continues, and at $\frac{x}{D_e} = 9.0$, the field has become almost like a single jet.

The velocity profiles at NPR 5 at different axial locations are shown in Figure 9.3.6. It can be seen that, at $\frac{x}{D_e} = 2.0$, the jets behave independently, and merging is yet to take place, unlike NPR 4 case. At $\frac{x}{D_e} = 3.0$, the peak value of the strapon jet is decreased, and the merging is on. As the jet proceeds downstream, the velocity peak value decreases, and the velocity at the merging location progressively increases. In this process, jets slightly lose their identity up to $\frac{x}{D_e} = 5.0$, but the jet is yet to become like a single jet. This tendency continues, and the jets confluence completely beyond $9D_e$, which is slightly downstream as compared to overexpanded case.

When the NPR is increased to 7, the velocity profiles of the combined jet are drastically different from NPR 4 and 5 cases (Figure 9.3.7). Unlike the

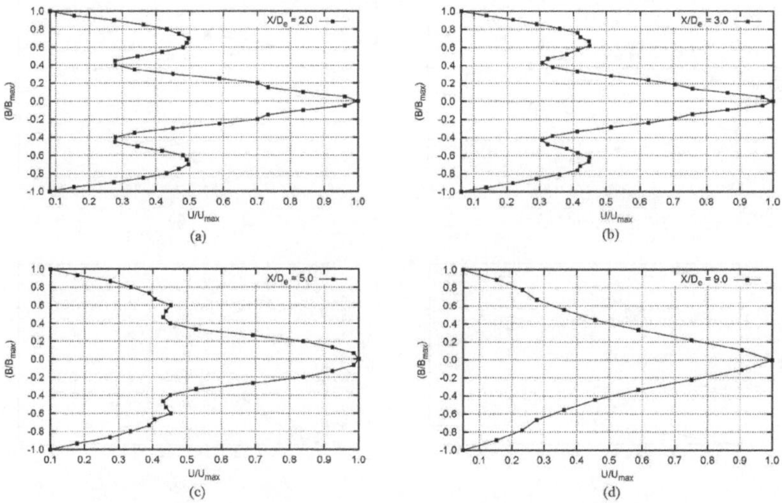

Figure 9.3.5 Velocity profiles of multijets at NPR 4 (overexpansion).

lower NPRs, at NPR 7, the jet retains its identity over a long distance as high as $9D_e$. However, the unfavorable pressure gradient is greatly reduced; as a result, the shocks in the core jet are not driven to increase pressure levels to the extent seen at NPR 4 and 5. The physics behind this behavior could be that the strapon provides a shield for the core jet propagation, preventing the core jet from regaining energy in succeeding cells when the strapon nozzles are flowing.

In the present case of strapon with outward canting, a combined effect of canting and differential overexpansion (higher for strapon jets and lower for a core jet) make the jets behave peculiarly as in the present case. The differential overexpansion compels the strapon to get deflected toward the main jet, whereas the outward canting forces the strapon to move away from the central jet. A closer look at the individual jet suggests that $\frac{U}{U_{max}}$ closer to the nozzle exit should be more than one for the strapon jet, but the actual values are much less than one. This behavior opens a new direction in the research of multijets. It is worthwhile to investigate deeply into this to understand the combined effect of canting and overexpansion.

9.3.3 MASS ENTRAINMENT

Mass entrainment, or simply entrainment, refers to the introduction of surrounding mass into the jet field by eddies near the jet periphery (i.e., entrained

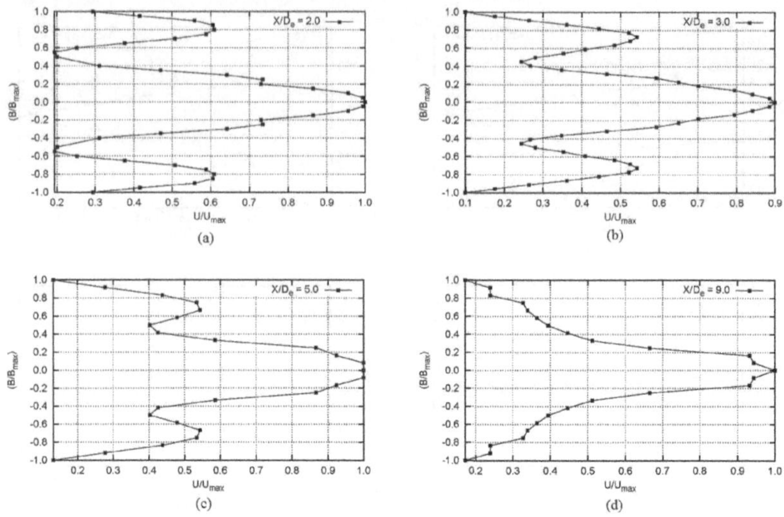

Figure 9.3.6 Velocity profiles of multijets at NPR 5 (correct expansion).

mass from zero momentum surrounding zone). When transported into the jet field, it acquires momentum from the jet flow, reducing the momentum of the original fluid since momentum is conserved. Through the momentum exchange mechanism, mass entrainment has a direct influence on jet decay – the stronger the entrainment, the faster the decay.

The mass flow rate, $\dot{m}_j$, at different cross-sections of the jet field is computed from the measured pitot pressure at different grid points. The difference between $\dot{m}_j$ and $\dot{m}_{exit}$, i.e., the net mass flow at the nozzle exit, is made non-dimensional with $\dot{m}_{exit}$, and the axial distance is made non-dimensional with D_e. The variation of $\frac{\dot{m}_j - \dot{m}_{exit}}{\dot{m}_{exit}}$ with $\frac{X}{D_e}$ for the core and strapon nozzles at NPR 4, 5, and 7 are shown in Figure 9.3.8. It is seen that at all axial locations, the entrainment increases with an increase of NPR. However, the mass entrainment gets augmented faster with an increase of NPR in the far field beyond $10D_e$ for NPR 4 and 5 and beyond $7D_e$ for NPR 7. This is because the shocks in the core jet become progressively stronger with the increase in NPR, making the jet subsonic early at higher NPR. Also, the cross-sectional area in the subsonic zone at higher NPR is larger than for lower NPRs. The combination of lower Mach number and the larger periphery available eases for higher entrainment in the far field. This causes the multijets to decay faster than the individual jets.

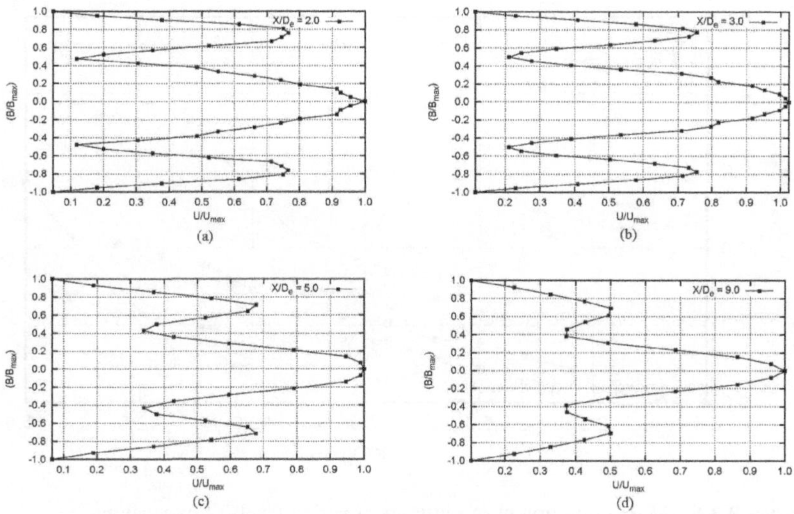

Figure 9.3.7 Velocity profiles of multijets at NPR 7 (underexpansion).

Listing 9.3 Fortran program for calculation of mass entrainment.

```
C       For calculating entrainment for all the nozzles.

        PARAMETER (N=100)
        DIMENSION X(N), DELTA'MASS(N)
        REAL NX, M(N), INITIAL'MASS
        OPEN (1, FILE = 'total mass')
        OPEN (2, FILE = 'entrain'mass')

        DO I = 1, 15
        READ (1, *) X(I), M(I)
        ENDDO
        INITIAL'MASS = 0.3176
        DO I = 1, 15
        DELTA'MASS(T) = (m(I)-INITIAL'MASS))/(INITIAL'MASS)
        WRITE (2,1000)X(I), DELTA'MASS(I)
        WRITE (*,*)X(I), DELTA'MASS(I)
        ENDDO
1000    FORMAT(2F12.4)
        STOP
        END
```

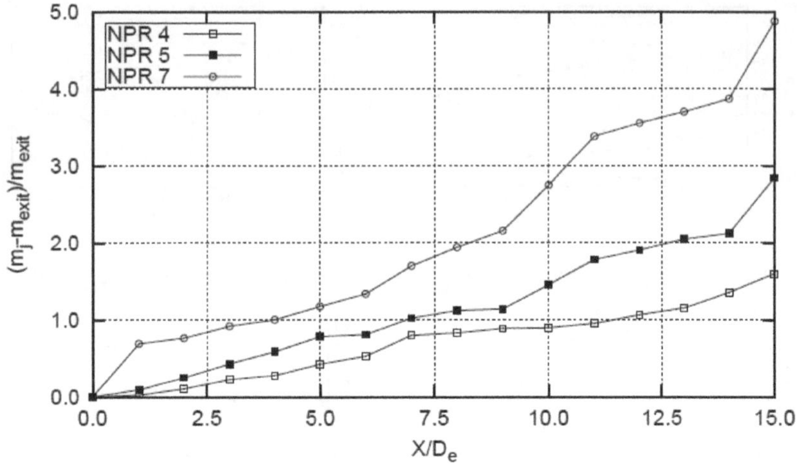

Figure 9.3.8 Mass entrainment of multijets at varied levels of expansion.

9.3.4 FLOW VISUALIZATION

Shadowgraphic images of the multijets, as shown in Figure 9.3.9(a)–(c), were acquired at NPRs 4, 5, and 7 to get insight into the wave pattern in the jets. The shock cells become longer as the NPR increases, a characteristic of a free jet. The figures illustrate that when NPR increases, the Mach disk in the first cell of the core jet shrinks. The wave pattern of the strapon jets is identical to that of the core jet. The first shock cell of the core jet is significantly altered in the presence of surrounding jets. Similarly, the existence of the central jet substantially modifies the shock cell structures of the strapon jets. These images support the claims based on the velocity profile and centerline decay plots.

9.4 SUMMARY

This chapter explored the mixing augmentation characteristics of a Mach 1.6 core in the presence of two Mach 2.5 jets with 6° outward canting arranged on either side of the core jet. The core jet decays faster in the presence of strapon jets, according to the jet velocity profile plots. A vast zone of subsonic flow exists shortly after the initial shock cross-over point. It could be owing to the shielding provided by the strapon jets to the core jet. Furthermore, the decay of the core jet in the presence of strapon jets is observed to be larger at all degrees of expansion when compared to the core jet alone. The presence of the core jet also improves strapon jet mixing. The existence of neighboring jets considerably modifies the waves in the individual jets.

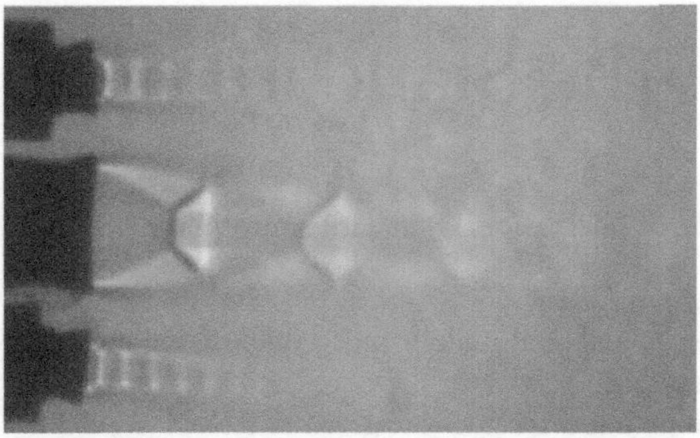

(a) At NPR 4 [core and strapon (overexpanded)]

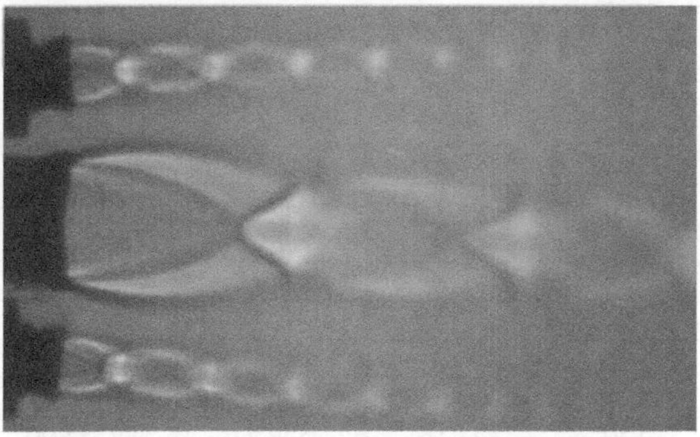

(b) At NPR 5 [core (correctly expanded), and strapon (overexpanded)]

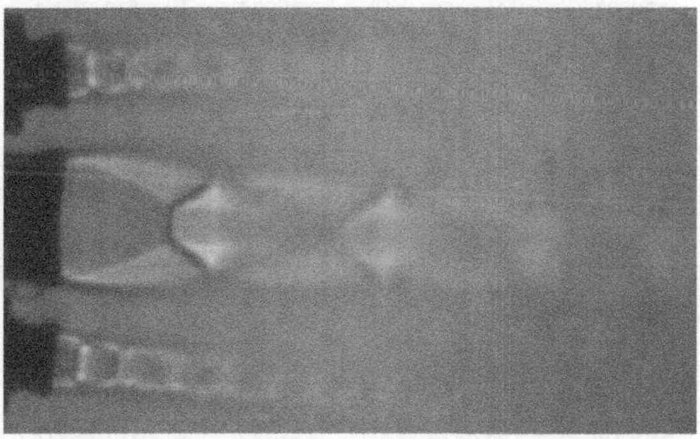

(c) At NPR 7 [core (underexpanded), and strapon (overexpanded)]

Figure 9.3.9 Shadowgraphic pictures of multijets.

A The Standard Atmosphere

The properties of International Standard Atmosphere (ISA) are tabulated in SI units.

- The geo-potential altitude h and geometric altitude h_G are measured in meters (m).
- The temperature T values are given in Kelvin (K).
- The speed of sound a is given in $^{m}/_{s}$.
- The pressure p is given in Pascals (Pa).
- The density ρ is shown in $^{kg}/_{m^3}$.

h (m)	h_G (m)	T (K)	a ($^{m}/_{s}$)	p (Pa)	ρ ($^{kg}/_{m^3}$)
49825	50217.73	270.65	329.8	77.64	0.001
0	0	288.15	340.29	101325	1.225
25	25	287.99	340.2	101025.03	1.222
50	50	287.83	340.1	100725.78	1.219
75	75	287.66	340.01	100427.25	1.216
100	100	287.5	339.91	100129.44	1.213
125	125	287.34	339.81	99832.34	1.21
150	150	287.18	339.72	99535.96	1.207
175	175	287.01	339.62	99240.29	1.205
200	200.01	286.85	339.53	98945.33	1.202
225	225.01	286.69	339.43	98651.08	1.199
250	250.01	286.53	339.33	98357.54	1.196
275	275.01	286.36	339.24	98064.7	1.193
300	300.01	286.2	339.14	97772.58	1.19
325	325.02	286.04	339.04	97481.16	1.187
350	350.02	285.88	338.95	97190.44	1.184
375	375.02	285.71	338.85	96900.42	1.182
400	400.03	285.55	338.76	96611.11	1.179
425	425.03	285.39	338.66	96322.5	1.176
450	450.03	285.23	338.56	96034.58	1.173
475	475.04	285.06	338.47	95747.36	1.17
500	500.04	284.9	338.37	95460.84	1.167
525	525.04	284.74	338.27	95175.01	1.164
550	550.05	284.58	338.18	94889.88	1.162
575	575.05	284.41	338.08	94605.44	1.159
600	600.06	284.25	337.98	94321.68	1.156
625	625.06	284.09	337.89	94038.62	1.153

DOI: 10.1201/9781003139447-A

h (m)	h_G (m)	T (K)	a (m/s)	p (Pa)	ρ (kg/m^3)
49825	50217.73	270.65	329.8	77.64	0.001
650	650.07	283.93	337.79	93756.25	1.15
675	675.07	283.76	337.69	93474.56	1.148
700	700.08	283.6	337.6	93193.56	1.145
725	725.08	283.44	337.5	92913.24	1.142
750	750.09	283.28	337.4	92633.61	1.139
775	775.09	283.11	337.31	92354.66	1.136
800	800.1	282.95	337.21	92076.39	1.134
825	825.11	282.79	337.11	91798.8	1.131
850	850.11	282.63	337.02	91521.88	1.128
875	875.12	282.46	336.92	91245.65	1.125
900	900.13	282.3	336.82	90970.09	1.123
925	925.13	282.14	336.73	90695.2	1.12
950	950.14	281.98	336.63	90420.98	1.117
975	975.15	281.81	336.53	90147.44	1.114
1000	1000.16	281.65	336.43	89874.57	1.112
1025	1025.16	281.49	336.34	89602.37	1.109
1050	1050.17	281.33	336.24	89330.83	1.106
1075	1075.18	281.16	336.14	89059.97	1.103
1100	1100.19	281	336.05	88789.76	1.101
1125	1125.2	280.84	335.95	88520.22	1.098
1150	1150.21	280.68	335.85	88251.35	1.095
1175	1175.22	280.51	335.75	87983.13	1.093
1200	1200.23	280.35	335.66	87715.58	1.09
1225	1225.24	280.19	335.56	87448.69	1.087
1250	1250.25	280.03	335.46	87182.45	1.085
1275	1275.26	279.86	335.36	86916.87	1.082
1300	1300.27	279.7	335.27	86651.95	1.079
1325	1325.28	279.54	335.17	86387.68	1.077
1350	1350.29	279.38	335.07	86124.06	1.074
1375	1375.3	279.21	334.98	85861.1	1.071
1400	1400.31	279.05	334.88	85598.78	1.069
1425	1425.32	278.89	334.78	85337.12	1.066
1450	1450.33	278.73	334.68	85076.1	1.063
1475	1475.34	278.56	334.58	84815.73	1.061
1500	1500.35	278.4	334.49	84556	1.058
1525	1525.37	278.24	334.39	84296.92	1.055
1550	1550.38	278.08	334.29	84038.49	1.053
1575	1575.39	277.91	334.19	83780.69	1.05
1600	1600.4	277.75	334.1	83523.54	1.048
1625	1625.41	277.59	334	83267.02	1.045
1650	1650.43	277.43	333.9	83011.15	1.042

h (m)	h_G (m)	T (K)	a ($^m/_s$)	p (Pa)	ρ ($^{kg}/_{m^3}$)
49825	50217.73	270.65	329.8	77.64	0.001
1675	1675.44	277.26	333.8	82755.91	1.04
1700	1700.45	277.1	333.71	82501.3	1.037
1725	1725.47	276.94	333.61	82247.33	1.035
1750	1750.48	276.78	333.51	81994	1.032
1775	1775.49	276.61	333.41	81741.3	1.029
1800	1800.51	276.45	333.31	81489.22	1.027
1825	1825.52	276.29	333.22	81237.78	1.024
1850	1850.54	276.13	333.12	80986.97	1.022
1875	1875.55	275.96	333.02	80736.78	1.019
1900	1900.57	275.8	332.92	80487.22	1.017
1925	1925.58	275.64	332.82	80238.28	1.014
1950	1950.6	275.48	332.73	79989.97	1.012
1975	1975.61	275.31	332.63	79742.28	1.009
2000	2000.63	275.15	332.53	79495.22	1.006
2025	2025.64	274.99	332.43	79248.77	1.004
2050	2050.66	274.83	332.33	79002.94	1.001
2075	2075.68	274.66	332.23	78757.73	0.999
2100	2100.69	274.5	332.14	78513.14	0.996
2125	2125.71	274.34	332.04	78269.16	0.994
2150	2150.73	274.18	331.94	78025.79	0.991
2175	2175.74	274.01	331.84	77783.04	0.989
2200	2200.76	273.85	331.74	77540.9	0.986
2225	2225.78	273.69	331.64	77299.37	0.984
2250	2250.79	273.53	331.55	77058.46	0.981
2275	2275.81	273.36	331.45	76818.14	0.979
2300	2300.83	273.2	331.35	76578.44	0.976
2325	2325.85	273.04	331.25	76339.34	0.974
2350	2350.87	272.88	331.15	76100.85	0.972
2375	2375.89	272.71	331.05	75862.96	0.969
2400	2400.9	272.55	330.95	75625.68	0.967
2425	2425.92	272.39	330.86	75388.99	0.964
2450	2450.94	272.23	330.76	75152.91	0.962
2475	2475.96	272.06	330.66	74917.42	0.959
2500	2500.98	271.9	330.56	74682.53	0.957
2525	2526	271.74	330.46	74448.24	0.954
2550	2551.02	271.58	330.36	74214.55	0.952
2575	2576.04	271.41	330.26	73981.45	0.95
2600	2601.06	271.25	330.16	73748.94	0.947
2625	2626.08	271.09	330.07	73517.02	0.945
2650	2651.1	270.93	329.97	73285.7	0.942
2675	2676.12	270.76	329.87	73054.96	0.94

h (m)	h_G (m)	T (K)	a (m/s)	p (Pa)	ρ (kg/m³)
49825	50217.73	270.65	329.8	77.64	0.001
2700	2701.14	270.6	329.77	72824.81	0.938
2725	2726.17	270.44	329.67	72595.25	0.935
2750	2751.19	270.28	329.57	72366.28	0.933
2775	2776.21	270.11	329.47	72137.89	0.93
2800	2801.23	269.95	329.37	71910.09	0.928
2825	2826.25	269.79	329.27	71682.87	0.926
2850	2851.28	269.63	329.17	71456.23	0.923
2875	2876.3	269.46	329.07	71230.17	0.921
2900	2901.32	269.3	328.98	71004.69	0.919
2925	2926.34	269.14	328.88	70779.79	0.916
2950	2951.37	268.98	328.78	70555.47	0.914
2975	2976.39	268.81	328.68	70331.72	0.911
3000	3001.41	268.65	328.58	70108.54	0.909
3025	3026.44	268.49	328.48	69885.95	0.907
3050	3051.46	268.33	328.38	69663.92	0.904
3075	3076.48	268.16	328.28	69442.46	0.902
3100	3101.51	268	328.18	69221.58	0.9
3125	3126.53	267.84	328.08	69001.26	0.897
3150	3151.56	267.68	327.98	68781.52	0.895
3175	3176.58	267.51	327.88	68562.34	0.893
3200	3201.61	267.35	327.78	68343.72	0.891
3225	3226.63	267.19	327.68	68125.67	0.888
3250	3251.66	267.03	327.58	67908.19	0.886
3275	3276.68	266.86	327.48	67691.26	0.884
3300	3301.71	266.7	327.38	67474.9	0.881
3325	3326.74	266.54	327.28	67259.1	0.879
3350	3351.76	266.38	327.18	67043.86	0.877
3375	3376.79	266.21	327.08	66829.17	0.875
3400	3401.82	266.05	326.98	66615.05	0.872
3425	3426.84	265.89	326.88	66401.47	0.87
3450	3451.87	265.73	326.78	66188.46	0.868
3475	3476.9	265.56	326.68	65975.99	0.865
3500	3501.92	265.4	326.58	65764.08	0.863
3525	3526.95	265.24	326.48	65552.73	0.861
3550	3551.98	265.08	326.38	65341.92	0.859
3575	3577.01	264.91	326.28	65131.66	0.857
3600	3602.04	264.75	326.18	64921.95	0.854
3625	3627.06	264.59	326.08	64712.78	0.852
3650	3652.09	264.43	325.98	64504.16	0.85
3675	3677.12	264.26	325.88	64296.09	0.848
3700	3702.15	264.1	325.78	64088.56	0.845

h (m)	h_G (m)	T (K)	a (m/s)	p (Pa)	ρ (kg/m^3)
49825	50217.73	270.65	329.8	77.64	0.001
3725	3727.18	263.94	325.68	63881.58	0.843
3750	3752.21	263.78	325.58	63675.13	0.841
3775	3777.24	263.61	325.48	63469.23	0.839
3800	3802.27	263.45	325.38	63263.86	0.837
3825	3827.3	263.29	325.28	63059.04	0.834
3850	3852.33	263.13	325.18	62854.75	0.832
3875	3877.36	262.96	325.08	62650.99	0.83
3900	3902.39	262.8	324.98	62447.78	0.828
3925	3927.42	262.64	324.88	62245.09	0.826
3950	3952.45	262.48	324.78	62042.94	0.823
3975	3977.48	262.31	324.68	61841.32	0.821
4000	4002.51	262.15	324.58	61640.24	0.819
4025	4027.54	261.99	324.48	61439.68	0.817
4050	4052.58	261.83	324.38	61239.65	0.815
4075	4077.61	261.66	324.28	61040.15	0.813
4100	4102.64	261.5	324.18	60841.17	0.811
4125	4127.67	261.34	324.08	60642.72	0.808
4150	4152.71	261.18	323.97	60444.8	0.806
4175	4177.74	261.01	323.87	60247.39	0.804
4200	4202.77	260.85	323.77	60050.52	0.802
4225	4227.8	260.69	323.67	59854.16	0.8
4250	4252.84	260.53	323.57	59658.32	0.798
4275	4277.87	260.36	323.47	59463	0.796
4300	4302.9	260.2	323.37	59268.2	0.794
4325	4327.94	260.04	323.27	59073.92	0.791
4350	4352.97	259.88	323.17	58880.15	0.789
4375	4378.01	259.71	323.07	58686.9	0.787
4400	4403.04	259.55	322.97	58494.16	0.785
4425	4428.08	259.39	322.86	58301.93	0.783
4450	4453.11	259.23	322.76	58110.22	0.781
4475	4478.15	259.06	322.66	57919.02	0.779
4500	4503.18	258.9	322.56	57728.32	0.777
4525	4528.22	258.74	322.46	57538.14	0.775
4550	4553.25	258.58	322.36	57348.46	0.773
4575	4578.29	258.41	322.26	57159.29	0.771
4600	4603.32	258.25	322.16	56970.63	0.769
4625	4628.36	258.09	322.05	56782.47	0.766
4650	4653.4	257.93	321.95	56594.81	0.764
4675	4678.43	257.76	321.85	56407.66	0.762
4700	4703.47	257.6	321.75	56221	0.76
4725	4728.51	257.44	321.65	56034.85	0.758

h (m)	h_G (m)	T (K)	a (m/s)	p (Pa)	ρ (kg/m³)
49825	50217.73	270.65	329.8	77.64	0.001
4750	4753.54	257.28	321.55	55849.2	0.756
4775	4778.58	257.11	321.45	55664.04	0.754
4800	4803.62	256.95	321.34	55479.39	0.752
4825	4828.66	256.79	321.24	55295.23	0.75
4850	4853.69	256.63	321.14	55111.56	0.748
4875	4878.73	256.46	321.04	54928.39	0.746
4900	4903.77	256.3	320.94	54745.71	0.744
4925	4928.81	256.14	320.83	54563.53	0.742
4950	4953.85	255.98	320.73	54381.83	0.74
4975	4978.89	255.81	320.63	54200.63	0.738
5000	5003.93	255.65	320.53	54019.91	0.736
5025	5028.97	255.49	320.43	53839.69	0.734
5050	5054.01	255.33	320.33	53659.95	0.732
5075	5079.05	255.16	320.22	53480.69	0.73
5100	5104.09	255	320.12	53301.92	0.728
5125	5129.13	254.84	320.02	53123.64	0.726
5150	5154.17	254.68	319.92	52945.84	0.724
5175	5179.21	254.51	319.82	52768.52	0.722
5200	5204.25	254.35	319.71	52591.68	0.72
5225	5229.29	254.19	319.61	52415.33	0.718
5250	5254.33	254.03	319.51	52239.45	0.716
5275	5279.37	253.86	319.41	52064.05	0.714
5300	5304.41	253.7	319.3	51889.13	0.713
5325	5329.45	253.54	319.2	51714.68	0.711
5350	5354.5	253.38	319.1	51540.71	0.709
5375	5379.54	253.21	319	51367.21	0.707
5400	5404.58	253.05	318.9	51194.19	0.705
5425	5429.62	252.89	318.79	51021.63	0.703
5450	5454.67	252.73	318.69	50849.55	0.701
5475	5479.71	252.56	318.59	50677.94	0.699
5500	5504.75	252.4	318.49	50506.8	0.697
5525	5529.8	252.24	318.38	50336.13	0.695
5550	5554.84	252.08	318.28	50165.92	0.693
5575	5579.88	251.91	318.18	49996.19	0.691
5600	5604.93	251.75	318.08	49826.91	0.689
5625	5629.97	251.59	317.97	49658.1	0.688
5650	5655.02	251.43	317.87	49489.76	0.686
5675	5680.06	251.26	317.77	49321.87	0.684
5700	5705.1	251.1	317.66	49154.45	0.682
5725	5730.15	250.94	317.56	48987.49	0.68
5750	5755.19	250.78	317.46	48820.99	0.678

h (m)	h_G (m)	T (K)	a ($^m/_s$)	p (Pa)	ρ ($^{kg}/_{m^3}$)
49825	50217.73	270.65	329.8	77.64	0.001
5775	5780.24	250.61	317.36	48654.94	0.676
5800	5805.28	250.45	317.25	48489.36	0.674
5825	5830.33	250.29	317.15	48324.23	0.673
5850	5855.38	250.13	317.05	48159.56	0.671
5875	5880.42	249.96	316.94	47995.34	0.669
5900	5905.47	249.8	316.84	47831.57	0.667
5925	5930.52	249.64	316.74	47668.26	0.665
5950	5955.56	249.48	316.63	47505.4	0.663
5975	5980.61	249.31	316.53	47342.99	0.662
6000	6005.66	249.15	316.43	47181.03	0.66
6025	6030.7	248.99	316.33	47019.52	0.658
6050	6055.75	248.83	316.22	46858.45	0.656
6075	6080.8	248.66	316.12	46697.84	0.654
6100	6105.85	248.5	316.02	46537.67	0.652
6125	6130.89	248.34	315.91	46377.94	0.651
6150	6155.94	248.18	315.81	46218.66	0.649
6175	6180.99	248.01	315.71	46059.82	0.647
6200	6206.04	247.85	315.6	45901.43	0.645
6225	6231.09	247.69	315.5	45743.48	0.643
6250	6256.14	247.53	315.39	45585.96	0.642
6275	6281.19	247.36	315.29	45428.89	0.64
6300	6306.24	247.2	315.19	45272.25	0.638
6325	6331.29	247.04	315.08	45116.06	0.636
6350	6356.34	246.88	314.98	44960.29	0.634
6375	6381.39	246.71	314.88	44804.97	0.633
6400	6406.44	246.55	314.77	44650.08	0.631
6425	6431.49	246.39	314.67	44495.62	0.629
6450	6456.54	246.23	314.57	44341.6	0.627
6475	6481.59	246.06	314.46	44188.01	0.626
6500	6506.64	245.9	314.36	44034.85	0.624
6525	6531.69	245.74	314.25	43882.12	0.622
6550	6556.74	245.58	314.15	43729.81	0.62
6575	6581.79	245.41	314.05	43577.94	0.619
6600	6606.84	245.25	313.94	43426.5	0.617
6625	6631.9	245.09	313.84	43275.48	0.615
6650	6656.95	244.93	313.73	43124.88	0.613
6675	6682	244.76	313.63	42974.71	0.612
6700	6707.05	244.6	313.53	42824.97	0.61
6725	6732.11	244.44	313.42	42675.65	0.608
6750	6757.16	244.28	313.32	42526.75	0.606
6775	6782.21	244.11	313.21	42378.27	0.605

h (m)	h_G (m)	T (K)	a (m/s)	p (Pa)	ρ (kg/m^3)
49825	50217.73	270.65	329.8	77.64	0.001
6800	6807.27	243.95	313.11	42230.21	0.603
6825	6832.32	243.79	313	42082.57	0.601
6850	6857.37	243.63	312.9	41935.34	0.6
6875	6882.43	243.46	312.8	41788.54	0.598
6900	6907.48	243.3	312.69	41642.15	0.596
6925	6932.54	243.14	312.59	41496.18	0.595
6950	6957.59	242.98	312.48	41350.62	0.593
6975	6982.64	242.81	312.38	41205.47	0.591
7000	7007.7	242.65	312.27	41060.74	0.59
7025	7032.75	242.49	312.17	40916.42	0.588
7050	7057.81	242.33	312.06	40772.51	0.586
7075	7082.87	242.16	311.96	40629.02	0.584
7100	7107.92	242	311.86	40485.93	0.583
7125	7132.98	241.84	311.75	40343.25	0.581
7150	7158.03	241.68	311.65	40200.97	0.579
7175	7183.09	241.51	311.54	40059.1	0.578
7200	7208.15	241.35	311.44	39917.64	0.576
7225	7233.2	241.19	311.33	39776.59	0.575
7250	7258.26	241.03	311.23	39635.93	0.573
7275	7283.32	240.86	311.12	39495.68	0.571
7300	7308.37	240.7	311.02	39355.84	0.57
7325	7333.43	240.54	310.91	39216.39	0.568
7350	7358.49	240.38	310.81	39077.34	0.566
7375	7383.55	240.21	310.7	38938.7	0.565
7400	7408.61	240.05	310.6	38800.45	0.563
7425	7433.66	239.89	310.49	38662.6	0.561
7450	7458.72	239.73	310.39	38525.14	0.56
7475	7483.78	239.56	310.28	38388.09	0.558
7500	7508.84	239.4	310.18	38251.42	0.557
7525	7533.9	239.24	310.07	38115.16	0.555
7550	7558.96	239.08	309.96	37979.28	0.553
7575	7584.02	238.91	309.86	37843.8	0.552
7600	7609.08	238.75	309.75	37708.71	0.55
7625	7634.14	238.59	309.65	37574.01	0.549
7650	7659.2	238.43	309.54	37439.7	0.547
7675	7684.26	238.26	309.44	37305.78	0.545
7700	7709.32	238.1	309.33	37172.24	0.544
7725	7734.38	237.94	309.23	37039.1	0.542
7750	7759.44	237.78	309.12	36906.34	0.541
7775	7784.5	237.61	309.02	36773.96	0.539
7800	7809.56	237.45	308.91	36641.98	0.538

h (m)	h_G (m)	T (K)	a (m/s)	p (Pa)	ρ (kg/m^3)
49825	50217.73	270.65	329.8	77.64	0.001
7825	7834.62	237.29	308.8	36510.37	0.536
7850	7859.68	237.13	308.7	36379.15	0.534
7875	7884.75	236.96	308.59	36248.31	0.533
7900	7909.81	236.8	308.49	36117.85	0.531
7925	7934.87	236.64	308.38	35987.77	0.53
7950	7959.93	236.48	308.27	35858.07	0.528
7975	7985	236.31	308.17	35728.75	0.527
8000	8010.06	236.15	308.06	35599.81	0.525
8025	8035.12	235.99	307.96	35471.25	0.524
8050	8060.18	235.83	307.85	35343.06	0.522
8075	8085.25	235.66	307.74	35215.24	0.521
8100	8110.31	235.5	307.64	35087.81	0.519
8125	8135.38	235.34	307.53	34960.74	0.518
8150	8160.44	235.18	307.43	34834.05	0.516
8175	8185.5	235.01	307.32	34707.73	0.514
8200	8210.57	234.85	307.21	34581.78	0.513
8225	8235.63	234.69	307.11	34456.2	0.511
8250	8260.7	234.53	307	34330.99	0.51
8275	8285.76	234.36	306.89	34206.15	0.508
8300	8310.83	234.2	306.79	34081.68	0.507
8325	8335.89	234.04	306.68	33957.57	0.505
8350	8360.96	233.88	306.58	33833.83	0.504
8375	8386.02	233.71	306.47	33710.46	0.502
8400	8411.09	233.55	306.36	33587.45	0.501
8425	8436.16	233.39	306.26	33464.8	0.5
8450	8461.22	233.23	306.15	33342.52	0.498
8475	8486.29	233.06	306.04	33220.6	0.497
8500	8511.36	232.9	305.94	33099.04	0.495
8525	8536.42	232.74	305.83	32977.84	0.494
8550	8561.49	232.58	305.72	32857	0.492
8575	8586.56	232.41	305.62	32736.52	0.491
8600	8611.62	232.25	305.51	32616.4	0.489
8625	8636.69	232.09	305.4	32496.63	0.488
8650	8661.76	231.93	305.29	32377.22	0.486
8675	8686.83	231.76	305.19	32258.17	0.485
8700	8711.9	231.6	305.08	32139.47	0.483
8725	8736.97	231.44	304.97	32021.13	0.482
8750	8762.03	231.28	304.87	31903.13	0.481
8775	8787.1	231.11	304.76	31785.49	0.479
8800	8812.17	230.95	304.65	31668.21	0.478
8825	8837.24	230.79	304.54	31551.27	0.476

h (m)	h_G (m)	T (K)	a ($^{m}/_{s}$)	p (Pa)	ρ ($^{kg}/_{m^3}$)
49825	50217.73	270.65	329.8	77.64	0.001
8850	8862.31	230.63	304.44	31434.68	0.475
8875	8887.38	230.46	304.33	31318.44	0.473
8900	8912.45	230.3	304.22	31202.55	0.472
8925	8937.52	230.14	304.12	31087.01	0.471
8950	8962.59	229.98	304.01	30971.81	0.469
8975	8987.66	229.81	303.9	30856.96	0.468
9000	9012.73	229.65	303.79	30742.46	0.466
9025	9037.8	229.49	303.69	30628.3	0.465
9050	9062.87	229.33	303.58	30514.48	0.464
9075	9087.95	229.16	303.47	30401.01	0.462
9100	9113.02	229	303.36	30287.87	0.461
9125	9138.09	228.84	303.26	30175.08	0.459
9150	9163.16	228.68	303.15	30062.63	0.458
9175	9188.23	228.51	303.04	29950.52	0.457
9200	9213.3	228.35	302.93	29838.75	0.455
9225	9238.38	228.19	302.82	29727.31	0.454
9250	9263.45	228.03	302.72	29616.21	0.452
9275	9288.52	227.86	302.61	29505.45	0.451
9300	9313.6	227.7	302.5	29395.03	0.45
9325	9338.67	227.54	302.39	29284.94	0.448
9350	9363.74	227.38	302.28	29175.18	0.447
9375	9388.82	227.21	302.18	29065.76	0.446
9400	9413.89	227.05	302.07	28956.67	0.444
9425	9438.96	226.89	301.96	28847.91	0.443
9450	9464.04	226.73	301.85	28739.48	0.442
9475	9489.11	226.56	301.74	28631.38	0.44
9500	9514.19	226.4	301.64	28523.62	0.439
9525	9539.26	226.24	301.53	28416.18	0.438
9550	9564.34	226.08	301.42	28309.07	0.436
9575	9589.41	225.91	301.31	28202.28	0.435
9600	9614.49	225.75	301.2	28095.82	0.434
9625	9639.56	225.59	301.09	27989.69	0.432
9650	9664.64	225.43	300.99	27883.88	0.431
9675	9689.71	225.26	300.88	27778.4	0.43
9700	9714.79	225.1	300.77	27673.24	0.428
9725	9739.87	224.94	300.66	27568.4	0.427
9750	9764.94	224.78	300.55	27463.89	0.426
9775	9790.02	224.61	300.44	27359.69	0.424
9800	9815.1	224.45	300.33	27255.82	0.423
9825	9840.17	224.29	300.23	27152.26	0.422
9850	9865.25	224.13	300.12	27049.03	0.42

h (m)	h_G (m)	T (K)	a (m/s)	p (Pa)	ρ (kg/m^3)
49825	50217.73	270.65	329.8	77.64	0.001
9875	9890.33	223.96	300.01	26946.11	0.419
9900	9915.41	223.8	299.9	26843.51	0.418
9925	9940.49	223.64	299.79	26741.23	0.417
9950	9965.56	223.48	299.68	26639.26	0.415
9975	9990.64	223.31	299.57	26537.61	0.414
10000	10015.72	223.15	299.46	26436.27	0.413
10025	10040.8	222.99	299.35	26335.24	0.411
10050	10065.88	222.83	299.25	26234.53	0.41
10075	10090.96	222.66	299.14	26134.13	0.409
10100	10116.04	222.5	299.03	26034.04	0.408
10125	10141.12	222.34	298.92	25934.26	0.406
10150	10166.2	222.18	298.81	25834.8	0.405
10175	10191.28	222.01	298.7	25735.64	0.404
10200	10216.36	221.85	298.59	25636.79	0.403
10225	10241.44	221.69	298.48	25538.24	0.401
10250	10266.52	221.53	298.37	25440.01	0.4
10275	10291.6	221.36	298.26	25342.08	0.399
10300	10316.68	221.2	298.15	25244.45	0.398
10325	10341.76	221.04	298.04	25147.13	0.396
10350	10366.84	220.88	297.93	25050.12	0.395
10375	10391.92	220.71	297.82	24953.4	0.394
10400	10417	220.55	297.71	24857	0.393
10425	10442.09	220.39	297.6	24760.89	0.391
10450	10467.17	220.23	297.49	24665.08	0.39
10475	10492.25	220.06	297.38	24569.57	0.389
10500	10517.33	219.9	297.27	24474.37	0.388
10525	10542.42	219.74	297.16	24379.46	0.387
10550	10567.5	219.58	297.05	24284.85	0.385
10575	10592.58	219.41	296.94	24190.54	0.384
10600	10617.67	219.25	296.83	24096.52	0.383
10625	10642.75	219.09	296.72	24002.8	0.382
10650	10667.83	218.93	296.61	23909.38	0.38
10675	10692.92	218.76	296.5	23816.25	0.379
10700	10718	218.6	296.39	23723.42	0.378
10725	10743.09	218.44	296.28	23630.87	0.377
10750	10768.17	218.28	296.17	23538.62	0.376
10775	10793.25	218.11	296.06	23446.67	0.374
10800	10818.34	217.95	295.95	23355	0.373
10825	10843.42	217.79	295.84	23263.62	0.372
10850	10868.51	217.63	295.73	23172.54	0.371
10875	10893.59	217.46	295.62	23081.74	0.37

h (m)	h_G (m)	T (K)	a (m/s)	p (Pa)	ρ (kg/m³)
49825	50217.73	270.65	329.8	77.64	0.001
10900	10918.68	217.3	295.51	22991.23	0.369
10925	10943.77	217.14	295.4	22901.01	0.367
10950	10968.85	216.98	295.29	22811.08	0.366
10975	10993.94	216.81	295.18	22721.43	0.365
11000	11019.03	216.65	295.07	22632.06	0.364
11025	11044.11	216.65	295.07	22543.02	0.362
11050	11069.2	216.65	295.07	22454.32	0.361
11075	11094.29	216.65	295.07	22365.98	0.36
11100	11119.37	216.65	295.07	22277.98	0.358
11125	11144.46	216.65	295.07	22190.33	0.357
11150	11169.55	216.65	295.07	22103.02	0.355
11175	11194.64	216.65	295.07	22016.06	0.354
11200	11219.72	216.65	295.07	21929.44	0.353
11225	11244.81	216.65	295.07	21843.16	0.351
11250	11269.9	216.65	295.07	21757.22	0.35
11275	11294.99	216.65	295.07	21671.62	0.348
11300	11320.08	216.65	295.07	21586.35	0.347
11325	11345.17	216.65	295.07	21501.42	0.346
11350	11370.26	216.65	295.07	21416.82	0.344
11375	11395.35	216.65	295.07	21332.56	0.343
11400	11420.44	216.65	295.07	21248.63	0.342
11425	11445.53	216.65	295.07	21165.03	0.34
11450	11470.62	216.65	295.07	21081.75	0.339
11475	11495.71	216.65	295.07	20998.81	0.338
11500	11520.8	216.65	295.07	20916.19	0.336
11525	11545.89	216.65	295.07	20833.9	0.335
11550	11570.98	216.65	295.07	20751.93	0.334
11575	11596.07	216.65	295.07	20670.28	0.332
11600	11621.16	216.65	295.07	20588.95	0.331
11625	11646.25	216.65	295.07	20507.95	0.33
11650	11671.34	216.65	295.07	20427.26	0.328
11675	11696.43	216.65	295.07	20346.89	0.327
11700	11721.53	216.65	295.07	20266.83	0.326
11725	11746.62	216.65	295.07	20187.1	0.325
11750	11771.71	216.65	295.07	20107.67	0.323
11775	11796.8	216.65	295.07	20028.56	0.322
11800	11821.9	216.65	295.07	19949.76	0.321
11825	11846.99	216.65	295.07	19871.27	0.32
11850	11872.08	216.65	295.07	19793.08	0.318
11875	11897.18	216.65	295.07	19715.21	0.317
11900	11922.27	216.65	295.07	19637.64	0.316

h (m)	h_G (m)	T (K)	a (m/s)	p (Pa)	ρ (kg/m^3)
49825	50217.73	270.65	329.8	77.64	0.001
11925	11947.36	216.65	295.07	19560.38	0.315
11950	11972.46	216.65	295.07	19483.42	0.313
11975	11997.55	216.65	295.07	19406.76	0.312
12000	12022.65	216.65	295.07	19330.41	0.311
12025	12047.74	216.65	295.07	19254.35	0.31
12050	12072.83	216.65	295.07	19178.6	0.308
12075	12097.93	216.65	295.07	19103.14	0.307
12100	12123.02	216.65	295.07	19027.98	0.306
12125	12148.12	216.65	295.07	18953.11	0.305
12150	12173.22	216.65	295.07	18878.54	0.304
12175	12198.31	216.65	295.07	18804.27	0.302
12200	12223.41	216.65	295.07	18730.28	0.301
12225	12248.5	216.65	295.07	18656.59	0.3
12250	12273.6	216.65	295.07	18583.19	0.299
12275	12298.7	216.65	295.07	18510.07	0.298
12300	12323.79	216.65	295.07	18437.24	0.296
12325	12348.89	216.65	295.07	18364.7	0.295
12350	12373.99	216.65	295.07	18292.45	0.294
12375	12399.08	216.65	295.07	18220.48	0.293
12400	12424.18	216.65	295.07	18148.79	0.292
12425	12449.28	216.65	295.07	18077.38	0.291
12450	12474.38	216.65	295.07	18006.26	0.277
12475	12499.48	216.65	295.07	17935.42	0.288
12500	12524.57	216.65	295.07	17864.85	0.287
12525	12549.67	216.65	295.07	17794.56	0.286
12550	12574.77	216.65	295.07	17724.55	0.285
12575	12599.87	216.65	295.07	17654.81	0.284
12600	12624.97	216.65	295.07	17585.35	0.283
12625	12650.07	216.65	295.07	17516.16	0.282
12650	12675.17	216.65	295.07	17447.25	0.281
12675	12700.27	216.65	295.07	17378.6	0.279
12700	12725.37	216.65	295.07	17310.22	0.278
12725	12750.47	216.65	295.07	17242.12	0.277
12750	12775.57	216.65	295.07	17174.28	0.276
12775	12800.67	216.65	295.07	17106.71	0.275
12800	12825.77	216.65	295.07	17039.4	0.274
12825	12850.87	216.65	295.07	16972.36	0.273
12850	12875.97	216.65	295.07	16905.59	0.272
12875	12901.07	216.65	295.07	16839.07	0.271
12900	12926.17	216.65	295.07	16772.82	0.27
12925	12951.27	216.65	295.07	16706.83	0.269

h (m)	h_G (m)	T (K)	a (m/s)	p (Pa)	ρ (kg/m³)
49825	50217.73	270.65	329.8	77.64	0.001
12950	12976.38	216.65	295.07	16641.1	0.268
12975	13001.48	216.65	295.07	16575.62	0.267
13000	13026.58	216.65	295.07	16510.41	0.265
13025	13051.68	216.65	295.07	16445.45	0.264
13050	13076.79	216.65	295.07	16380.74	0.263
13075	13101.89	216.65	295.07	16316.29	0.261
13100	13126.99	216.65	295.07	16252.1	0.261
13125	13152.09	216.65	295.07	16188.15	0.26
13150	13177.2	216.65	295.07	16124.46	0.259
13175	13202.3	216.65	295.07	16061.02	0.258
13200	13227.41	216.65	295.07	15997.83	0.257
13225	13252.51	216.65	295.07	15934.89	0.256
13250	13277.61	216.65	295.07	15872.19	0.255
13275	13302.72	216.65	295.07	15809.74	0.254
13300	13327.82	216.65	295.07	15747.54	0.253
13325	13352.93	216.65	295.07	15685.58	0.252
13350	13378.03	216.65	295.07	15623.87	0.251
13375	13403.14	216.65	295.07	15562.4	0.25
13400	13428.24	216.65	295.07	15501.17	0.249
13425	13453.35	216.65	295.07	15440.18	0.248
13450	13478.45	216.65	295.07	15379.43	0.247
13475	13503.56	216.65	295.07	15318.92	0.246
13500	13528.67	216.65	295.07	15258.65	0.245
13525	13553.77	216.65	295.07	15198.62	0.244
13550	13578.88	216.65	295.07	15138.82	0.243
13575	13603.99	216.65	295.07	15079.26	0.242
13600	13629.09	216.65	295.07	15019.93	0.242
13625	13654.2	216.65	295.07	14960.83	0.241
13650	13679.31	216.65	295.07	14901.97	0.24
13675	13704.42	216.65	295.07	14843.34	0.239
13700	13729.52	216.65	295.07	14784.94	0.238
13725	13754.63	216.65	295.07	14726.77	0.237
13750	13779.74	216.65	295.07	14668.83	0.236
13775	13804.85	216.65	295.07	14611.11	0.235
13800	13829.96	216.65	295.07	14553.63	0.234
13825	13855.07	216.65	295.07	14496.36	0.233
13850	13880.17	216.65	295.07	14439.33	0.232
13875	13905.28	216.65	295.07	14382.52	0.231
13900	13930.39	216.65	295.07	14325.93	0.23
13925	13955.5	216.65	295.07	14269.57	0.229
13950	13980.61	216.65	295.07	14213.42	0.229

h (m)	h_G (m)	T (K)	a (m/s)	p (Pa)	ρ (kg/m³)
49825	50217.73	270.65	329.8	77.64	0.001
13975	14005.72	216.65	295.07	14157.5	0.228
14000	14030.83	216.65	295.07	14101.8	0.227
14025	14055.94	216.65	295.07	14046.32	0.226
14050	14081.05	216.65	295.07	13991.05	0.225
14075	14106.16	216.65	295.07	13936.01	0.224
14100	14131.27	216.65	295.07	13881.17	0.223
14125	14156.39	216.65	295.07	13826.56	0.222
14150	14181.5	216.65	295.07	13772.16	0.221
14175	14206.61	216.65	295.07	13717.97	0.221
14200	14231.72	216.65	295.07	13664	0.22
14225	14256.83	216.65	295.07	13610.24	0.219
14250	14281.94	216.65	295.07	13556.69	0.218
14275	14307.06	216.65	295.07	13503.35	0.217
14300	14332.17	216.65	295.07	13450.23	0.216
14325	14357.28	216.65	295.07	13397.31	0.215
14350	14382.39	216.65	295.07	13344.6	0.215
14375	14407.51	216.65	295.07	13292.09	0.214
14400	14432.62	216.65	295.07	13239.79	0.213
14425	14457.73	216.65	295.07	13187.7	0.212
14450	14482.85	216.65	295.07	13135.82	0.211
14475	14507.96	216.65	295.07	13084.13	0.21
14500	14533.08	216.65	295.07	13032.66	0.21
14525	14558.19	216.65	295.07	12981.38	0.209
14550	14583.31	216.65	295.07	12930.31	0.208
14575	14608.42	216.65	295.07	12879.43	0.207
14600	14633.53	216.65	295.07	12828.76	0.206
14625	14658.65	216.65	295.07	12778.28	0.205
14650	14683.77	216.65	295.07	12728.01	0.205
14675	14708.88	216.65	295.07	12677.93	0.204
14700	14734	216.65	295.07	12628.05	0.203
14725	14759.11	216.65	295.07	12578.37	0.202
14750	14784.23	216.65	295.07	12528.88	0.201
14775	14809.34	216.65	295.07	12479.58	0.201
14800	14834.46	216.65	295.07	12430.48	0.2
14825	14859.58	216.65	295.07	12381.57	0.199
14850	14884.69	216.65	295.07	12332.86	0.198
14875	14909.81	216.65	295.07	12284.34	0.198
14900	14934.93	216.65	295.07	12236.01	0.197
14925	14960.05	216.65	295.07	12187.86	0.196
14950	14985.16	216.65	295.07	12139.91	0.195
14975	15010.28	216.65	295.07	12092.15	0.194

h (m)	h_G (m)	T (K)	a (m/s)	p (Pa)	ρ (kg/m³)
49825	50217.73	270.65	329.8	77.64	0.001
15000	15035.4	216.65	295.07	12044.57	0.194
15025	15060.52	216.65	295.07	11997.18	0.193
15050	15085.64	216.65	295.07	11949.98	0.192
15075	15110.75	216.65	295.07	11902.96	0.191
15100	15135.87	216.65	295.07	11856.13	0.191
15125	15160.99	216.65	295.07	11809.48	0.19
15150	15186.11	216.65	295.07	11763.02	0.189
15175	15211.23	216.65	295.07	11716.74	0.188
15200	15236.35	216.65	295.07	11670.64	0.188
15225	15261.47	216.65	295.07	11624.72	0.187
15250	15286.59	216.65	295.07	11578.99	0.186
15275	15311.71	216.65	295.07	11533.43	0.185
15300	15336.83	216.65	295.07	11488.05	0.185
15325	15361.95	216.65	295.07	11442.85	0.184
15350	15387.07	216.65	295.07	11397.83	0.183
15375	15412.19	216.65	295.07	11352.99	0.183
15400	15437.32	216.65	295.07	11308.32	0.182
15425	15462.44	216.65	295.07	11263.83	0.181
15450	15487.56	216.65	295.07	11219.51	0.18
15475	15512.68	216.65	295.07	11175.37	0.18
15500	15537.8	216.65	295.07	11131.4	0.179
15525	15562.92	216.65	295.07	11087.6	0.178
15550	15588.05	216.65	295.07	11043.98	0.178
15575	15613.17	216.65	295.07	11000.53	0.177
15600	15638.29	216.65	295.07	10957.25	0.176
15625	15663.41	216.65	295.07	10914.14	0.175
15650	15688.54	216.65	295.07	10871.19	0.175
15675	15713.66	216.65	295.07	10828.42	0.174
15700	15738.78	216.65	295.07	10785.82	0.173
15725	15763.91	216.65	295.07	10743.38	0.173
15750	15789.03	216.65	295.07	10701.11	0.172
15775	15814.16	216.65	295.07	10659.01	0.171
15800	15839.28	216.65	295.07	10617.07	0.171
15825	15864.41	216.65	295.07	10575.3	0.17
15850	15889.53	216.65	295.07	10533.69	0.169
15875	15914.66	216.65	295.07	10492.25	0.169
15900	15939.78	216.65	295.07	10450.97	0.168
15925	15964.91	216.65	295.07	10409.85	0.167
15950	15990.03	216.65	295.07	10368.89	0.167
15975	16015.16	216.65	295.07	10328.09	0.166
16000	16040.28	216.65	295.07	10287.46	0.165

h (m)	h_G (m)	T (K)	a (m/s)	p (Pa)	ρ (kg/m^3)
49825	50217.73	270.65	329.8	77.64	0.001
16025	16065.41	216.65	295.07	10246.98	0.165
16050	16090.54	216.65	295.07	10206.67	0.164
16075	16115.66	216.65	295.07	10166.51	0.163
16100	16140.79	216.65	295.07	10126.51	0.163
16125	16165.92	216.65	295.07	10086.67	0.162
16150	16191.04	216.65	295.07	10046.98	0.162
16175	16216.17	216.65	295.07	10007.45	0.161
16200	16241.3	216.65	295.07	9968.08	0.16
16225	16266.43	216.65	295.07	9928.86	0.16
16250	16291.55	216.65	295.07	9889.8	0.159
16275	16316.68	216.65	295.07	9850.88	0.158
16300	16341.81	216.65	295.07	9812.13	0.158
16325	16366.94	216.65	295.07	9773.52	0.157
16350	16392.07	216.65	295.07	9735.07	0.157
16375	16417.2	216.65	295.07	9696.77	0.156
16400	16442.33	216.65	295.07	9658.61	0.155
16425	16467.45	216.65	295.07	9620.61	0.155
16450	16492.58	216.65	295.07	9582.76	0.154
16475	16517.71	216.65	295.07	9545.06	0.153
16500	16542.84	216.65	295.07	9507.5	0.153
16525	16567.97	216.65	295.07	9470.1	0.152
16550	16593.1	216.65	295.07	9432.84	0.152
16575	16618.23	216.65	295.07	9395.72	0.151
16600	16643.37	216.65	295.07	9358.76	0.15
16625	16668.5	216.65	295.07	9321.94	0.15
16650	16693.63	216.65	295.07	9285.26	0.149
16675	16718.76	216.65	295.07	9248.73	0.149
16700	16743.89	216.65	295.07	9212.34	0.148
16725	16769.02	216.65	295.07	9176.09	0.148
16750	16794.15	216.65	295.07	9139.99	0.147
16775	16819.29	216.65	295.07	9104.03	0.146
16800	16844.42	216.65	295.07	9068.21	0.146
16825	16869.55	216.65	295.07	9032.53	0.145
16850	16894.68	216.65	295.07	8996.99	0.145
16875	16919.82	216.65	295.07	8961.6	0.144
16900	16944.95	216.65	295.07	8926.34	0.144
16925	16970.08	216.65	295.07	8891.22	0.143
16950	16995.22	216.65	295.07	8856.23	0.142
16975	17020.35	216.65	295.07	8821.39	0.142
17000	17045.48	216.65	295.07	8786.68	0.141
17025	17070.62	216.65	295.07	8752.11	0.141

h (m)	h_G (m)	T (K)	a (m/s)	p (Pa)	ρ (kg/m^3)
49825	50217.73	270.65	329.8	77.64	0.001
17050	17095.75	216.65	295.07	8717.68	0.14
17075	17120.89	216.65	295.07	8683.38	0.14
17100	17146.02	216.65	295.07	8649.21	0.139
17125	17171.16	216.65	295.07	8615.18	0.139
17150	17196.29	216.65	295.07	8581.29	0.138
17175	17221.43	216.65	295.07	8547.52	0.137
17200	17246.56	216.65	295.07	8513.89	0.137
17225	17271.7	216.65	295.07	8480.4	0.136
17250	17296.83	216.65	295.07	8447.03	0.136
17275	17321.97	216.65	295.07	8413.8	0.135
17300	17347.1	216.65	295.07	8380.69	0.135
17325	17372.24	216.65	295.07	8347.72	0.134
17350	17397.38	216.65	295.07	8314.88	0.134
17375	17422.51	216.65	295.07	8282.16	0.133
17400	17447.65	216.65	295.07	8249.58	0.133
17425	17472.79	216.65	295.07	8217.12	0.132
17450	17497.93	216.65	295.07	8184.79	0.132
17475	17523.06	216.65	295.07	8152.59	0.131
17500	17548.2	216.65	295.07	8120.51	0.131
17525	17573.34	216.65	295.07	8088.56	0.13
17550	17598.48	216.65	295.07	8056.74	0.13
17575	17623.62	216.65	295.07	8025.04	0.129
17600	17648.75	216.65	295.07	7993.46	0.129
17625	17673.89	216.65	295.07	7962.01	0.128
17650	17699.03	216.65	295.07	7930.69	0.128
17675	17724.17	216.65	295.07	7899.48	0.127
17700	17749.31	216.65	295.07	7868.4	0.127
17725	17774.45	216.65	295.07	7837.45	0.126
17750	17799.59	216.65	295.07	7806.61	0.126
17775	17824.73	216.65	295.07	7775.9	0.125
17800	17849.87	216.65	295.07	7745.3	0.125
17825	17875.01	216.65	295.07	7714.83	0.124
17850	17900.15	216.65	295.07	7684.47	0.124
17875	17925.29	216.65	295.07	7654.24	0.123
17900	17950.43	216.65	295.07	7624.13	0.123
17925	17975.57	216.65	295.07	7594.13	0.122
17950	18000.72	216.65	295.07	7564.25	0.122
17975	18025.86	216.65	295.07	7534.49	0.121
18000	18051	216.65	295.07	7504.84	0.121
18025	18076.14	216.65	295.07	7475.32	0.12
18050	18101.28	216.65	295.07	7445.91	0.12

h (m)	h_G (m)	T (K)	a (m/s)	p (Pa)	ρ (kg/m^3)
49825	50217.73	270.65	329.8	77.64	0.001
18075	18126.43	216.65	295.07	7416.61	0.119
18100	18151.57	216.65	295.07	7387.43	0.119
18125	18176.71	216.65	295.07	7358.36	0.118
18150	18201.85	216.65	295.07	7329.41	0.118
18175	18227	216.65	295.07	7300.58	0.117
18200	18252.14	216.65	295.07	7271.85	0.117
18225	18277.28	216.65	295.07	7243.24	0.116
18250	18302.43	216.65	295.07	7214.74	0.116
18275	18327.57	216.65	295.07	7186.36	0.116
18300	18352.72	216.65	295.07	7158.08	0.115
18325	18377.86	216.65	295.07	7129.92	0.115
18350	18403.01	216.65	295.07	7101.87	0.114
18375	18428.15	216.65	295.07	7073.93	0.114
18400	18453.29	216.65	295.07	7046.09	0.113
18425	18478.44	216.65	295.07	7018.37	0.113
18450	18503.59	216.65	295.07	6990.76	0.112
18475	18528.73	216.65	295.07	6963.25	0.112
18500	18553.88	216.65	295.07	6935.86	0.112
18525	18579.02	216.65	295.07	6908.57	0.111
18550	18604.17	216.65	295.07	6881.39	0.111
18575	18629.31	216.65	295.07	6854.31	0.11
18600	18654.46	216.65	295.07	6827.34	0.11
18625	18679.61	216.65	295.07	6800.48	0.109
18650	18704.75	216.65	295.07	6773.73	0.109
18675	18729.9	216.65	295.07	6747.08	0.108
18700	18755.05	216.65	295.07	6720.53	0.108
18725	18780.2	216.65	295.07	6694.09	0.108
18750	18805.34	216.65	295.07	6667.75	0.107
18775	18830.49	216.65	295.07	6641.52	0.107
18800	18855.64	216.65	295.07	6615.39	0.106
18825	18880.79	216.65	295.07	6589.36	0.106
18850	18905.94	216.65	295.07	6563.43	0.106
18875	18931.09	216.65	295.07	6537.61	0.105
18900	18956.23	216.65	295.07	6511.89	0.105
18925	18981.38	216.65	295.07	6486.27	0.104
18950	19006.53	216.65	295.07	6460.75	0.104
18975	19031.68	216.65	295.07	6435.33	0.103
19000	19056.83	216.65	295.07	6410.01	0.103
19025	19081.98	216.65	295.07	6384.79	0.103
19050	19107.13	216.65	295.07	6359.67	0.102
19075	19132.28	216.65	295.07	6334.64	0.102

h (m)	h_G (m)	T (K)	a (m/s)	p (Pa)	ρ (kg/m^3)
49825	50217.73	270.65	329.8	77.64	0.001
19100	19157.43	216.65	295.07	6309.72	0.101
19125	19182.58	216.65	295.07	6284.9	0.101
19150	19207.73	216.65	295.07	6260.17	0.101
19175	19232.89	216.65	295.07	6235.54	0.1
19200	19258.04	216.65	295.07	6211	0.1
19225	19283.19	216.65	295.07	6186.57	0.099
19250	19308.34	216.65	295.07	6162.23	0.099
19275	19333.49	216.65	295.07	6137.98	0.099
19300	19358.64	216.65	295.07	6113.83	0.098
19325	19383.8	216.65	295.07	6089.78	0.098
19350	19408.95	216.65	295.07	6065.82	0.098
19375	19434.1	216.65	295.07	6041.95	0.097
19400	19459.25	216.65	295.07	6018.18	0.097
19425	19484.41	216.65	295.07	5994.5	0.096
19450	19509.56	216.65	295.07	5970.92	0.096
19475	19534.71	216.65	295.07	5947.43	0.096
19500	19559.87	216.65	295.07	5924.03	0.095
19525	19585.02	216.65	295.07	5900.72	0.095
19550	19610.18	216.65	295.07	5877.5	0.095
19575	19635.33	216.65	295.07	5854.38	0.094
19600	19660.48	216.65	295.07	5831.34	0.094
19625	19685.64	216.65	295.07	5808.4	0.093
19650	19710.79	216.65	295.07	5785.55	0.093
19675	19735.95	216.65	295.07	5762.78	0.093
19700	19761.1	216.65	295.07	5740.11	0.092
19725	19786.26	216.65	295.07	5717.53	0.092
19750	19811.42	216.65	295.07	5695.03	0.092
19775	19836.57	216.65	295.07	5672.62	0.091
19800	19861.73	216.65	295.07	5650.31	0.091
19825	19886.88	216.65	295.07	5628.07	0.09
19850	19912.04	216.65	295.07	5605.93	0.09
19875	19937.2	216.65	295.07	5583.88	0.09
19900	19962.35	216.65	295.07	5561.91	0.089
19925	19987.51	216.65	295.07	5540.02	0.089
19950	20012.67	216.65	295.07	5518.23	0.089
19975	20037.82	216.65	295.07	5496.51	0.088
20000	20062.98	216.65	295.07	5474.89	0.088
20025	20088.14	216.68	295.09	5453.35	0.088
20050	20113.3	216.7	295.1	5431.9	0.087
20075	20138.46	216.73	295.12	5410.53	0.087
20100	20163.61	216.75	295.14	5389.25	0.087

h (m)	h_G (m)	T (K)	a (m/s)	p (Pa)	ρ (kg/m^3)
49825	50217.73	270.65	329.8	77.64	0.001
20125	20188.77	216.78	295.15	5368.06	0.086
20150	20213.93	216.8	295.17	5346.95	0.086
20175	20239.09	216.83	295.19	5325.93	0.086
20200	20264.25	216.85	295.21	5304.99	0.085
20225	20289.41	216.88	295.22	5284.14	0.085
20250	20314.57	216.9	295.24	5263.38	0.085
20275	20339.73	216.93	295.26	5242.69	0.084
20300	20364.89	216.95	295.27	5222.09	0.084
20325	20390.05	216.98	295.29	5201.58	0.084
20350	20415.21	217	295.31	5181.14	0.083
20375	20440.37	217.03	295.32	5160.79	0.083
20400	20465.53	217.05	295.34	5140.52	0.083
20425	20490.69	217.08	295.36	5120.34	0.082
20450	20515.85	217.1	295.38	5100.23	0.082
20475	20541.01	217.13	295.39	5080.21	0.082
20500	20566.18	217.15	295.41	5060.26	0.081
20525	20591.34	217.18	295.43	5040.4	0.081
20550	20616.5	217.2	295.44	5020.62	0.081
20575	20641.66	217.23	295.46	5000.92	0.08
20600	20666.82	217.25	295.48	4981.29	0.08
20625	20691.99	217.28	295.49	4961.75	0.08
20650	20717.15	217.3	295.51	4942.29	0.079
20675	20742.31	217.33	295.53	4922.9	0.079
20700	20767.48	217.35	295.55	4903.59	0.079
20725	20792.64	217.38	295.56	4884.36	0.078
20750	20817.8	217.4	295.58	4865.21	0.078
20775	20842.97	217.43	295.6	4846.14	0.078
20800	20868.13	217.45	295.61	4827.14	0.077
20825	20893.29	217.48	295.63	4808.22	0.077
20850	20918.46	217.5	295.65	4789.37	0.077
20875	20943.62	217.53	295.66	4770.6	0.076
20900	20968.79	217.55	295.68	4751.91	0.076
20925	20993.95	217.58	295.7	4733.29	0.076
20950	21019.12	217.6	295.72	4714.75	0.075
20975	21044.28	217.63	295.73	4696.28	0.075
21000	21069.45	217.65	295.75	4677.89	0.075
21025	21094.61	217.68	295.77	4659.57	0.075
21050	21119.78	217.7	295.78	4641.32	0.074
21075	21144.95	217.73	295.8	4623.15	0.074
21100	21170.11	217.75	295.82	4605.05	0.074
21125	21195.28	217.78	295.83	4587.02	0.073

h (m)	h_G (m)	T (K)	*a* (m/s)	*p* (Pa)	ρ (kg/m³)
49825	50217.73	270.65	329.8	77.64	0.001
21150	21220.45	217.8	295.85	4569.07	0.073
21175	21245.61	217.83	295.87	4551.19	0.073
21200	21270.78	217.85	295.89	4533.38	0.072
21225	21295.95	217.88	295.9	4515.64	0.072
21250	21321.12	217.9	295.92	4497.98	0.072
21275	21346.28	217.93	295.94	4480.38	0.072
21300	21371.45	217.95	295.95	4462.86	0.071
21325	21396.62	217.98	295.97	4445.41	0.071
21350	21421.79	218	295.99	4428.02	0.071
21375	21446.96	218.03	296	4410.71	0.07
21400	21472.12	218.05	296.02	4393.47	0.07
21425	21497.29	218.08	296.04	4376.29	0.07
21450	21522.46	218.1	296.06	4359.19	0.07
21475	21547.63	218.13	296.07	4342.15	0.069
21500	21572.8	218.15	296.09	4325.18	0.069
21525	21597.97	218.18	296.11	4308.28	0.069
21550	21623.14	218.2	296.12	4291.45	0.069
21575	21648.31	218.23	296.14	4274.69	0.068
21600	21673.48	218.25	296.16	4257.99	0.068
21625	21698.65	218.28	296.17	4241.36	0.068
21650	21723.82	218.3	296.19	4224.8	0.067
21675	21748.99	218.33	296.21	4208.3	0.067
21700	21774.16	218.35	296.23	4191.87	0.067
21725	21799.34	218.38	296.24	4175.51	0.067
21750	21824.51	218.4	296.26	4159.21	0.066
21775	21849.68	218.43	296.28	4142.98	0.066
21800	21874.85	218.45	296.29	4126.81	0.066
21825	21900.02	218.48	296.31	4110.71	0.066
21850	21925.19	218.5	296.33	4094.67	0.065
21875	21950.37	218.53	296.34	4078.7	0.065
21900	21975.54	218.55	296.36	4062.79	0.065
21925	22000.71	218.58	296.38	4046.95	0.065
21950	22025.89	218.6	296.39	4031.16	0.064
21975	22051.06	218.63	296.41	4015.45	0.064
22000	22076.23	218.65	296.43	3999.79	0.064
22025	22101.41	218.68	296.45	3984.2	0.063
22050	22126.58	218.7	296.46	3968.67	0.063
22075	22151.75	218.73	296.48	3953.2	0.063
22100	22176.93	218.75	296.5	3937.79	0.063
22125	22202.1	218.78	296.51	3922.45	0.062
22150	22227.28	218.8	296.53	3907.17	0.062

h (m)	h_G (m)	T (K)	a (m/s)	p (Pa)	ρ (kg/m³)
49825	50217.73	270.65	329.8	77.64	0.001
22175	22252.45	218.83	296.55	3891.95	0.062
22200	22277.63	218.85	296.56	3876.79	0.062
22225	22302.8	218.88	296.58	3861.69	0.061
22250	22327.98	218.9	296.6	3846.65	0.061
22275	22353.15	218.93	296.61	3831.67	0.061
22300	22378.33	218.95	296.63	3816.75	0.061
22325	22403.51	218.98	296.65	3801.9	0.06
22350	22428.68	219	296.67	3787.1	0.06
22375	22453.86	219.03	296.68	3772.36	0.06
22400	22479.03	219.05	296.7	3757.68	0.06
22425	22504.21	219.08	296.72	3743.05	0.06
22450	22529.39	219.1	296.73	3728.49	0.059
22475	22554.57	219.13	296.75	3713.99	0.059
22500	22579.74	219.15	296.77	3699.54	0.059
22525	22604.92	219.18	296.78	3685.15	0.059
22550	22630.1	219.2	296.8	3670.82	0.058
22575	22655.28	219.23	296.82	3656.54	0.058
22600	22680.45	219.25	296.83	3642.33	0.058
22625	22705.63	219.28	296.85	3628.17	0.058
22650	22730.81	219.3	296.87	3614.06	0.057
22675	22755.99	219.33	296.89	3600.02	0.057
22700	22781.17	219.35	296.9	3586.03	0.057
22725	22806.35	219.38	296.92	3572.09	0.057
22750	22831.53	219.4	296.94	3558.21	0.056
22775	22856.71	219.43	296.95	3544.39	0.056
22800	22881.89	219.45	296.97	3530.62	0.056
22825	22907.07	219.48	296.99	3516.91	0.056
22850	22932.25	219.5	297	3503.25	0.056
22875	22957.43	219.53	297.02	3489.64	0.055
22900	22982.61	219.55	297.04	3476.09	0.055
22925	23007.79	219.58	297.05	3462.6	0.055
22950	23032.97	219.6	297.07	3449.16	0.055
22975	23058.15	219.63	297.09	3435.77	0.054
23000	23083.33	219.65	297.11	3422.43	0.054
23025	23108.51	219.68	297.12	3409.15	0.054
23050	23133.7	219.7	297.14	3395.93	0.054
23075	23158.88	219.73	297.16	3382.75	0.054
23100	23184.06	219.75	297.17	3369.63	0.053
23125	23209.24	219.78	297.19	3356.56	0.053
23150	23234.43	219.8	297.21	3343.54	0.053
23175	23259.61	219.83	297.22	3330.57	0.053

h (m)	h_G (m)	T (K)	a (m/s)	p (Pa)	ρ (kg/m³)
49825	50217.73	270.65	329.8	77.64	0.001
23200	23284.79	219.85	297.24	3317.66	0.053
23225	23309.97	219.88	297.26	3304.8	0.052
23250	23335.16	219.9	297.27	3291.98	0.052
23275	23360.34	219.93	297.29	3279.22	0.052
23300	23385.53	219.95	297.31	3266.51	0.052
23325	23410.71	219.98	297.33	3253.86	0.052
23350	23435.89	220	297.34	3241.25	0.051
23375	23461.08	220.03	297.36	3228.69	0.051
23400	23486.26	220.05	297.38	3216.18	0.051
23425	23511.45	220.08	297.39	3203.72	0.051
23450	23536.63	220.1	297.41	3191.32	0.051
23475	23561.82	220.13	297.43	3178.96	0.05
23500	23587	220.15	297.44	3166.65	0.05
23525	23612.19	220.18	297.46	3154.39	0.05
23550	23637.37	220.2	297.48	3142.17	0.05
23575	23662.56	220.23	297.49	3130.01	0.05
23600	23687.75	220.25	297.51	3117.9	0.049
23625	23712.93	220.28	297.53	3105.83	0.049
23650	23738.12	220.3	297.54	3093.81	0.049
23675	23763.31	220.33	297.56	3081.84	0.049
23700	23788.49	220.35	297.58	3069.92	0.049
23725	23813.68	220.38	297.6	3058.04	0.048
23750	23838.87	220.4	297.61	3046.21	0.048
23775	23864.05	220.43	297.63	3034.43	0.048
23800	23889.24	220.45	297.65	3022.7	0.048
23825	23914.43	220.48	297.66	3011.01	0.048
23850	23939.62	220.5	297.68	2999.37	0.047
23875	23964.81	220.53	297.7	2987.78	0.047
23900	23990	220.55	297.71	2976.23	0.047
23925	24015.18	220.58	297.73	2964.73	0.047
23950	24040.37	220.6	297.75	2953.27	0.047
23975	24065.56	220.63	297.76	2941.86	0.046
24000	24090.75	220.65	297.78	2930.49	0.046
24025	24115.94	220.68	297.8	2919.17	0.046
24050	24141.13	220.7	297.81	2907.9	0.046
24075	24166.32	220.73	297.83	2896.67	0.046
24100	24191.51	220.75	297.85	2885.48	0.046
24125	24216.7	220.78	297.87	2874.34	0.045
24150	24241.89	220.8	297.88	2863.24	0.045
24175	24267.08	220.83	297.9	2852.19	0.045
24200	24292.27	220.85	297.92	2841.18	0.045

h (m)	h_G (m)	T (K)	a (m/s)	p (Pa)	ρ (kg/m³)
49825	50217.73	270.65	329.8	77.64	0.001
24225	24317.46	220.88	297.93	2830.21	0.045
24250	24342.66	220.9	297.95	2819.29	0.044
24275	24367.85	220.93	297.97	2808.41	0.044
24300	24393.04	220.95	297.98	2797.58	0.044
24325	24418.23	220.98	298	2786.78	0.044
24350	24443.42	221	298.02	2776.03	0.044
24375	24468.62	221.03	298.03	2765.33	0.044
24400	24493.81	221.05	298.05	2754.66	0.043
24425	24519	221.08	298.07	2744.04	0.043
24450	24544.19	221.1	298.08	2733.46	0.043
24475	24569.39	221.13	298.1	2722.92	0.043
24500	24594.58	221.15	298.12	2712.42	0.043
24525	24619.77	221.18	298.14	2701.97	0.043
24550	24644.97	221.2	298.15	2691.56	0.042
24575	24670.16	221.23	298.17	2681.19	0.042
24600	24695.35	221.25	298.19	2670.85	0.042
24625	24720.55	221.28	298.2	2660.57	0.042
24650	24745.74	221.3	298.22	2650.32	0.042
24675	24770.94	221.33	298.24	2640.11	0.042
24700	24796.13	221.35	298.25	2629.94	0.041
24725	24821.33	221.38	298.27	2619.81	0.041
24750	24846.52	221.4	298.29	2609.73	0.041
24775	24871.72	221.43	298.3	2599.68	0.041
24800	24896.91	221.45	298.32	2589.67	0.041
24825	24922.11	221.48	298.34	2579.7	0.041
24850	24947.31	221.5	298.35	2569.77	0.04
24875	24972.5	221.53	298.37	2559.88	0.04
24900	24997.7	221.55	298.39	2550.03	0.04
24925	25022.9	221.58	298.4	2540.22	0.04
24950	25048.09	221.6	298.42	2530.45	0.04
24975	25073.29	221.63	298.44	2520.72	0.04
25000	25098.49	221.65	298.46	2511.02	0.039
25025	25123.68	221.68	298.47	2501.37	0.039
25050	25148.88	221.7	298.49	2491.75	0.039
25075	25174.08	221.73	298.51	2482.17	0.039
25100	25199.28	221.75	298.52	2472.63	0.039
25125	25224.48	221.78	298.54	2463.12	0.039
25150	25249.67	221.8	298.56	2453.65	0.039
25175	25274.87	221.83	298.57	2444.22	0.038
25200	25300.07	221.85	298.59	2434.83	0.038
25225	25325.27	221.88	298.61	2425.48	0.038

h (m)	h_G (m)	T (K)	a (m/s)	p (Pa)	ρ (kg/m³)
49825	50217.73	270.65	329.8	77.64	0.001
25250	25350.47	221.9	298.62	2416.16	0.038
25275	25375.67	221.93	298.64	2406.88	0.038
25300	25400.87	221.95	298.66	2397.63	0.038
25325	25426.07	221.98	298.67	2388.43	0.037
25350	25451.27	222	298.69	2379.25	0.037
25375	25476.47	222.03	298.71	2370.12	0.037
25400	25501.67	222.05	298.72	2361.02	0.037
25425	25526.87	222.08	298.74	2351.96	0.037
25450	25552.07	222.1	298.76	2342.93	0.037
25475	25577.27	222.13	298.77	2333.94	0.037
25500	25602.47	222.15	298.79	2324.98	0.036
25525	25627.68	222.18	298.81	2316.06	0.036
25550	25652.88	222.2	298.83	2307.17	0.036
25575	25678.08	222.23	298.84	2298.32	0.036
25600	25703.28	222.25	298.86	2289.51	0.036
25625	25728.48	222.28	298.88	2280.73	0.036
25650	25753.69	222.3	298.89	2271.98	0.036
25675	25778.89	222.33	298.91	2263.27	0.035
25700	25804.09	222.35	298.93	2254.59	0.035
25725	25829.29	222.38	298.94	2245.95	0.035
25750	25854.5	222.4	298.96	2237.34	0.035
25775	25879.7	222.43	298.98	2228.76	0.035
25800	25904.9	222.45	298.99	2220.22	0.035
25825	25930.11	222.48	299.01	2211.71	0.035
25850	25955.31	222.5	299.03	2203.24	0.034
25875	25980.52	222.53	299.04	2194.8	0.034
25900	26005.72	222.55	299.06	2186.39	0.034
25925	26030.93	222.58	299.08	2178.02	0.034
25950	26056.13	222.6	299.09	2169.68	0.034
25975	26081.34	222.63	299.11	2161.37	0.034
26000	26106.54	222.65	299.13	2153.09	0.034
26025	26131.75	222.68	299.14	2144.85	0.034
26050	26156.95	222.7	299.16	2136.64	0.033
26075	26182.16	222.73	299.18	2128.46	0.033
26100	26207.36	222.75	299.19	2120.32	0.033
26125	26232.57	222.78	299.21	2112.2	0.033
26150	26257.78	222.8	299.23	2104.12	0.033
26175	26282.98	222.83	299.25	2096.07	0.033
26200	26308.19	222.85	299.26	2088.05	0.033
26225	26333.4	222.88	299.28	2080.07	0.033
26250	26358.6	222.9	299.3	2072.11	0.032

h (m)	h_G (m)	T (K)	a (m/s)	p (Pa)	ρ (kg/m³)
49825	50217.73	270.65	329.8	77.64	0.001
26275	26383.81	222.93	299.31	2064.19	0.032
26300	26409.02	222.95	299.33	2056.29	0.032
26325	26434.23	222.98	299.35	2048.43	0.032
26350	26459.43	223	299.36	2040.6	0.032
26375	26484.64	223.03	299.38	2032.8	0.032
26400	26509.85	223.05	299.4	2025.03	0.032
26425	26535.06	223.08	299.41	2017.29	0.032
26450	26560.27	223.1	299.43	2009.58	0.031
26475	26585.48	223.13	299.45	2001.91	0.031
26500	26610.69	223.15	299.46	1994.26	0.031
26525	26635.9	223.18	299.48	1986.64	0.031
26550	26661.11	223.2	299.5	1979.05	0.031
26575	26686.32	223.23	299.51	1971.49	0.031
26600	26711.53	223.25	299.53	1963.97	0.031
26625	26736.74	223.28	299.55	1956.47	0.031
26650	26761.95	223.3	299.56	1949	0.03
26675	26787.16	223.33	299.58	1941.56	0.03
26700	26812.37	223.35	299.6	1934.15	0.03
26725	26837.58	223.38	299.61	1926.77	0.03
26750	26862.79	223.4	299.63	1919.41	0.03
26775	26888	223.43	299.65	1912.09	0.03
26800	26913.21	223.45	299.66	1904.8	0.03
26825	26938.42	223.48	299.68	1897.53	0.03
26850	26963.64	223.5	299.7	1890.29	0.029
26875	26988.85	223.53	299.71	1883.08	0.029
26900	27014.06	223.55	299.73	1875.9	0.029
26925	27039.27	223.58	299.75	1868.75	0.029
26950	27064.49	223.6	299.77	1861.62	0.029
26975	27089.7	223.63	299.78	1854.53	0.029
27000	27114.91	223.65	299.8	1847.46	0.029
27025	27140.13	223.68	299.82	1840.42	0.029
27050	27165.34	223.7	299.83	1833.4	0.029
27075	27190.55	223.73	299.85	1826.42	0.028
27100	27215.77	223.75	299.87	1819.46	0.028
27125	27240.98	223.78	299.88	1812.53	0.028
27150	27266.19	223.8	299.9	1805.62	0.028
27175	27291.41	223.83	299.92	1798.74	0.028
27200	27316.62	223.85	299.93	1791.89	0.028
27225	27341.84	223.88	299.95	1785.07	0.028
27250	27367.05	223.9	299.97	1778.27	0.028
27275	27392.27	223.93	299.98	1771.5	0.028

h (m)	h_G (m)	T (K)	a (m/s)	p (Pa)	ρ (kg/m^3)
49825	50217.73	270.65	329.8	77.64	0.001
27300	27417.49	223.95	300	1764.76	0.027
27325	27442.7	223.98	300.02	1758.04	0.027
27350	27467.92	224	300.03	1751.35	0.027
27375	27493.13	224.03	300.05	1744.69	0.027
27400	27518.35	224.05	300.07	1738.05	0.027
27425	27543.57	224.08	300.08	1731.44	0.027
27450	27568.78	224.1	300.1	1724.85	0.027
27475	27594	224.13	300.12	1718.29	0.027
27500	27619.22	224.15	300.13	1711.75	0.027
27525	27644.43	224.18	300.15	1705.25	0.026
27550	27669.65	224.2	300.17	1698.76	0.026
27575	27694.87	224.23	300.18	1692.3	0.026
27600	27720.09	224.25	300.2	1685.87	0.026
27625	27745.31	224.28	300.22	1679.46	0.026
27650	27770.52	224.3	300.23	1673.08	0.026
27675	27795.74	224.33	300.25	1666.72	0.026
27700	27820.96	224.35	300.27	1660.39	0.026
27725	27846.18	224.38	300.28	1654.08	0.026
27750	27871.4	224.4	300.3	1647.79	0.026
27775	27896.62	224.43	300.32	1641.53	0.025
27800	27921.84	224.45	300.33	1635.3	0.025
27825	27947.06	224.48	300.35	1629.09	0.025
27850	27972.28	224.5	300.37	1622.9	0.025
27875	27997.5	224.53	300.38	1616.74	0.025
27900	28022.72	224.55	300.4	1610.6	0.025
27925	28047.94	224.58	300.42	1604.49	0.025
27950	28073.16	224.6	300.43	1598.4	0.025
27975	28098.38	224.63	300.45	1592.33	0.025
28000	28123.6	224.65	300.47	1586.29	0.025
28025	28148.82	224.68	300.48	1580.27	0.025
28050	28174.04	224.7	300.5	1574.28	0.024
28075	28199.27	224.73	300.52	1568.3	0.024
28100	28224.49	224.75	300.53	1562.35	0.024
28125	28249.71	224.78	300.55	1556.43	0.024
28150	28274.93	224.8	300.57	1550.53	0.024
28175	28300.15	224.83	300.59	1544.65	0.024
28200	28325.38	224.85	300.6	1538.79	0.024
28225	28350.6	224.88	300.62	1532.96	0.024
28250	28375.82	224.9	300.64	1527.15	0.024
28275	28401.05	224.93	300.65	1521.36	0.024
28300	28426.27	224.95	300.67	1515.59	0.023

h (m)	h_G (m)	T (K)	a (m/s)	p (Pa)	ρ (kg/m³)
49825	50217.73	270.65	329.8	77.64	0.001
28325	28451.49	224.98	300.69	1509.85	0.023
28350	28476.72	225	300.7	1504.13	0.023
28375	28501.94	225.03	300.72	1498.43	0.023
28400	28527.17	225.05	300.74	1492.75	0.023
28425	28552.39	225.08	300.75	1487.1	0.023
28450	28577.61	225.1	300.77	1481.47	0.023
28475	28602.84	225.13	300.79	1475.86	0.023
28500	28628.06	225.15	300.8	1470.27	0.023
28525	28653.29	225.18	300.82	1464.7	0.023
28550	28678.52	225.2	300.84	1459.16	0.023
28575	28703.74	225.23	300.85	1453.64	0.022
28600	28728.97	225.25	300.87	1448.13	0.022
28625	28754.19	225.28	300.89	1442.65	0.022
28650	28779.42	225.3	300.9	1437.19	0.022
28675	28804.65	225.33	300.92	1431.76	0.022
28700	28829.87	225.35	300.94	1426.34	0.022
28725	28855.1	225.38	300.95	1420.95	0.022
28750	28880.33	225.4	300.97	1415.57	0.022
28775	28905.55	225.43	300.99	1410.22	0.022
28800	28930.78	225.45	301	1404.89	0.022
28825	28956.01	225.48	301.02	1399.57	0.022
28850	28981.24	225.5	301.04	1394.28	0.022
28875	29006.46	225.53	301.05	1389.01	0.021
28900	29031.69	225.55	301.07	1383.76	0.021
28925	29056.92	225.58	301.09	1378.53	0.021
28950	29082.15	225.6	301.1	1373.32	0.021
28975	29107.38	225.63	301.12	1368.13	0.021
29000	29132.61	225.65	301.14	1362.96	0.021
29025	29157.84	225.68	301.15	1357.82	0.021
29050	29183.07	225.7	301.17	1352.69	0.021
29075	29208.3	225.73	301.19	1347.58	0.021
29100	29233.53	225.75	301.2	1342.49	0.021
29125	29258.76	225.78	301.22	1337.42	0.021
29150	29283.99	225.8	301.24	1332.37	0.021
29175	29309.22	225.83	301.25	1327.34	0.02
29200	29334.45	225.85	301.27	1322.33	0.02
29225	29359.68	225.88	301.29	1317.34	0.02
29250	29384.91	225.9	301.3	1312.37	0.02
29275	29410.14	225.93	301.32	1307.42	0.02
29300	29435.37	225.95	301.34	1302.48	0.02
29325	29460.6	225.98	301.35	1297.57	0.02

h (m)	h_G (m)	T (K)	a (m/s)	p (Pa)	ρ (kg/m^3)
49825	50217.73	270.65	329.8	77.64	0.001
29350	29485.84	226	301.37	1292.68	0.02
29375	29511.07	226.03	301.39	1287.8	0.02
29400	29536.3	226.05	301.4	1282.94	0.02
29425	29561.53	226.08	301.42	1278.1	0.02
29450	29586.77	226.1	301.44	1273.29	0.02
29475	29612	226.13	301.45	1268.49	0.02
29500	29637.23	226.15	301.47	1263.7	0.019
29525	29662.46	226.18	301.49	1258.94	0.019
29550	29687.7	226.2	301.5	1254.2	0.019
29575	29712.93	226.23	301.52	1249.47	0.019
29600	29738.17	226.25	301.54	1244.76	0.019
29625	29763.4	226.28	301.55	1240.07	0.019
29650	29788.63	226.3	301.57	1235.4	0.019
29675	29813.87	226.33	301.59	1230.75	0.019
29700	29839.1	226.35	301.6	1226.11	0.019
29725	29864.34	226.38	301.62	1221.49	0.019
29750	29889.57	226.4	301.64	1216.89	0.019
29775	29914.81	226.43	301.65	1212.31	0.019
29800	29940.04	226.45	301.67	1207.75	0.019
29825	29965.28	226.48	301.69	1203.2	0.019
29850	29990.51	226.5	301.7	1198.67	0.018
29875	30015.75	226.53	301.72	1194.16	0.018
29900	30040.99	226.55	301.74	1189.67	0.018
29925	30066.22	226.58	301.75	1185.19	0.018
29950	30091.46	226.6	301.77	1180.73	0.018
29975	30116.7	226.63	301.79	1176.29	0.018
30000	30141.93	226.65	301.8	1171.87	0.018
30025	30167.17	226.68	301.82	1167.46	0.018
30050	30192.41	226.7	301.84	1163.07	0.018
30075	30217.65	226.73	301.85	1158.7	0.018
30100	30242.88	226.75	301.87	1154.34	0.018
30125	30268.12	226.78	301.89	1150	0.018
30150	30293.36	226.8	301.9	1145.68	0.018
30175	30318.6	226.83	301.92	1141.37	0.018
30200	30343.84	226.85	301.94	1137.08	0.017
30225	30369.08	226.88	301.95	1132.81	0.017
30250	30394.31	226.9	301.97	1128.55	0.017
30275	30419.55	226.93	301.99	1124.31	0.017
30300	30444.79	226.95	302	1120.09	0.017
30325	30470.03	226.98	302.02	1115.88	0.017
30350	30495.27	227	302.04	1111.69	0.017

h (m)	h_G (m)	T (K)	a ($^m/_s$)	p (Pa)	ρ ($^{kg}/_{m^3}$)
49825	50217.73	270.65	329.8	77.64	0.001
30375	30520.51	227.03	302.05	1107.52	0.017
30400	30545.75	227.05	302.07	1103.36	0.017
30425	30570.99	227.08	302.09	1099.22	0.017
30450	30596.23	227.1	302.1	1095.09	0.017
30475	30621.47	227.13	302.12	1090.98	0.017
30500	30646.72	227.15	302.14	1086.88	0.017
30525	30671.96	227.18	302.15	1082.81	0.017
30550	30697.2	227.2	302.17	1078.74	0.017
30575	30722.44	227.23	302.19	1074.7	0.016
30600	30747.68	227.25	302.2	1070.66	0.016
30625	30772.92	227.28	302.22	1066.65	0.016
30650	30798.17	227.3	302.24	1062.65	0.016
30675	30823.41	227.33	302.25	1058.66	0.016
30700	30848.65	227.35	302.27	1054.69	0.016
30725	30873.89	227.38	302.28	1050.74	0.016
30750	30899.14	227.4	302.3	1046.8	0.016
30775	30924.38	227.43	302.32	1042.87	0.016
30800	30949.62	227.45	302.33	1038.97	0.016
30825	30974.87	227.48	302.35	1035.07	0.016
30850	31000.11	227.5	302.37	1031.19	0.016
30875	31025.35	227.53	302.38	1027.33	0.016
30900	31050.6	227.55	302.4	1023.48	0.016
30925	31075.84	227.58	302.42	1019.65	0.016
30950	31101.09	227.6	302.43	1015.83	0.016
30975	31126.33	227.63	302.45	1012.02	0.015
31000	31151.58	227.65	302.47	1008.23	0.015
31025	31176.82	227.68	302.48	1004.46	0.015
31050	31202.07	227.7	302.5	1000.7	0.015
31075	31227.31	227.73	302.52	996.95	0.015
31100	31252.56	227.75	302.53	993.22	0.015
31125	31277.81	227.78	302.55	989.5	0.015
31150	31303.05	227.8	302.57	985.8	0.015
31175	31328.3	227.83	302.58	982.11	0.015
31200	31353.54	227.85	302.6	978.43	0.015
31225	31378.79	227.88	302.62	974.77	0.015
31250	31404.04	227.9	302.63	971.13	0.015
31275	31429.29	227.93	302.65	967.49	0.015
31300	31454.53	227.95	302.67	963.88	0.015
31325	31479.78	227.98	302.68	960.27	0.015
31350	31505.03	228	302.7	956.68	0.015
31375	31530.28	228.03	302.72	953.1	0.015

h (m)	h_G (m)	T (K)	a (m/s)	p (Pa)	ρ (kg/m^3)
49825	50217.73	270.65	329.8	77.64	0.001
31400	31555.52	228.05	302.73	949.54	0.015
31425	31580.77	228.08	302.75	945.99	0.014
31450	31606.02	228.1	302.77	942.46	0.014
31475	31631.27	228.13	302.78	938.93	0.014
31500	31656.52	228.15	302.8	935.43	0.014
31525	31681.77	228.18	302.82	931.93	0.014
31550	31707.02	228.2	302.83	928.45	0.014
31575	31732.27	228.23	302.85	924.98	0.014
31600	31757.52	228.25	302.87	921.53	0.014
31625	31782.77	228.28	302.88	918.08	0.014
31650	31808.02	228.3	302.9	914.66	0.014
31675	31833.27	228.33	302.92	911.24	0.014
31700	31858.52	228.35	302.93	907.84	0.014
31725	31883.77	228.38	302.95	904.45	0.014
31750	31909.02	228.4	302.97	901.07	0.014
31775	31934.27	228.43	302.98	897.71	0.014
31800	31959.52	228.45	303	894.36	0.014
31825	31984.77	228.48	303.02	891.02	0.014
31850	32010.03	228.5	303.03	887.7	0.014
31875	32035.28	228.53	303.05	884.39	0.013
31900	32060.53	228.55	303.06	881.09	0.013
31925	32085.78	228.58	303.08	877.8	0.013
31950	32111.03	228.6	303.1	874.53	0.013
31975	32136.29	228.63	303.11	871.27	0.013
32000	32161.54	228.65	303.13	868.02	0.013
32025	32186.79	228.72	303.18	864.78	0.013
32050	32212.05	228.79	303.22	861.56	0.013
32075	32237.3	228.86	303.27	858.35	0.013
32100	32262.55	228.93	303.32	855.15	0.013
32125	32287.81	229	303.36	851.97	0.013
32150	32313.06	229.07	303.41	848.8	0.013
32175	32338.32	229.14	303.46	845.64	0.013
32200	32363.57	229.21	303.5	842.49	0.013
32225	32388.83	229.28	303.55	839.36	0.013
32250	32414.08	229.35	303.59	836.24	0.013
32275	32439.34	229.42	303.64	833.13	0.013
32300	32464.59	229.49	303.69	830.04	0.013
32325	32489.85	229.56	303.73	826.96	0.013
32350	32515.1	229.63	303.78	823.88	0.012
32375	32540.36	229.7	303.83	820.83	0.012
32400	32565.61	229.77	303.87	817.78	0.012

h (m)	h_G (m)	T (K)	a (m/s)	p (Pa)	ρ (kg/m^3)
49825	50217.73	270.65	329.8	77.64	0.001
32425	32590.87	229.84	303.92	814.75	0.012
32450	32616.13	229.91	303.97	811.73	0.012
32475	32641.38	229.98	304.01	808.72	0.012
32500	32666.64	230.05	304.06	805.72	0.012
32525	32691.9	230.12	304.1	802.73	0.012
32550	32717.15	230.19	304.15	799.76	0.012
32575	32742.41	230.26	304.2	796.8	0.012
32600	32767.67	230.33	304.24	793.85	0.012
32625	32792.93	230.4	304.29	790.91	0.012
32650	32818.19	230.47	304.34	787.99	0.012
32675	32843.44	230.54	304.38	785.07	0.012
32700	32868.7	230.61	304.43	782.17	0.012
32725	32893.96	230.68	304.47	779.28	0.012
32750	32919.22	230.75	304.52	776.4	0.012
32775	32944.48	230.82	304.57	773.53	0.012
32800	32969.74	230.89	304.61	770.67	0.012
32825	32995	230.96	304.66	767.83	0.012
32850	33020.26	231.03	304.7	764.99	0.012
32875	33045.52	231.1	304.75	762.17	0.011
32900	33070.78	231.17	304.8	759.36	0.011
32925	33096.04	231.24	304.84	756.56	0.011
32950	33121.3	231.31	304.89	753.77	0.011
32975	33146.56	231.38	304.94	750.99	0.011
33000	33171.82	231.45	304.98	748.23	0.011
33025	33197.08	231.52	305.03	745.47	0.011
33050	33222.34	231.59	305.07	742.73	0.011
33075	33247.6	231.66	305.12	739.99	0.011
33100	33272.87	231.73	305.17	737.27	0.011
33125	33298.13	231.8	305.21	734.56	0.011
33150	33323.39	231.87	305.26	731.86	0.011
33175	33348.65	231.94	305.3	729.17	0.011
33200	33373.92	232.01	305.35	726.49	0.011
33225	33399.18	232.08	305.4	723.82	0.011
33250	33424.44	232.15	305.44	721.16	0.011
33275	33449.7	232.22	305.49	718.51	0.011
33300	33474.97	232.29	305.53	715.88	0.011
33325	33500.23	232.36	305.58	713.25	0.011
33350	33525.49	232.43	305.63	710.63	0.011
33375	33550.76	232.5	305.67	708.03	0.011
33400	33576.02	232.57	305.72	705.43	0.011
33425	33601.29	232.64	305.76	702.85	0.011

h (m)	h_G (m)	T (K)	a (m/s)	p (Pa)	ρ (kg/m³)
49825	50217.73	270.65	329.8	77.64	0.001
33450	33626.55	232.71	305.81	700.27	0.01
33475	33651.82	232.78	305.86	697.7	0.01
33500	33677.08	232.85	305.9	695.15	0.01
33525	33702.35	232.92	305.95	692.61	0.01
33550	33727.61	232.99	305.99	690.07	0.01
33575	33752.88	233.06	306.04	687.55	0.01
33600	33778.14	233.13	306.09	685.03	0.01
33625	33803.41	233.2	306.13	682.53	0.01
33650	33828.67	233.27	306.18	680.03	0.01
33675	33853.94	233.34	306.22	677.55	0.01
33700	33879.21	233.41	306.27	675.07	0.01
33725	33904.47	233.48	306.32	672.61	0.01
33750	33929.74	233.55	306.36	670.15	0.01
33775	33955.01	233.62	306.41	667.71	0.01
33800	33980.28	233.69	306.45	665.27	0.01
33825	34005.54	233.76	306.5	662.84	0.01
33850	34030.81	233.83	306.55	660.43	0.01
33875	34056.08	233.9	306.59	658.02	0.01
33900	34081.35	233.97	306.64	655.62	0.01
33925	34106.61	234.04	306.68	653.23	0.01
33950	34131.88	234.11	306.73	650.85	0.01
33975	34157.15	234.18	306.78	648.48	0.01
34000	34182.42	234.25	306.82	646.12	0.01
34025	34207.69	234.32	306.87	643.77	0.01
34050	34232.96	234.39	306.91	641.43	0.01
34075	34258.23	234.46	306.96	639.1	0.009
34100	34283.5	234.53	307	636.77	0.009
34125	34308.77	234.6	307.05	634.46	0.009
34150	34334.04	234.67	307.1	632.15	0.009
34175	34359.31	234.74	307.14	629.86	0.009
34200	34384.58	234.81	307.19	627.57	0.009
34225	34409.85	234.88	307.23	625.29	0.009
34250	34435.12	234.95	307.28	623.02	0.009
34275	34460.39	235.02	307.32	620.76	0.009
34300	34485.66	235.09	307.37	618.51	0.009
34325	34510.93	235.16	307.42	616.27	0.009
34350	34536.21	235.23	307.46	614.03	0.009
34375	34561.48	235.3	307.51	611.81	0.009
34400	34586.75	235.37	307.55	609.59	0.009
34425	34612.02	235.44	307.6	607.39	0.009
34450	34637.29	235.51	307.64	605.19	0.009

h (m)	h_G (m)	T (K)	a ($^m/_s$)	p (Pa)	ρ ($^{kg}/_{m^3}$)
49825	50217.73	270.65	329.8	77.64	0.001
34475	34662.57	235.58	307.69	603	0.009
34500	34687.84	235.65	307.74	600.81	0.009
34525	34713.11	235.72	307.78	598.64	0.009
34550	34738.39	235.79	307.83	596.48	0.009
34575	34763.66	235.86	307.87	594.32	0.009
34600	34788.93	235.93	307.92	592.17	0.009
34625	34814.21	236	307.96	590.03	0.009
34650	34839.48	236.07	308.01	587.9	0.009
34675	34864.76	236.14	308.06	585.78	0.009
34700	34890.03	236.21	308.1	583.66	0.009
34725	34915.31	236.28	308.15	581.56	0.009
34750	34940.58	236.35	308.19	579.46	0.009
34775	34965.86	236.42	308.24	577.37	0.009
34800	34991.13	236.49	308.28	575.29	0.008
34825	35016.41	236.56	308.33	573.21	0.008
34850	35041.68	236.63	308.38	571.15	0.008
34875	35066.96	236.7	308.42	569.09	0.008
34900	35092.23	236.77	308.47	567.04	0.008
34925	35117.51	236.84	308.51	565	0.008
34950	35142.79	236.91	308.56	562.97	0.008
34975	35168.06	236.98	308.6	560.94	0.008
35000	35193.34	237.05	308.65	558.92	0.008
35025	35218.62	237.12	308.69	556.91	0.008
35050	35243.89	237.19	308.74	554.91	0.008
35075	35269.17	237.26	308.79	552.92	0.008
35100	35294.45	237.33	308.83	550.93	0.008
35125	35319.73	237.4	308.88	548.95	0.008
35150	35345.01	237.47	308.92	546.98	0.008
35175	35370.28	237.54	308.97	545.02	0.008
35200	35395.56	237.61	309.01	543.06	0.008
35225	35420.84	237.68	309.06	541.11	0.008
35250	35446.12	237.75	309.1	539.17	0.008
35275	35471.4	237.82	309.15	537.24	0.008
35300	35496.68	237.89	309.2	535.31	0.008
35325	35521.96	237.96	309.24	533.4	0.008
35350	35547.24	238.03	309.29	531.49	0.008
35375	35572.52	238.1	309.33	529.58	0.008
35400	35597.8	238.17	309.38	527.69	0.008
35425	35623.08	238.24	309.42	525.8	0.008
35450	35648.36	238.31	309.47	523.92	0.008
35475	35673.64	238.38	309.51	522.04	0.008

h (m)	h_G (m)	T (K)	a (m/s)	p (Pa)	ρ (kg/m³)
49825	50217.73	270.65	329.8	77.64	0.001
35500	35698.92	238.45	309.56	520.18	0.008
35525	35724.2	238.52	309.6	518.32	0.008
35550	35749.48	238.59	309.65	516.46	0.008
35575	35774.76	238.66	309.7	514.62	0.008
35600	35800.04	238.73	309.74	512.78	0.007
35625	35825.33	238.8	309.79	510.95	0.007
35650	35850.61	238.87	309.83	509.13	0.007
35675	35875.89	238.94	309.88	507.31	0.007
35700	35901.17	239.01	309.92	505.5	0.007
35725	35926.46	239.08	309.97	503.7	0.007
35750	35951.74	239.15	310.01	501.9	0.007
35775	35977.02	239.22	310.06	500.11	0.007
35800	36002.3	239.29	310.1	498.33	0.007
35825	36027.59	239.36	310.15	496.55	0.007
35850	36052.87	239.43	310.19	494.78	0.007
35875	36078.16	239.5	310.24	493.02	0.007
35900	36103.44	239.57	310.29	491.27	0.007
35925	36128.72	239.64	310.33	489.52	0.007
35950	36154.01	239.71	310.38	487.78	0.007
35975	36179.29	239.78	310.42	486.04	0.007
36000	36204.58	239.85	310.47	484.32	0.007
36025	36229.86	239.92	310.51	482.6	0.007
36050	36255.15	239.99	310.56	480.88	0.007
36075	36280.43	240.06	310.6	479.17	0.007
36100	36305.72	240.13	310.65	477.47	0.007
36125	36331	240.2	310.69	475.78	0.007
36150	36356.29	240.27	310.74	474.09	0.007
36175	36381.58	240.34	310.78	472.41	0.007
36200	36406.86	240.41	310.83	470.73	0.007
36225	36432.15	240.48	310.87	469.06	0.007
36250	36457.44	240.55	310.92	467.4	0.007
36275	36482.72	240.62	310.96	465.74	0.007
36300	36508.01	240.69	311.01	464.09	0.007
36325	36533.3	240.76	311.06	462.45	0.007
36350	36558.59	240.83	311.1	460.81	0.007
36375	36583.87	240.9	311.15	459.18	0.007
36400	36609.16	240.97	311.19	457.56	0.007
36425	36634.45	241.04	311.24	455.94	0.007
36450	36659.74	241.11	311.28	454.32	0.007
36475	36685.03	241.18	311.33	452.72	0.007
36500	36710.32	241.25	311.37	451.12	0.007

h (m)	h_G (m)	T (K)	a (m/s)	p (Pa)	ρ (kg/m^3)
49825	50217.73	270.65	329.8	77.64	0.001
36525	36735.61	241.32	311.42	449.52	0.006
36550	36760.89	241.39	311.46	447.94	0.006
36575	36786.18	241.46	311.51	446.35	0.006
36600	36811.47	241.53	311.55	444.78	0.006
36625	36836.76	241.6	311.6	443.21	0.006
36650	36862.05	241.67	311.64	441.64	0.006
36675	36887.34	241.74	311.69	440.09	0.006
36700	36912.63	241.81	311.73	438.53	0.006
36725	36937.93	241.88	311.78	436.99	0.006
36750	36963.22	241.95	311.82	435.45	0.006
36775	36988.51	242.02	311.87	433.91	0.006
36800	37013.8	242.09	311.91	432.39	0.006
36825	37039.09	242.16	311.96	430.86	0.006
36850	37064.38	242.23	312	429.35	0.006
36875	37089.67	242.3	312.05	427.84	0.006
36900	37114.97	242.37	312.09	426.33	0.006
36925	37140.26	242.44	312.14	424.83	0.006
36950	37165.55	242.51	312.18	423.34	0.006
36975	37190.84	242.58	312.23	421.85	0.006
37000	37216.14	242.65	312.27	420.37	0.006
37025	37241.43	242.72	312.32	418.89	0.006
37050	37266.72	242.79	312.36	417.42	0.006
37075	37292.01	242.86	312.41	415.95	0.006
37100	37317.31	242.93	312.45	414.49	0.006
37125	37342.6	243	312.5	413.04	0.006
37150	37367.9	243.07	312.54	411.59	0.006
37175	37393.19	243.14	312.59	410.15	0.006
37200	37418.48	243.21	312.63	408.71	0.006
37225	37443.78	243.28	312.68	407.28	0.006
37250	37469.07	243.35	312.72	405.85	0.006
37275	37494.37	243.42	312.77	404.43	0.006
37300	37519.66	243.49	312.81	403.01	0.006
37325	37544.96	243.56	312.86	401.6	0.006
37350	37570.26	243.63	312.9	400.19	0.006
37375	37595.55	243.7	312.95	398.79	0.006
37400	37620.85	243.77	312.99	397.4	0.006
37425	37646.14	243.84	313.04	396.01	0.006
37450	37671.44	243.91	313.08	394.62	0.006
37475	37696.74	243.98	313.13	393.25	0.006
37500	37722.03	244.05	313.17	391.87	0.006
37525	37747.33	244.12	313.22	390.5	0.006

h (m)	h_G (m)	T (K)	a (m/s)	p (Pa)	ρ (kg/m³)
49825	50217.73	270.65	329.8	77.64	0.001
37550	37772.63	244.19	313.26	389.14	0.006
37575	37797.93	244.26	313.31	387.78	0.006
37600	37823.22	244.33	313.35	386.43	0.006
37625	37848.52	244.4	313.4	385.08	0.005
37650	37873.82	244.47	313.44	383.74	0.005
37675	37899.12	244.54	313.49	382.4	0.005
37700	37924.42	244.61	313.53	381.06	0.005
37725	37949.71	244.68	313.58	379.74	0.005
37750	37975.01	244.75	313.62	378.41	0.005
37775	38000.31	244.82	313.67	377.1	0.005
37800	38025.61	244.89	313.71	375.78	0.005
37825	38050.91	244.96	313.76	374.47	0.005
37850	38076.21	245.03	313.8	373.17	0.005
37875	38101.51	245.1	313.85	371.87	0.005
37900	38126.81	245.17	313.89	370.58	0.005
37925	38152.11	245.24	313.94	369.29	0.005
37950	38177.41	245.31	313.98	368.01	0.005
37975	38202.71	245.38	314.03	366.73	0.005
38000	38228.01	245.45	314.07	365.45	0.005
38025	38253.31	245.52	314.11	364.19	0.005
38050	38278.61	245.59	314.16	362.92	0.005
38075	38303.92	245.66	314.2	361.66	0.005
38100	38329.22	245.73	314.25	360.41	0.005
38125	38354.52	245.8	314.29	359.16	0.005
38150	38379.82	245.87	314.34	357.91	0.005
38175	38405.12	245.94	314.38	356.67	0.005
38200	38430.43	246.01	314.43	355.43	0.005
38225	38455.73	246.08	314.47	354.2	0.005
38250	38481.03	246.15	314.52	352.97	0.005
38275	38506.33	246.22	314.56	351.75	0.005
38300	38531.64	246.29	314.61	350.53	0.005
38325	38556.94	246.36	314.65	349.32	0.005
38350	38582.24	246.43	314.7	348.11	0.005
38375	38607.55	246.5	314.74	346.91	0.005
38400	38632.85	246.57	314.79	345.71	0.005
38425	38658.16	246.64	314.83	344.51	0.005
38450	38683.46	246.71	314.88	343.32	0.005
38475	38708.77	246.78	314.92	342.14	0.005
38500	38734.07	246.85	314.96	340.95	0.005
38525	38759.38	246.92	315.01	339.78	0.005
38550	38784.68	246.99	315.05	338.6	0.005

h (m)	h_G (m)	T (K)	a (m/s)	p (Pa)	ρ (kg/m³)
49825	50217.73	270.65	329.8	77.64	0.001
38575	38809.99	247.06	315.1	337.43	0.005
38600	38835.29	247.13	315.14	336.27	0.005
38625	38860.6	247.2	315.19	335.11	0.005
38650	38885.9	247.27	315.23	333.95	0.005
38675	38911.21	247.34	315.28	332.8	0.005
38700	38936.52	247.41	315.32	331.66	0.005
38725	38961.82	247.48	315.37	330.51	0.005
38750	38987.13	247.55	315.41	329.38	0.005
38775	39012.44	247.62	315.46	328.24	0.005
38800	39037.74	247.69	315.5	327.11	0.005
38825	39063.05	247.76	315.54	325.98	0.005
38850	39088.36	247.83	315.59	324.86	0.005
38875	39113.67	247.9	315.63	323.75	0.005
38900	39138.97	247.97	315.68	322.63	0.005
38925	39164.28	248.04	315.72	321.52	0.005
38950	39189.59	248.11	315.77	320.42	0.004
38975	39214.9	248.18	315.81	319.32	0.004
39000	39240.21	248.25	315.86	318.22	0.004
39025	39265.52	248.32	315.9	317.13	0.004
39050	39290.83	248.39	315.95	316.04	0.004
39075	39316.14	248.46	315.99	314.95	0.004
39100	39341.45	248.53	316.03	313.87	0.004
39125	39366.76	248.6	316.08	312.8	0.004
39150	39392.07	248.67	316.12	311.72	0.004
39175	39417.38	248.74	316.17	310.66	0.004
39200	39442.69	248.81	316.21	309.59	0.004
39225	39468	248.88	316.26	308.53	0.004
39250	39493.31	248.95	316.3	307.47	0.004
39275	39518.62	249.02	316.35	306.42	0.004
39300	39543.93	249.09	316.39	305.37	0.004
39325	39569.24	249.16	316.43	304.33	0.004
39350	39594.55	249.23	316.48	303.29	0.004
39375	39619.86	249.3	316.52	302.25	0.004
39400	39645.18	249.37	316.57	301.21	0.004
39425	39670.49	249.44	316.61	300.18	0.004
39450	39695.8	249.51	316.66	299.16	0.004
39475	39721.11	249.58	316.7	298.14	0.004
39500	39746.43	249.65	316.75	297.12	0.004
39525	39771.74	249.72	316.79	296.1	0.004
39550	39797.05	249.79	316.83	295.09	0.004
39575	39822.37	249.86	316.88	294.09	0.004

h (m)	h_G (m)	T (K)	a (m/s)	p (Pa)	ρ (kg/m^3)
49825	50217.73	270.65	329.8	77.64	0.001
39600	39847.68	249.93	316.92	293.08	0.004
39625	39872.99	250	316.97	292.08	0.004
39650	39898.31	250.07	317.01	291.09	0.004
39675	39923.62	250.14	317.06	290.09	0.004
39700	39948.94	250.21	317.1	289.11	0.004
39725	39974.25	250.28	317.15	288.12	0.004
39750	39999.57	250.35	317.19	287.14	0.004
39775	40024.88	250.42	317.23	286.16	0.004
39800	40050.2	250.49	317.28	285.19	0.004
39825	40075.51	250.56	317.32	284.22	0.004
39850	40100.83	250.63	317.37	283.25	0.004
39875	40126.14	250.7	317.41	282.29	0.004
39900	40151.46	250.77	317.46	281.33	0.004
39925	40176.77	250.84	317.5	280.37	0.004
39950	40202.09	250.91	317.54	279.42	0.004
39975	40227.41	250.98	317.59	278.47	0.004
40000	40252.72	251.05	317.63	277.52	0.004
40025	40278.04	251.12	317.68	276.58	0.004
40050	40303.36	251.19	317.72	275.64	0.004
40075	40328.68	251.26	317.77	274.7	0.004
40100	40353.99	251.33	317.81	273.77	0.004
40125	40379.31	251.4	317.85	272.84	0.004
40150	40404.63	251.47	317.9	271.92	0.004
40175	40429.95	251.54	317.94	271	0.004
40200	40455.27	251.61	317.99	270.08	0.004
40225	40480.58	251.68	318.03	269.16	0.004
40250	40505.9	251.75	318.08	268.25	0.004
40275	40531.22	251.82	318.12	267.34	0.004
40300	40556.54	251.89	318.16	266.44	0.004
40325	40581.86	251.96	318.21	265.54	0.004
40350	40607.18	252.03	318.25	264.64	0.004
40375	40632.5	252.1	318.3	263.74	0.004
40400	40657.82	252.17	318.34	262.85	0.004
40425	40683.14	252.24	318.38	261.96	0.004
40450	40708.46	252.31	318.43	261.08	0.004
40475	40733.78	252.38	318.47	260.19	0.004
40500	40759.1	252.45	318.52	259.32	0.004
40525	40784.42	252.52	318.56	258.44	0.004
40550	40809.74	252.59	318.61	257.57	0.004
40575	40835.07	252.66	318.65	256.7	0.004
40600	40860.39	252.73	318.69	255.83	0.004

h (m)	h_G (m)	T (K)	a (m/s)	p (Pa)	ρ (kg/m³)
49825	50217.73	270.65	329.8	77.64	0.001
40625	40885.71	252.8	318.74	254.97	0.004
40650	40911.03	252.87	318.78	254.11	0.004
40675	40936.35	252.94	318.83	253.25	0.003
40700	40961.68	253.01	318.87	252.4	0.003
40725	40987	253.08	318.91	251.55	0.003
40750	41012.32	253.15	318.96	250.7	0.003
40775	41037.64	253.22	319	249.86	0.003
40800	41062.97	253.29	319.05	249.02	0.003
40825	41088.29	253.36	319.09	248.18	0.003
40850	41113.62	253.43	319.13	247.34	0.003
40875	41138.94	253.5	319.18	246.51	0.003
40900	41164.26	253.57	319.22	245.68	0.003
40925	41189.59	253.64	319.27	244.86	0.003
40950	41214.91	253.71	319.31	244.03	0.003
40975	41240.24	253.78	319.36	243.21	0.003
41000	41265.56	253.85	319.4	242.4	0.003
41025	41290.89	253.92	319.44	241.58	0.003
41050	41316.21	253.99	319.49	240.77	0.003
41075	41341.54	254.06	319.53	239.96	0.003
41100	41366.86	254.13	319.58	239.16	0.003
41125	41392.19	254.2	319.62	238.35	0.003
41150	41417.51	254.27	319.66	237.56	0.003
41175	41442.84	254.34	319.71	236.76	0.003
41200	41468.17	254.41	319.75	235.97	0.003
41225	41493.49	254.48	319.8	235.17	0.003
41250	41518.82	254.55	319.84	234.39	0.003
41275	41544.15	254.62	319.88	233.6	0.003
41300	41569.47	254.69	319.93	232.82	0.003
41325	41594.8	254.76	319.97	232.04	0.003
41350	41620.13	254.83	320.02	231.26	0.003
41375	41645.46	254.9	320.06	230.49	0.003
41400	41670.78	254.97	320.1	229.72	0.003
41425	41696.11	255.04	320.15	228.95	0.003
41450	41721.44	255.11	320.19	228.19	0.003
41475	41746.77	255.18	320.23	227.42	0.003
41500	41772.1	255.25	320.28	226.66	0.003
41525	41797.43	255.32	320.32	225.91	0.003
41550	41822.76	255.39	320.37	225.15	0.003
41575	41848.09	255.46	320.41	224.4	0.003
41600	41873.42	255.53	320.45	223.65	0.003
41625	41898.75	255.6	320.5	222.91	0.003

h (m)	h_G (m)	T (K)	a (m/s)	p (Pa)	ρ (kg/m³)
49825	50217.73	270.65	329.8	77.64	0.001
41650	41924.08	255.67	320.54	222.16	0.003
41675	41949.41	255.74	320.59	221.42	0.003
41700	41974.74	255.81	320.63	220.68	0.003
41725	42000.07	255.88	320.67	219.95	0.003
41750	42025.4	255.95	320.72	219.21	0.003
41775	42050.73	256.02	320.76	218.48	0.003
41800	42076.06	256.09	320.81	217.76	0.003
41825	42101.39	256.16	320.85	217.03	0.003
41850	42126.72	256.23	320.89	216.31	0.003
41875	42152.05	256.3	320.94	215.59	0.003
41900	42177.39	256.37	320.98	214.87	0.003
41925	42202.72	256.44	321.02	214.16	0.003
41950	42228.05	256.51	321.07	213.45	0.003
41975	42253.38	256.58	321.11	212.74	0.003
42000	42278.72	256.65	321.16	212.03	0.003
42025	42304.05	256.72	321.2	211.33	0.003
42050	42329.38	256.79	321.24	210.62	0.003
42075	42354.72	256.86	321.29	209.92	0.003
42100	42380.05	256.93	321.33	209.23	0.003
42125	42405.38	257	321.37	208.53	0.003
42150	42430.72	257.07	321.42	207.84	0.003
42175	42456.05	257.14	321.46	207.15	0.003
42200	42481.39	257.21	321.51	206.47	0.003
42225	42506.72	257.28	321.55	205.78	0.003
42250	42532.06	257.35	321.59	205.1	0.003
42275	42557.39	257.42	321.64	204.42	0.003
42300	42582.73	257.49	321.68	203.74	0.003
42325	42608.06	257.56	321.72	203.07	0.003
42350	42633.4	257.63	321.77	202.4	0.003
42375	42658.73	257.7	321.81	201.73	0.003
42400	42684.07	257.77	321.86	201.06	0.003
42425	42709.41	257.84	321.9	200.39	0.003
42450	42734.74	257.91	321.94	199.73	0.003
42475	42760.08	257.98	321.99	199.07	0.003
42500	42785.42	258.05	322.03	198.41	0.003
42525	42810.75	258.12	322.07	197.76	0.003
42550	42836.09	258.19	322.12	197.1	0.003
42575	42861.43	258.26	322.16	196.45	0.003
42600	42886.76	258.33	322.21	195.81	0.003
42625	42912.1	258.4	322.25	195.16	0.003
42650	42937.44	258.47	322.29	194.52	0.003

h (m)	h_G (m)	T (K)	a (m/s)	p (Pa)	ρ (kg/m^3)
49825	50217.73	270.65	329.8	77.64	0.001
42675	42962.78	258.54	322.34	193.87	0.003
42700	42988.12	258.61	322.38	193.23	0.003
42725	43013.46	258.68	322.42	192.6	0.003
42750	43038.79	258.75	322.47	191.96	0.003
42775	43064.13	258.82	322.51	191.33	0.003
42800	43089.47	258.89	322.55	190.7	0.003
42825	43114.81	258.96	322.6	190.07	0.003
42850	43140.15	259.03	322.64	189.45	0.003
42875	43165.49	259.1	322.69	188.82	0.003
42900	43190.83	259.17	322.73	188.2	0.003
42925	43216.17	259.24	322.77	187.58	0.003
42950	43241.51	259.31	322.82	186.97	0.003
42975	43266.85	259.38	322.86	186.35	0.003
43000	43292.19	259.45	322.9	185.74	0.002
43025	43317.53	259.52	322.95	185.13	0.002
43050	43342.88	259.59	322.99	184.52	0.002
43075	43368.22	259.66	323.03	183.91	0.002
43100	43393.56	259.73	323.08	183.31	0.002
43125	43418.9	259.8	323.12	182.71	0.002
43150	43444.24	259.87	323.16	182.11	0.002
43175	43469.58	259.94	323.21	181.51	0.002
43200	43494.93	260.01	323.25	180.92	0.002
43225	43520.27	260.08	323.29	180.32	0.002
43250	43545.61	260.15	323.34	179.73	0.002
43275	43570.96	260.22	323.38	179.14	0.002
43300	43596.3	260.29	323.43	178.56	0.002
43325	43621.64	260.36	323.47	177.97	0.002
43350	43646.99	260.43	323.51	177.39	0.002
43375	43672.33	260.5	323.56	176.81	0.002
43400	43697.67	260.57	323.6	176.23	0.002
43425	43723.02	260.64	323.64	175.65	0.002
43450	43748.36	260.71	323.69	175.08	0.002
43475	43773.71	260.78	323.73	174.5	0.002
43500	43799.05	260.85	323.77	173.93	0.002
43525	43824.4	260.92	323.82	173.37	0.002
43550	43849.74	260.99	323.86	172.8	0.002
43575	43875.09	261.06	323.9	172.23	0.002
43600	43900.43	261.13	323.95	171.67	0.002
43625	43925.78	261.2	323.99	171.11	0.002
43650	43951.12	261.27	324.03	170.55	0.002
43675	43976.47	261.34	324.08	170	0.002

h (m)	h_G (m)	T (K)	a (m/s)	p (Pa)	ρ (kg/m³)
49825	50217.73	270.65	329.8	77.64	0.001
43700	44001.82	261.41	324.12	169.44	0.002
43725	44027.16	261.48	324.16	168.89	0.002
43750	44052.51	261.55	324.21	168.34	0.002
43775	44077.86	261.62	324.25	167.79	0.002
43800	44103.21	261.69	324.29	167.24	0.002
43825	44128.55	261.76	324.34	166.7	0.002
43850	44153.9	261.83	324.38	166.16	0.002
43875	44179.25	261.9	324.42	165.61	0.002
43900	44204.6	261.97	324.47	165.08	0.002
43925	44229.94	262.04	324.51	164.54	0.002
43950	44255.29	262.11	324.55	164	0.002
43975	44280.64	262.18	324.6	163.47	0.002
44000	44305.99	262.25	324.64	162.94	0.002
44025	44331.34	262.32	324.68	162.41	0.002
44050	44356.69	262.39	324.73	161.88	0.002
44075	44382.04	262.46	324.77	161.35	0.002
44100	44407.39	262.53	324.81	160.83	0.002
44125	44432.74	262.6	324.86	160.31	0.002
44150	44458.09	262.67	324.9	159.79	0.002
44175	44483.44	262.74	324.94	159.27	0.002
44200	44508.79	262.81	324.99	158.75	0.002
44225	44534.14	262.88	325.03	158.24	0.002
44250	44559.49	262.95	325.07	157.72	0.002
44275	44584.84	263.02	325.12	157.21	0.002
44300	44610.19	263.09	325.16	156.7	0.002
44325	44635.54	263.16	325.2	156.19	0.002
44350	44660.89	263.23	325.25	155.69	0.002
44375	44686.25	263.3	325.29	155.18	0.002
44400	44711.6	263.37	325.33	154.68	0.002
44425	44736.95	263.44	325.38	154.18	0.002
44450	44762.3	263.51	325.42	153.68	0.002
44475	44787.66	263.58	325.46	153.18	0.002
44500	44813.01	263.65	325.51	152.69	0.002
44525	44838.36	263.72	325.55	152.2	0.002
44550	44863.72	263.79	325.59	151.7	0.002
44575	44889.07	263.86	325.64	151.21	0.002
44600	44914.42	263.93	325.68	150.72	0.002
44625	44939.78	264	325.72	150.24	0.002
44650	44965.13	264.07	325.77	149.75	0.002
44675	44990.48	264.14	325.81	149.27	0.002
44700	45015.84	264.21	325.85	148.79	0.002

h (m)	h_G (m)	T (K)	a (m/s)	p (Pa)	ρ (kg/m^3)
49825	50217.73	270.65	329.8	77.64	0.001
44725	45041.19	264.28	325.89	148.31	0.002
44750	45066.55	264.35	325.94	147.83	0.002
44775	45091.9	264.42	325.98	147.35	0.002
44800	45117.26	264.49	326.02	146.88	0.002
44825	45142.61	264.56	326.07	146.4	0.002
44850	45167.97	264.63	326.11	145.93	0.002
44875	45193.33	264.7	326.15	145.46	0.002
44900	45218.68	264.77	326.2	144.99	0.002
44925	45244.04	264.84	326.24	144.53	0.002
44950	45269.39	264.91	326.28	144.06	0.002
44975	45294.75	264.98	326.33	143.6	0.002
45000	45320.11	265.05	326.37	143.13	0.002
45025	45345.46	265.12	326.41	142.67	0.002
45050	45370.82	265.19	326.46	142.22	0.002
45075	45396.18	265.26	326.5	141.76	0.002
45100	45421.54	265.33	326.54	141.3	0.002
45125	45446.89	265.4	326.58	140.85	0.002
45150	45472.25	265.47	326.63	140.4	0.002
45175	45497.61	265.54	326.67	139.95	0.002
45200	45522.97	265.61	326.71	139.5	0.002
45225	45548.33	265.68	326.76	139.05	0.002
45250	45573.69	265.75	326.8	138.6	0.002
45275	45599.05	265.82	326.84	138.16	0.002
45300	45624.41	265.89	326.89	137.71	0.002
45325	45649.76	265.96	326.93	137.27	0.002
45350	45675.12	266.03	326.97	136.83	0.002
45375	45700.48	266.1	327.01	136.39	0.002
45400	45725.84	266.17	327.06	135.96	0.002
45425	45751.2	266.24	327.1	135.52	0.002
45450	45776.56	266.31	327.14	135.09	0.002
45475	45801.93	266.38	327.19	134.65	0.002
45500	45827.29	266.45	327.23	134.22	0.002
45525	45852.65	266.52	327.27	133.79	0.002
45550	45878.01	266.59	327.32	133.37	0.002
45575	45903.37	266.66	327.36	132.94	0.002
45600	45928.73	266.73	327.4	132.51	0.002
45625	45954.09	266.8	327.44	132.09	0.002
45650	45979.46	266.87	327.49	131.67	0.002
45675	46004.82	266.94	327.53	131.25	0.002
45700	46030.18	267.01	327.57	130.83	0.002
45725	46055.54	267.08	327.62	130.41	0.002

h (m)	h_G (m)	T (K)	a (m/s)	p (Pa)	ρ (kg/m^3)
49825	50217.73	270.65	329.8	77.64	0.001
45750	46080.91	267.15	327.66	130	0.002
45775	46106.27	267.22	327.7	129.58	0.002
45800	46131.63	267.29	327.75	129.17	0.002
45825	46157	267.36	327.79	128.75	0.002
45850	46182.36	267.43	327.83	128.34	0.002
45875	46207.72	267.5	327.87	127.94	0.002
45900	46233.09	267.57	327.92	127.53	0.002
45925	46258.45	267.64	327.96	127.12	0.002
45950	46283.82	267.71	328	126.72	0.002
45975	46309.18	267.78	328.05	126.31	0.002
46000	46334.55	267.85	328.09	125.91	0.002
46025	46359.91	267.92	328.13	125.51	0.002
46050	46385.28	267.99	328.17	125.11	0.002
46075	46410.64	268.06	328.22	124.71	0.002
46100	46436.01	268.13	328.26	124.32	0.002
46125	46461.37	268.2	328.3	123.92	0.002
46150	46486.74	268.27	328.35	123.53	0.002
46175	46512.11	268.34	328.39	123.13	0.002
46200	46537.47	268.41	328.43	122.74	0.002
46225	46562.84	268.48	328.47	122.35	0.002
46250	46588.21	268.55	328.52	121.96	0.002
46275	46613.57	268.62	328.56	121.58	0.002
46300	46638.94	268.69	328.6	121.19	0.002
46325	46664.31	268.76	328.65	120.81	0.002
46350	46689.67	268.83	328.69	120.42	0.002
46375	46715.04	268.9	328.73	120.04	0.002
46400	46740.41	268.97	328.77	119.66	0.002
46425	46765.78	269.04	328.82	119.28	0.002
46450	46791.15	269.11	328.86	118.9	0.002
46475	46816.52	269.18	328.9	118.53	0.002
46500	46841.88	269.25	328.94	118.15	0.002
46525	46867.25	269.32	328.99	117.78	0.002
46550	46892.62	269.39	329.03	117.4	0.002
46575	46917.99	269.46	329.07	117.03	0.002
46600	46943.36	269.53	329.12	116.66	0.002
46625	46968.73	269.6	329.16	116.29	0.002
46650	46994.1	269.67	329.2	115.93	0.001
46675	47019.47	269.74	329.24	115.56	0.001
46700	47044.84	269.81	329.29	115.19	0.001
46725	47070.21	269.88	329.33	114.83	0.001
46750	47095.58	269.95	329.37	114.47	0.001

h (m)	h_G (m)	T (K)	a (m/s)	p (Pa)	ρ (kg/m³)
49825	50217.73	270.65	329.8	77.64	0.001
46775	47120.96	270.02	329.41	114.11	0.001
46800	47146.33	270.09	329.46	113.74	0.001
46825	47171.7	270.16	329.5	113.39	0.001
46850	47197.07	270.23	329.54	113.03	0.001
46875	47222.44	270.3	329.59	112.67	0.001
46900	47247.81	270.37	329.63	112.32	0.001
46925	47273.19	270.44	329.67	111.96	0.001
46950	47298.56	270.51	329.71	111.61	0.001
46975	47323.93	270.58	329.76	111.26	0.001
47000	47349.3	270.65	329.8	110.91	0.001
47025	47374.68	270.65	329.8	110.56	0.001
47050	47400.05	270.65	329.8	110.21	0.001
47075	47425.42	270.65	329.8	109.86	0.001
47100	47450.8	270.65	329.8	109.52	0.001
47125	47476.17	270.65	329.8	109.17	0.001
47150	47501.55	270.65	329.8	108.83	0.001
47175	47526.92	270.65	329.8	108.48	0.001
47200	47552.29	270.65	329.8	108.14	0.001
47225	47577.67	270.65	329.8	107.8	0.001
47250	47603.04	270.65	329.8	107.46	0.001
47275	47628.42	270.65	329.8	107.12	0.001
47300	47653.79	270.65	329.8	106.79	0.001
47325	47679.17	270.65	329.8	106.45	0.001
47350	47704.55	270.65	329.8	106.11	0.001
47375	47729.92	270.65	329.8	105.78	0.001
47400	47755.3	270.65	329.8	105.45	0.001
47425	47780.67	270.65	329.8	105.11	0.001
47450	47806.05	270.65	329.8	104.78	0.001
47475	47831.43	270.65	329.8	104.45	0.001
47500	47856.8	270.65	329.8	104.12	0.001
47525	47882.18	270.65	329.8	103.79	0.001
47550	47907.56	270.65	329.8	103.47	0.001
47575	47932.94	270.65	329.8	103.14	0.001
47600	47958.31	270.65	329.8	102.82	0.001
47625	47983.69	270.65	329.8	102.49	0.001
47650	48009.07	270.65	329.8	102.17	0.001
47675	48034.45	270.65	329.8	101.85	0.001
47700	48059.83	270.65	329.8	101.53	0.001
47725	48085.21	270.65	329.8	101.21	0.001
47750	48110.58	270.65	329.8	100.89	0.001
47775	48135.96	270.65	329.8	100.57	0.001

h (m)	h_G (m)	T (K)	a (m/s)	p (Pa)	ρ (kg/m³)
49825	50217.73	270.65	329.8	77.64	0.001
47800	48161.34	270.65	329.8	100.25	0.001
47825	48186.72	270.65	329.8	99.94	0.001
47850	48212.1	270.65	329.8	99.62	0.001
47875	48237.48	270.65	329.8	99.31	0.001
47900	48262.86	270.65	329.8	99	0.001
47925	48288.24	270.65	329.8	98.68	0.001
47950	48313.62	270.65	329.8	98.37	0.001
47975	48339	270.65	329.8	98.06	0.001
48000	48364.38	270.65	329.8	97.75	0.001
48025	48389.77	270.65	329.8	97.45	0.001
48050	48415.15	270.65	329.8	97.14	0.001
48075	48440.53	270.65	329.8	96.83	0.001
48100	48465.91	270.65	329.8	96.53	0.001
48125	48491.29	270.65	329.8	96.22	0.001
48150	48516.67	270.65	329.8	95.92	0.001
48175	48542.06	270.65	329.8	95.62	0.001
48200	48567.44	270.65	329.8	95.32	0.001
48225	48592.82	270.65	329.8	95.02	0.001
48250	48618.2	270.65	329.8	94.72	0.001
48275	48643.59	270.65	329.8	94.42	0.001
48300	48668.97	270.65	329.8	94.12	0.001
48325	48694.35	270.65	329.8	93.83	0.001
48350	48719.74	270.65	329.8	93.53	0.001
48375	48745.12	270.65	329.8	93.24	0.001
48400	48770.51	270.65	329.8	92.94	0.001
48425	48795.89	270.65	329.8	92.65	0.001
48450	48821.27	270.65	329.8	92.36	0.001
48475	48846.66	270.65	329.8	92.07	0.001
48500	48872.04	270.65	329.8	91.78	0.001
48525	48897.43	270.65	329.8	91.49	0.001
48550	48922.81	270.65	329.8	91.2	0.001
48575	48948.2	270.65	329.8	90.91	0.001
48600	48973.59	270.65	329.8	90.62	0.001
48625	48998.97	270.65	329.8	90.34	0.001
48650	49024.36	270.65	329.8	90.05	0.001
48675	49049.74	270.65	329.8	89.77	0.001
48700	49075.13	270.65	329.8	89.49	0.001
48725	49100.52	270.65	329.8	89.21	0.001
48750	49125.9	270.65	329.8	88.92	0.001
48775	49151.29	270.65	329.8	88.64	0.001
48800	49176.68	270.65	329.8	88.37	0.001

h (m)	h_G (m)	T (K)	a (m/s)	p (Pa)	ρ (kg/m^3)
49825	50217.73	270.65	329.8	77.64	0.001
48825	49202.07	270.65	329.8	88.09	0.001
48850	49227.45	270.65	329.8	87.81	0.001
48875	49252.84	270.65	329.8	87.53	0.001
48900	49278.23	270.65	329.8	87.26	0.001
48925	49303.62	270.65	329.8	86.98	0.001
48950	49329.01	270.65	329.8	86.71	0.001
48975	49354.4	270.65	329.8	86.43	0.001
49000	49379.78	270.65	329.8	86.16	0.001
49025	49405.17	270.65	329.8	85.89	0.001
49050	49430.56	270.65	329.8	85.62	0.001
49075	49455.95	270.65	329.8	85.35	0.001
49100	49481.34	270.65	329.8	85.08	0.001
49125	49506.73	270.65	329.8	84.81	0.001
49150	49532.12	270.65	329.8	84.55	0.001
49175	49557.51	270.65	329.8	84.28	0.001
49200	49582.9	270.65	329.8	84.01	0.001
49225	49608.29	270.65	329.8	83.75	0.001
49250	49633.69	270.65	329.8	83.49	0.001
49275	49659.08	270.65	329.8	83.22	0.001
49300	49684.47	270.65	329.8	82.96	0.001
49325	49709.86	270.65	329.8	82.7	0.001
49350	49735.25	270.65	329.8	82.44	0.001
49375	49760.64	270.65	329.8	82.18	0.001
49400	49786.04	270.65	329.8	81.92	0.001
49425	49811.43	270.65	329.8	81.66	0.001
49450	49836.82	270.65	329.8	81.4	0.001
49475	49862.21	270.65	329.8	81.15	0.001
49500	49887.61	270.65	329.8	80.89	0.001
49525	49913	270.65	329.8	80.64	0.001
49550	49938.39	270.65	329.8	80.38	0.001
49575	49963.79	270.65	329.8	80.13	0.001
49600	49989.18	270.65	329.8	79.88	0.001
49625	50014.57	270.65	329.8	79.63	0.001
49650	50039.97	270.65	329.8	79.38	0.001
49675	50065.36	270.65	329.8	79.13	0.001
49700	50090.76	270.65	329.8	78.88	0.001
49725	50116.15	270.65	329.8	78.63	0.001
49750	50141.55	270.65	329.8	78.38	0.001
49775	50166.94	270.65	329.8	78.13	0.001
49800	50192.34	270.65	329.8	77.89	0.001
49850	50243.13	270.65	329.8	77.4	0.001

h (m)	h_G (m)	T (K)	a (m/s)	p (Pa)	ρ (kg/m³)
49825	50217.73	270.65	329.8	77.64	0.001
49875	50268.52	270.65	329.8	77.15	0.001
49900	50293.92	270.65	329.8	76.91	0.001
49925	50319.32	270.65	329.8	76.67	0.001
49950	50344.71	270.65	329.8	76.43	0.001
49975	50370.11	270.65	329.8	76.18	0.001
50000	50395.51	270.65	329.8	75.94	0.001

B Isentropic Table ($\gamma = 1.4$)

M	$\frac{p}{p_0}$	$\frac{T}{T_0}$	$\frac{\rho}{\rho_0}$	$\frac{A}{A^*}$	$\frac{a}{a_0}$	M*	μ	ν
0.00	1.0000	1.0000	1.0000	∞	1.0000	0.0000	-	-
0.01	0.9999	1.0000	0.9999	57.8738	1.0000	0.0110	-	-
0.02	0.9997	0.9999	0.9998	28.9421	1.0000	0.0219	-	-
0.03	0.9994	0.9998	0.9996	19.3005	0.9999	0.0329	-	-
0.04	0.9989	0.9997	0.9992	14.4815	0.9998	0.0438	-	-
0.05	0.9983	0.9995	0.9988	11.5914	0.9998	0.0548	-	-
0.06	0.9975	0.9993	0.9982	9.6659	0.9996	0.0657	-	-
0.07	0.9966	0.9990	0.9976	8.2915	0.9995	0.0766	-	-
0.08	0.9955	0.9987	0.9968	7.2616	0.9994	0.0876	-	-
0.09	0.9944	0.9984	0.9960	6.4613	0.9992	0.0985	-	-
0.10	0.9930	0.9980	0.9950	5.8218	0.9990	0.1094	-	-
0.11	0.9916	0.9976	0.9940	5.2992	0.9988	0.1204	-	-
0.12	0.9900	0.9971	0.9928	4.8643	0.9986	0.1313	-	-
0.13	0.9883	0.9966	0.9916	4.4969	0.9983	0.1422	-	-
0.14	0.9864	0.9961	0.9903	4.1824	0.9980	0.1531	-	-
0.15	0.9844	0.9955	0.9888	3.9103	0.9978	0.1639	-	-
0.16	0.9823	0.9949	0.9873	3.6727	0.9974	0.1748	-	-
0.17	0.9800	0.9943	0.9857	3.4635	0.9971	0.1857	-	-
0.18	0.9776	0.9936	0.9840	3.2779	0.9968	0.1965	-	-
0.19	0.9751	0.9928	0.9822	3.1123	0.9964	0.2074	-	-
0.20	0.9725	0.9921	0.9803	2.9635	0.9960	0.2182	-	-
0.21	0.9697	0.9913	0.9783	2.8293	0.9956	0.2290	-	-
0.22	0.9668	0.9904	0.9762	2.7076	0.9952	0.2398	-	-
0.23	0.9638	0.9895	0.9740	2.5968	0.9948	0.2506	-	-
0.24	0.9607	0.9886	0.9718	2.4956	0.9943	0.2614	-	-
0.25	0.9575	0.9877	0.9694	2.4027	0.9938	0.2722	-	-
0.26	0.9541	0.9867	0.9670	2.3173	0.9933	0.2829	-	-
0.27	0.9506	0.9856	0.9645	2.2385	0.9928	0.2936	-	-
0.28	0.9470	0.9846	0.9619	2.1656	0.9923	0.3043	-	-
0.29	0.9433	0.9835	0.9592	2.0979	0.9917	0.3150	-	-
0.30	0.9395	0.9823	0.9564	2.0351	0.9911	0.3257	-	-
0.31	0.9355	0.9811	0.9535	1.9765	0.9905	0.3364	-	-
0.32	0.9315	0.9799	0.9506	1.9219	0.9899	0.3470	-	-
0.33	0.9274	0.9787	0.9476	1.8707	0.9893	0.3576	-	-
0.34	0.9231	0.9774	0.9445	1.8229	0.9886	0.3682	-	-
0.35	0.9188	0.9761	0.9413	1.7780	0.9880	0.3788	-	-
0.36	0.9143	0.9747	0.9380	1.7358	0.9873	0.3893	-	-

DOI: 10.1201/9781003139447-B

M	$\frac{p}{p_0}$	$\frac{T}{T_0}$	$\frac{\rho}{\rho_0}$	$\frac{A}{A^*}$	$\frac{a}{a_0}$	M*	μ	ν
0.37	0.9098	0.9733	0.9347	1.6961	0.9866	0.3999	-	-
0.38	0.9052	0.9719	0.9313	1.6587	0.9859	0.4104	-	-
0.39	0.9004	0.9705	0.9278	1.6234	0.9851	0.4209	-	-
0.40	0.8956	0.9690	0.9243	1.5901	0.9844	0.4313	-	-
0.41	0.8907	0.9675	0.9207	1.5587	0.9836	0.4418	-	-
0.42	0.8857	0.9659	0.9170	1.5289	0.9828	0.4522	-	-
0.43	0.8807	0.9643	0.9132	1.5007	0.9820	0.4626	-	-
0.44	0.8755	0.9627	0.9094	1.4740	0.9812	0.4729	-	-
0.45	0.8703	0.9611	0.9055	1.4487	0.9803	0.4833	-	-
0.46	0.8650	0.9594	0.9016	1.4246	0.9795	0.4936	-	-
0.47	0.8596	0.9577	0.8976	1.4018	0.9786	0.5038	-	-
0.48	0.8541	0.9559	0.8935	1.3801	0.9777	0.5141	-	-
0.49	0.8486	0.9542	0.8894	1.3595	0.9768	0.5243	-	-
0.50	0.8430	0.9524	0.8852	1.3398	0.9759	0.5345	-	-
0.51	0.8374	0.9506	0.8809	1.3212	0.9750	0.5447	-	-
0.52	0.8317	0.9487	0.8766	1.3034	0.9740	0.5548	-	-
0.53	0.8259	0.9468	0.8723	1.2865	0.9730	0.5649	-	-
0.54	0.8201	0.9449	0.8679	1.2703	0.9721	0.5750	-	-
0.55	0.8142	0.9430	0.8634	1.2549	0.9711	0.5851	-	-
0.56	0.8082	0.9410	0.8589	1.2403	0.9700	0.5951	-	-
0.57	0.8022	0.9390	0.8544	1.2263	0.9690	0.6051	-	-
0.58	0.7962	0.9370	0.8498	1.2130	0.9680	0.6150	-	-
0.59	0.7901	0.9349	0.8451	1.2003	0.9669	0.6249	-	-
0.60	0.7840	0.9328	0.8405	1.1882	0.9658	0.6348	-	-
0.61	0.7778	0.9307	0.8357	1.1767	0.9647	0.6447	-	-
0.62	0.7716	0.9286	0.8310	1.1656	0.9636	0.6545	-	-
0.63	0.7654	0.9265	0.8262	1.1552	0.9625	0.6643	-	-
0.64	0.7591	0.9243	0.8213	1.1451	0.9614	0.6740	-	-
0.65	0.7528	0.9221	0.8164	1.1356	0.9603	0.6837	-	-
0.66	0.7465	0.9199	0.8115	1.1265	0.9591	0.6934	-	-
0.67	0.7401	0.9176	0.8066	1.1179	0.9579	0.7031	-	-
0.68	0.7338	0.9153	0.8016	1.1097	0.9567	0.7127	-	-
0.69	0.7274	0.9131	0.7966	1.1018	0.9555	0.7223	-	-
0.70	0.7209	0.9107	0.7916	1.0944	0.9543	0.7318	-	-
0.71	0.7145	0.9084	0.7865	1.0873	0.9531	0.7413	-	-
0.72	0.7080	0.9061	0.7814	1.0806	0.9519	0.7508	-	-
0.73	0.7016	0.9037	0.7763	1.0742	0.9506	0.7602	-	-
0.74	0.6951	0.9013	0.7712	1.0681	0.9494	0.7696	-	-
0.75	0.6886	0.8989	0.7660	1.0624	0.9481	0.7789	-	-
0.76	0.6821	0.8964	0.7609	1.0570	0.9468	0.7883	-	-
0.77	0.6756	0.894	0.7557	1.0519	0.9455	0.7975	-	-
0.78	0.6691	0.8915	0.7505	1.0471	0.9442	0.8068	-	-
0.79	0.6625	0.8890	0.7452	1.0425	0.9429	0.8160	-	-

M	$\frac{p}{p_0}$	$\frac{T}{T_0}$	$\frac{\rho}{\rho_0}$	$\frac{A}{A^*}$	$\frac{a}{a_0}$	M*	μ	ν
0.80	0.6560	0.8865	0.7400	1.0382	0.9416	0.8251	-	-
0.81	0.6495	0.8840	0.7347	1.0342	0.9402	0.8343	-	-
0.82	0.6430	0.8815	0.7295	1.0305	0.9389	0.8433	-	-
0.83	0.6365	0.8789	0.7242	1.0270	0.9375	0.8524	-	-
0.84	0.6300	0.8763	0.7189	1.0237	0.9361	0.8614	-	-
0.85	0.6235	0.8737	0.7136	1.0207	0.9347	0.8704	-	-
0.86	0.6170	0.8711	0.7083	1.0179	0.9333	0.8793	-	-
0.87	0.6106	0.8685	0.7030	1.0153	0.9319	0.8882	-	-
0.88	0.6041	0.8659	0.6977	1.0129	0.9305	0.8970	-	-
0.89	0.5977	0.8632	0.6924	1.0108	0.9291	0.9058	-	-
0.90	0.5913	0.8606	0.6870	1.0089	0.9277	0.9146	-	-
0.91	0.5849	0.8579	0.6817	1.0071	0.9262	0.9233	-	-
0.92	0.5785	0.8552	0.6764	1.0056	0.9248	0.9320	-	-
0.93	0.5721	0.8525	0.6711	1.0043	0.9233	0.9406	-	-
0.94	0.5658	0.8498	0.6658	1.0031	0.9219	0.9493	-	-
0.95	0.5595	0.8471	0.6604	1.0021	0.9204	0.9578	-	-
0.96	0.5532	0.8444	0.6551	1.0014	0.9189	0.9663	-	-
0.97	0.5469	0.8416	0.6498	1.0008	0.9174	0.9748	-	-
0.98	0.5407	0.8389	0.6445	1.0003	0.9159	0.9832	-	-
0.99	0.5345	0.8361	0.6392	1.0001	0.9144	0.9916	-	-
1.00	0.5283	0.8333	0.6339	1.0000	0.9129	1.0000	90.0000	0.0000
1.01	0.5221	0.8306	0.6287	1.0001	0.9113	1.0083	81.9307	0.0447
1.02	0.5160	0.8278	0.6234	1.0003	0.9098	1.0166	78.6351	0.1257
1.03	0.5099	0.8250	0.6181	1.0007	0.9083	1.0248	76.1376	0.2294
1.04	0.5039	0.8222	0.6129	1.0013	0.9067	1.0330	74.0576	0.3510
1.05	0.4979	0.8193	0.6077	1.0020	0.9052	1.0411	72.2472	0.4874
1.06	0.4919	0.8165	0.6024	1.0029	0.9036	1.0492	70.6300	0.6367
1.07	0.4860	0.8137	0.5972	1.0039	0.9020	1.0573	69.1603	0.7973
1.08	0.4800	0.8108	0.5920	1.0051	0.9005	1.0653	67.8084	0.9680
1.09	0.4742	0.8080	0.5869	1.0064	0.8989	1.0733	66.5534	1.1479
1.10	0.4684	0.8052	0.5817	1.0079	0.8973	1.0812	65.3800	1.3362
1.11	0.4626	0.8023	0.5766	1.0095	0.8957	1.0891	64.2767	1.5321
1.12	0.4568	0.7994	0.5714	1.0113	0.8941	1.097	63.2345	1.735
1.13	0.4511	0.7966	0.5663	1.0132	0.8925	1.1048	62.2461	1.9445
1.14	0.4455	0.7937	0.5612	1.0153	0.8909	1.1126	61.3056	2.16
1.15	0.4398	0.7908	0.5562	1.0175	0.8893	1.1203	60.4082	2.381
1.16	0.4343	0.7879	0.5511	1.0198	0.8877	1.128	59.5497	2.6073
1.17	0.4287	0.7851	0.5461	1.0222	0.886	1.1356	58.7267	2.8385
1.18	0.4232	0.7822	0.5411	1.0248	0.8844	1.1432	57.9362	3.0742
1.19	0.4178	0.7793	0.5361	1.0276	0.8828	1.1508	57.1756	3.3142
1.20	0.4124	0.7764	0.5311	1.0304	0.8811	1.1583	56.4427	3.5582
1.21	0.4070	0.7735	0.5262	1.0334	0.8795	1.1658	55.7354	3.8060
1.22	0.4017	0.7706	0.5213	1.0366	0.8778	1.1732	55.0520	4.0572

M	$\frac{p}{p_0}$	$\frac{T}{T_0}$	$\frac{\rho}{\rho_0}$	$\frac{A}{A^*}$	$\frac{a}{a_0}$	M^*	μ	ν
1.23	0.3964	0.7677	0.5164	1.0398	0.8762	1.1806	54.3909	4.3117
1.24	0.3912	0.7648	0.5115	1.0432	0.8745	1.1879	53.7507	4.5694
1.25	0.3861	0.7619	0.5067	1.0468	0.8729	1.1952	53.1301	4.8299
1.26	0.3809	0.7590	0.5019	1.0504	0.8712	1.2025	52.5280	5.0931
1.27	0.3759	0.7561	0.4971	1.0542	0.8695	1.2097	51.9434	5.3590
1.28	0.3708	0.7532	0.4923	1.0581	0.8679	1.2169	51.3752	5.6272
1.29	0.3658	0.7503	0.4876	1.0621	0.8662	1.2240	50.8226	5.8977
1.30	0.3609	0.7474	0.4829	1.0663	0.8645	1.2311	50.2849	6.1703
1.31	0.3560	0.7445	0.4782	1.0706	0.8628	1.2382	49.7612	6.4449
1.32	0.3512	0.7416	0.4736	1.0750	0.8611	1.2452	49.2510	6.7213
1.33	0.3464	0.7387	0.4690	1.0796	0.8595	1.2522	48.7535	6.9995
1.34	0.3417	0.7358	0.4644	1.0842	0.8578	1.2591	48.2682	7.2794
1.35	0.3370	0.7329	0.4598	1.0890	0.8561	1.2660	47.7946	7.5607
1.36	0.3323	0.7300	0.4553	1.0940	0.8544	1.2729	47.3321	7.8435
1.37	0.3277	0.7271	0.4508	1.0990	0.8527	1.2797	46.8803	8.1276
1.38	0.3232	0.7242	0.4463	1.1042	0.8510	1.2864	46.4387	8.4130
1.39	0.3187	0.7213	0.4418	1.1095	0.8493	1.2932	46.0070	8.6995
1.40	0.3142	0.7184	0.4374	1.1149	0.8476	1.2999	45.5847	8.9870
1.41	0.3098	0.7155	0.4330	1.1205	0.8459	1.3065	45.1715	9.2756
1.42	0.3055	0.7126	0.4287	1.1262	0.8442	1.3131	44.7670	9.5650
1.43	0.3012	0.7097	0.4244	1.1320	0.8425	1.3197	44.3709	9.8553
1.44	0.2969	0.7069	0.4201	1.1379	0.8407	1.3262	43.9830	10.1463
1.45	0.2927	0.7040	0.4158	1.1440	0.8390	1.3327	43.6028	10.4381
1.46	0.2886	0.7011	0.4116	1.1501	0.8373	1.3392	43.2302	10.7305
1.47	0.2845	0.6982	0.4074	1.1565	0.8356	1.3456	42.8649	11.0234
1.48	0.2804	0.6954	0.4032	1.1629	0.8339	1.3520	42.5067	11.3169
1.49	0.2764	0.6925	0.3991	1.1695	0.8322	1.3583	42.1552	11.6109
1.50	0.2724	0.6897	0.3950	1.1762	0.8305	1.3646	41.8103	11.9052
1.51	0.2685	0.6868	0.3909	1.1830	0.8287	1.3708	41.4718	12.1999
1.52	0.2646	0.6840	0.3869	1.1899	0.8270	1.3770	41.1395	12.4949
1.53	0.2608	0.6811	0.3829	1.1970	0.8253	1.3832	40.8132	12.7901
1.54	0.2570	0.6783	0.3789	1.2042	0.8236	1.3894	40.4927	13.0856
1.55	0.2533	0.6754	0.3750	1.2116	0.8219	1.3955	40.1778	13.3812
1.56	0.2496	0.6726	0.3710	1.2190	0.8201	1.4015	39.8684	13.6769
1.57	0.2459	0.6698	0.3672	1.2266	0.8184	1.4075	39.5642	13.9728
1.58	0.2423	0.6670	0.3633	1.2344	0.8167	1.4135	39.2653	14.2686
1.59	0.2388	0.6642	0.3595	1.2422	0.8150	1.4195	38.9713	14.5645
1.60	0.2353	0.6614	0.3557	1.2502	0.8133	1.4254	38.6822	14.8603
1.61	0.2318	0.6586	0.3520	1.2584	0.8115	1.4313	38.3978	15.1561
1.62	0.2284	0.6558	0.3483	1.2666	0.8098	1.4371	38.1181	15.4518
1.63	0.2250	0.6530	0.3446	1.2750	0.8081	1.4429	37.8428	15.7473
1.64	0.2217	0.6502	0.3409	1.2836	0.8064	1.4487	37.5719	16.0427
1.65	0.2184	0.6475	0.3373	1.2922	0.8046	1.4544	37.3052	16.3379

M	$\frac{p}{p_0}$	$\frac{T}{T_0}$	$\frac{\rho}{\rho_0}$	$\frac{A}{A^*}$	$\frac{a}{a_0}$	M^*	μ	ν
1.66	0.2152	0.6447	0.3337	1.3010	0.8029	1.4601	37.0427	16.6328
1.67	0.2119	0.6419	0.3302	1.3100	0.8012	1.4657	36.7842	16.9275
1.68	0.2088	0.6392	0.3266	1.3190	0.7995	1.4713	36.5296	17.2220
1.69	0.2057	0.6364	0.3232	1.3283	0.7978	1.4769	36.2789	17.5161
1.70	0.2026	0.6337	0.3197	1.3376	0.7961	1.4825	36.0319	17.8099
1.71	0.1996	0.6310	0.3163	1.3471	0.7943	1.4880	35.7885	18.1033
1.72	0.1966	0.6283	0.3129	1.3567	0.7926	1.4935	35.5488	18.3964
1.73	0.1936	0.6256	0.3095	1.3665	0.7909	1.4989	35.3125	18.6891
1.74	0.1907	0.6229	0.3062	1.3764	0.7892	1.5043	35.0795	18.9813
1.75	0.1878	0.6202	0.3029	1.3865	0.7875	1.5097	34.8499	19.2732
1.76	0.1850	0.6175	0.2996	1.3967	0.7858	1.5150	34.6236	19.5645
1.77	0.1822	0.6148	0.2964	1.4070	0.7841	1.5203	34.4003	19.8554
1.78	0.1794	0.6121	0.2931	1.4175	0.7824	1.5256	34.1802	20.1458
1.79	0.1767	0.6095	0.2900	1.4282	0.7807	1.5308	33.9631	20.4357
1.80	0.1740	0.6068	0.2868	1.4390	0.7790	1.5360	33.7490	20.7250
1.81	0.1714	0.6041	0.2837	1.4499	0.7773	1.5411	33.5378	21.0138
1.82	0.1688	0.6015	0.2806	1.4610	0.7756	1.5463	33.3293	21.3021
1.83	0.1662	0.5989	0.2776	1.4723	0.7739	1.5514	33.1237	21.5898
1.84	0.1637	0.5963	0.2745	1.4836	0.7722	1.5564	32.9207	21.8768
1.85	0.1612	0.5936	0.2715	1.4952	0.7705	1.5614	32.7205	22.1633
1.86	0.1587	0.5910	0.2686	1.5069	0.7688	1.5664	32.5228	22.4491
1.87	0.1563	0.5885	0.2656	1.5187	0.7671	1.5714	32.3276	22.7344
1.88	0.1539	0.5859	0.2627	1.5308	0.7654	1.5763	32.1349	23.0189
1.89	0.1516	0.5833	0.2598	1.5429	0.7637	1.5812	31.9447	23.3029
1.90	0.1492	0.5807	0.2570	1.5553	0.7621	1.5861	31.7569	23.5861
1.91	0.1470	0.5782	0.2542	1.5677	0.7604	1.5909	31.5714	23.8687
1.92	0.1447	0.5756	0.2514	1.5804	0.7587	1.5957	31.3882	24.1506
1.93	0.1425	0.5731	0.2486	1.5932	0.7570	1.6005	31.2072	24.4318
1.94	0.1403	0.5705	0.2459	1.6062	0.7553	1.6052	31.0285	24.7122
1.95	0.1381	0.5680	0.2432	1.6193	0.7537	1.6099	30.8519	24.9920
1.96	0.1360	0.5655	0.2405	1.6326	0.7520	1.6146	30.6774	25.2710
1.97	0.1339	0.5630	0.2378	1.6461	0.7503	1.6192	30.5051	25.5493
1.98	0.1318	0.5605	0.2352	1.6597	0.7487	1.6239	30.3347	25.8269
1.99	0.1298	0.5580	0.2326	1.6735	0.7470	1.6284	30.1664	26.1037
2.00	0.1278	0.5556	0.2300	1.6875	0.7454	1.6330	30.0000	26.3797
2.01	0.1258	0.5531	0.2275	1.7016	0.7437	1.6375	29.8356	26.6550
2.02	0.1239	0.5506	0.2250	1.7160	0.7420	1.6420	29.6730	26.9295
2.03	0.1220	0.5482	0.2225	1.7305	0.7404	1.6465	29.5123	27.2032
2.04	0.1201	0.5458	0.2200	1.7451	0.7388	1.6509	29.3535	27.4762
2.05	0.1182	0.5433	0.2176	1.7600	0.7371	1.6553	29.1964	27.7483
2.06	0.1164	0.5409	0.2152	1.7750	0.7355	1.6597	29.0411	28.0197
2.07	0.1146	0.5385	0.2128	1.7902	0.7338	1.6640	28.8875	28.2903
2.08	0.1128	0.5361	0.2104	1.8056	0.7322	1.6683	28.7357	28.5600

M	$\frac{p}{p_0}$	$\frac{T}{T_0}$	$\frac{\rho}{\rho_0}$	$\frac{A}{A^*}$	$\frac{a}{a_0}$	M^*	μ	ν
2.09	0.1111	0.5337	0.2081	1.8212	0.7306	1.6726	28.5855	28.8289
2.10	0.1094	0.5313	0.2058	1.8369	0.7289	1.6769	28.4369	29.0971
2.11	0.1077	0.5290	0.2035	1.8529	0.7273	1.6811	28.2900	29.3644
2.12	0.1060	0.5266	0.2013	1.8690	0.7257	1.6853	28.1446	29.6308
2.13	0.1043	0.5243	0.1990	1.8853	0.7241	1.6895	28.0008	29.8965
2.14	0.1027	0.5219	0.1968	1.9018	0.7225	1.6936	27.8585	30.1613
2.15	0.1011	0.5196	0.1946	1.9185	0.7208	1.6977	27.7178	30.4252
2.16	0.0996	0.5173	0.1925	1.9354	0.7192	1.7018	27.5785	30.6884
2.17	0.0980	0.5150	0.1903	1.9525	0.7176	1.7059	27.4407	30.9507
2.18	0.0965	0.5127	0.1882	1.9698	0.7160	1.7099	27.3043	31.2121
2.19	0.0950	0.5104	0.1861	1.9873	0.7144	1.7139	27.1693	31.4727
2.20	0.0935	0.5081	0.1841	2.0050	0.7128	1.7179	27.0357	31.7325
2.21	0.0921	0.5059	0.1820	2.0229	0.7112	1.7219	26.9035	31.9914
2.22	0.0906	0.5036	0.1800	2.0409	0.7097	1.7258	26.7726	32.2494
2.23	0.0892	0.5014	0.1780	2.0592	0.7081	1.7297	26.6430	32.5066
2.24	0.0878	0.4991	0.1760	2.0777	0.7065	1.7336	26.5148	32.7629
2.25	0.0865	0.4969	0.1740	2.0964	0.7049	1.7374	26.3878	33.0184
2.26	0.0851	0.4947	0.1721	2.1153	0.7033	1.7412	26.2621	33.2730
2.27	0.0838	0.4925	0.1702	2.1345	0.7018	1.7450	26.1376	33.5267
2.28	0.0825	0.4903	0.1683	2.1538	0.7002	1.7488	26.0144	33.7796
2.29	0.0812	0.4881	0.1664	2.1734	0.6986	1.7526	25.8923	34.0316
2.30	0.0800	0.4859	0.1646	2.1931	0.6971	1.7563	25.7715	34.2828
2.31	0.0787	0.4837	0.1628	2.2131	0.6955	1.7600	25.6518	34.5330
2.32	0.0775	0.4816	0.1609	2.2333	0.6940	1.7637	25.5332	34.7825
2.33	0.0763	0.4794	0.1592	2.2537	0.6924	1.7673	25.4158	35.0310
2.34	0.0751	0.4773	0.1574	2.2744	0.6909	1.7709	25.2995	35.2787
2.35	0.0740	0.4752	0.1556	2.2953	0.6893	1.7745	25.1844	35.5255
2.36	0.0728	0.4731	0.1539	2.3164	0.6878	1.7781	25.0702	35.7714
2.37	0.0717	0.4709	0.1522	2.3377	0.6863	1.7817	24.9572	36.0165
2.38	0.0706	0.4688	0.1505	2.3593	0.6847	1.7852	24.8452	36.2607
2.39	0.0695	0.4668	0.1488	2.3811	0.6832	1.7887	24.7343	36.5040
2.40	0.0684	0.4647	0.1472	2.4031	0.6817	1.7922	24.6243	36.7465
2.41	0.0673	0.4626	0.1456	2.4254	0.6802	1.7956	24.5154	36.9881
2.42	0.0663	0.4606	0.1439	2.4479	0.6786	1.7991	24.4075	37.2288
2.43	0.0653	0.4585	0.1424	2.4706	0.6771	1.8025	24.3005	37.4687
2.44	0.0643	0.4565	0.1408	2.4936	0.6756	1.8059	24.1946	37.7077
2.45	0.0633	0.4544	0.1392	2.5168	0.6741	1.8092	24.0895	37.9458
2.46	0.0623	0.4524	0.1377	2.5403	0.6726	1.8126	23.9854	38.1831
2.47	0.0613	0.4504	0.1362	2.5640	0.6711	1.8159	23.8823	38.4195
2.48	0.0604	0.4484	0.1346	2.5880	0.6696	1.8192	23.7800	38.6550
2.49	0.0594	0.4464	0.1332	2.6122	0.6682	1.8225	23.6787	38.8897
2.50	0.0585	0.4444	0.1317	2.6367	0.6667	1.8257	23.5782	39.1235
2.51	0.0576	0.4425	0.1302	2.6615	0.6652	1.8290	23.4786	39.3565

M	$\frac{p}{p_0}$	$\frac{T}{T_0}$	$\frac{\rho}{\rho_0}$	$\frac{A}{A^*}$	$\frac{a}{a_0}$	M^*	μ	ν
2.52	0.0567	0.4405	0.1288	2.6864	0.6637	1.8322	23.3799	39.5886
2.53	0.0559	0.4386	0.1274	2.7117	0.6622	1.8354	23.2820	39.8198
2.54	0.0550	0.4366	0.1260	2.7372	0.6608	1.8386	23.1850	40.0502
2.55	0.0542	0.4347	0.1246	2.7630	0.6593	1.8417	23.0888	40.2798
2.56	0.0533	0.4328	0.1232	2.7891	0.6578	1.8448	22.9934	40.5084
2.57	0.0525	0.4309	0.1218	2.8154	0.6564	1.8479	22.8988	40.7363
2.58	0.0517	0.4289	0.1205	2.8420	0.6549	1.8510	22.8051	40.9633
2.59	0.0509	0.4271	0.1192	2.8688	0.6535	1.8541	22.7121	41.1894
2.60	0.0501	0.4252	0.1179	2.8960	0.6521	1.8571	22.6199	41.4147
2.61	0.0493	0.4233	0.1166	2.9234	0.6506	1.8602	22.5284	41.6391
2.62	0.0486	0.4214	0.1153	2.9511	0.6492	1.8632	22.4378	41.8627
2.63	0.0478	0.4196	0.1140	2.9791	0.6477	1.8662	22.3478	42.0855
2.64	0.0471	0.4177	0.1128	3.0073	0.6463	1.8691	22.2586	42.3074
2.65	0.0464	0.4159	0.1115	3.0359	0.6449	1.8721	22.1702	42.5285
2.66	0.0457	0.4141	0.1103	3.0647	0.6435	1.8750	22.0824	42.7487
2.67	0.0450	0.4122	0.1091	3.0938	0.6421	1.8779	21.9954	42.9682
2.68	0.0443	0.4104	0.1079	3.1233	0.6406	1.8808	21.9091	43.1867
2.69	0.0436	0.4086	0.1067	3.1530	0.6392	1.8837	21.8234	43.4045
2.70	0.0430	0.4068	0.1056	3.1830	0.6378	1.8865	21.7385	43.6214
2.71	0.0423	0.4051	0.1044	3.2133	0.6364	1.8894	21.6542	43.8376
2.72	0.0417	0.4033	0.1033	3.2439	0.6350	1.8922	21.5706	44.0528
2.73	0.0410	0.4015	0.1022	3.2749	0.6337	1.8950	21.4876	44.2673
2.74	0.0404	0.3998	0.1010	3.3061	0.6323	1.8978	21.4053	44.4810
2.75	0.0398	0.3980	0.0999	3.3377	0.6309	1.9005	21.3237	44.6938
2.76	0.0392	0.3963	0.0989	3.3695	0.6295	1.9033	21.2427	44.9058
2.77	0.0386	0.3945	0.0978	3.4017	0.6281	1.9060	21.1623	45.1170
2.78	0.0380	0.3928	0.0967	3.4342	0.6268	1.9087	21.0825	45.3275
2.79	0.0374	0.3911	0.0957	3.4670	0.6254	1.9114	21.0034	45.5371
2.80	0.0368	0.3894	0.0946	3.5001	0.6240	1.9140	20.9248	45.7459
2.81	0.0363	0.3877	0.0936	3.5336	0.6227	1.9167	20.8469	45.9539
2.82	0.0357	0.3860	0.0926	3.5674	0.6213	1.9193	20.7696	46.1611
2.83	0.0352	0.3844	0.0916	3.6015	0.6200	1.9219	20.6928	46.3675
2.84	0.0347	0.3827	0.0906	3.6359	0.6186	1.9246	20.6166	46.5731
2.85	0.0341	0.3810	0.0896	3.6707	0.6173	1.9271	20.5410	46.7779
2.86	0.0336	0.3794	0.0886	3.7058	0.6159	1.9297	20.4659	46.9819
2.87	0.0331	0.3777	0.0877	3.7413	0.6146	1.9323	20.3915	47.1852
2.88	0.0326	0.3761	0.0867	3.7771	0.6133	1.9348	20.3175	47.3877
2.89	0.0321	0.3745	0.0858	3.8133	0.6119	1.9373	20.2441	47.5894
2.90	0.0317	0.3729	0.0849	3.8498	0.6106	1.9398	20.1713	47.7903
2.91	0.0312	0.3712	0.084	3.8866	0.6093	1.9423	20.0990	47.9904
2.92	0.0307	0.3696	0.0831	3.9238	0.6080	1.9448	20.0272	48.1898
2.93	0.0302	0.3681	0.0822	3.9614	0.6067	1.9472	19.9559	48.3884
2.94	0.0298	0.3665	0.0813	3.9993	0.6054	1.9497	19.8852	48.5862

M	$\frac{p}{p_0}$	$\frac{T}{T_0}$	$\frac{\rho}{\rho_0}$	$\frac{A}{A^*}$	$\frac{a}{a_0}$	M^*	μ	ν
2.95	0.0293	0.3649	0.0804	4.0376	0.6041	1.9521	19.8149	48.7833
2.96	0.0289	0.3633	0.0796	4.0762	0.6028	1.9545	19.7452	48.9796
2.97	0.0285	0.3618	0.0787	4.1153	0.6015	1.9569	19.6760	49.1752
2.98	0.0281	0.3602	0.0779	4.1547	0.6002	1.9593	19.6072	49.3700
2.99	0.0276	0.3587	0.0770	4.1944	0.5989	1.9616	19.5390	49.5640
3.00	0.0272	0.3571	0.0762	4.2346	0.5976	1.9640	19.4712	49.7573
3.01	0.0268	0.3556	0.0754	4.2751	0.5963	1.9663	19.4039	49.9499
3.02	0.0264	0.3541	0.0746	4.3160	0.5951	1.9686	19.3371	50.1417
3.03	0.0260	0.3526	0.0738	4.3573	0.5938	1.9709	19.2708	50.3327
3.04	0.0256	0.3511	0.0730	4.3989	0.5925	1.9732	19.2049	50.5231
3.05	0.0253	0.3496	0.0723	4.4410	0.5913	1.9755	19.1395	50.7127
3.06	0.0249	0.3481	0.0715	4.4835	0.5900	1.9777	19.0745	50.9015
3.07	0.0245	0.3466	0.0707	4.5263	0.5887	1.9800	19.0100	51.0897
3.08	0.0242	0.3452	0.0700	4.5696	0.5875	1.9822	18.9459	51.2771
3.09	0.0238	0.3437	0.0692	4.6132	0.5862	1.9844	18.8823	51.4637
3.10	0.0234	0.3422	0.0685	4.6573	0.5850	1.9866	18.8191	51.6497
3.11	0.0231	0.3408	0.0678	4.7018	0.5838	1.9888	18.7563	51.8349
3.12	0.0228	0.3393	0.0671	4.7467	0.5825	1.9910	18.6940	52.0195
3.13	0.0224	0.3379	0.0664	4.7920	0.5813	1.9931	18.6320	52.2033
3.14	0.0221	0.3365	0.0657	4.8377	0.5801	1.9953	18.5705	52.3864
3.15	0.0218	0.3351	0.0650	4.8838	0.5788	1.9974	18.5094	52.5688
3.16	0.0215	0.3337	0.0643	4.9304	0.5776	1.9995	18.4487	52.7505
3.17	0.0211	0.3323	0.0636	4.9774	0.5764	2.0016	18.3884	52.9315
3.18	0.0208	0.3309	0.0630	5.0248	0.5752	2.0037	18.3286	53.1118
3.19	0.0205	0.3295	0.0623	5.0727	0.5740	2.0058	18.2691	53.2914
3.20	0.0202	0.3281	0.0617	5.1209	0.5728	2.0079	18.2100	53.4703
3.21	0.0199	0.3267	0.0610	5.1697	0.5716	2.0099	18.1513	53.6485
3.22	0.0196	0.3253	0.0604	5.2189	0.5704	2.0119	18.0929	53.8260
3.23	0.0194	0.3240	0.0597	5.2685	0.5692	2.0140	18.0350	54.0029
3.24	0.0191	0.3226	0.0591	5.3186	0.5680	2.0160	17.9774	54.1791
3.25	0.0188	0.3213	0.0585	5.3691	0.5668	2.0180	17.9202	54.3546
3.26	0.0185	0.3199	0.0579	5.4201	0.5656	2.0200	17.8634	54.5294
3.27	0.0183	0.3186	0.0573	5.4715	0.5645	2.0220	17.8069	54.7035
3.28	0.0180	0.3173	0.0567	5.5234	0.5633	2.0239	17.7508	54.8770
3.29	0.0177	0.3160	0.0561	5.5758	0.5621	2.0259	17.6951	55.0498
3.30	0.0175	0.3147	0.0555	5.6286	0.5609	2.0278	17.6397	55.2219
3.31	0.0172	0.3134	0.0550	5.6820	0.5598	2.0297	17.5847	55.3934
3.32	0.0170	0.3121	0.0544	5.7357	0.5586	2.0317	17.5300	55.5642
3.33	0.0167	0.3108	0.0538	5.7900	0.5575	2.0336	17.4757	55.7344
3.34	0.0165	0.3095	0.0533	5.8448	0.5563	2.0355	17.4217	55.9039
3.35	0.0163	0.3082	0.0527	5.9000	0.5552	2.0373	17.3680	56.0728
3.36	0.0160	0.3069	0.0522	5.9558	0.554	2.0392	17.3147	56.2410
3.37	0.0158	0.3057	0.0517	6.0120	0.5529	2.0411	17.2617	56.4086

M	$\frac{p}{p_0}$	$\frac{T}{T_0}$	$\frac{\rho}{\rho_0}$	$\frac{A}{A^*}$	$\frac{a}{a_0}$	M^*	μ	ν
3.38	0.0156	0.3044	0.0511	6.0687	0.5517	2.0429	17.2090	56.5755
3.39	0.0153	0.3032	0.0506	6.1260	0.5506	2.0447	17.1567	56.7418
3.40	0.0151	0.3019	0.0501	6.1837	0.5495	2.0466	17.1046	56.9075
3.41	0.0149	0.3007	0.0496	6.2419	0.5484	2.0484	17.0529	57.0725
3.42	0.0147	0.2995	0.0491	6.3007	0.5472	2.0502	17.0016	57.2369
3.43	0.0145	0.2982	0.0486	6.3600	0.5461	2.0520	16.9505	57.4007
3.44	0.0143	0.2970	0.0481	6.4197	0.5450	2.0537	16.8997	57.5638
3.45	0.0141	0.2958	0.0476	6.4801	0.5439	2.0555	16.8493	57.7263
3.46	0.0139	0.2946	0.0471	6.5409	0.5428	2.0573	16.7991	57.8882
3.47	0.0137	0.2934	0.0466	6.6023	0.5417	2.059	16.7493	58.0495
3.48	0.0135	0.2922	0.0462	6.6642	0.5406	2.0607	16.6998	58.2102
3.49	0.0133	0.2910	0.0457	6.7266	0.5395	2.0625	16.6505	58.3703
3.5	0.0131	0.2899	0.0452	6.7896	0.5384	2.0642	16.6016	58.5297
3.51	0.0129	0.2887	0.0448	6.8531	0.5373	2.0659	16.5529	58.6886
3.52	0.0127	0.2875	0.0443	6.9172	0.5362	2.0676	16.5045	58.8468
3.53	0.0126	0.2864	0.0439	6.9819	0.5351	2.0693	16.4564	59.0045
3.54	0.0124	0.2852	0.0434	7.0470	0.5340	2.0709	16.4086	59.1615
3.55	0.0122	0.2841	0.0430	7.1128	0.5330	2.0726	16.3611	59.318
3.56	0.0120	0.2829	0.0426	7.1791	0.5319	2.0743	16.3139	59.4738
3.57	0.0119	0.2818	0.0421	7.2460	0.5308	2.0759	16.2669	59.6291
3.58	0.0117	0.2806	0.0417	7.3134	0.5298	2.0775	16.2202	59.7838
3.59	0.0115	0.2795	0.0413	7.3815	0.5287	2.0792	16.1738	59.9379
3.60	0.0114	0.2784	0.0409	7.4501	0.5276	2.0808	16.1276	60.0914
3.61	0.0112	0.2773	0.0405	7.5193	0.5266	2.0824	16.0817	60.2444
3.62	0.0111	0.2762	0.0401	7.5891	0.5255	2.0840	16.0361	60.3967
3.63	0.0109	0.2751	0.0397	7.6595	0.5245	2.0856	15.9908	60.5485
3.64	0.0108	0.2740	0.0393	7.7304	0.5234	2.0871	15.9457	60.6997
3.65	0.0106	0.2729	0.0389	7.8020	0.5224	2.0887	15.9008	60.8504
3.66	0.0105	0.2718	0.0385	7.8742	0.5213	2.0903	15.8562	61.0005
3.67	0.0103	0.2707	0.0381	7.9470	0.5203	2.0918	15.8119	61.1500
3.68	0.0102	0.2697	0.0378	8.0204	0.5193	2.0933	15.7678	61.2990
3.69	0.0100	0.2686	0.0374	8.0944	0.5183	2.0949	15.7240	61.4474
3.70	0.0099	0.2675	0.0370	8.1690	0.5172	2.0964	15.6804	61.5952
3.71	0.0098	0.2665	0.0367	8.2443	0.5162	2.0979	15.6370	61.7425
3.72	0.0096	0.2654	0.0363	8.3202	0.5152	2.0994	15.5939	61.8893
3.73	0.0095	0.2644	0.0359	8.3967	0.5142	2.1009	15.5510	62.0355
3.74	0.0094	0.2633	0.0356	8.4739	0.5132	2.1024	15.5084	62.1811
3.75	0.0092	0.2623	0.0352	8.5517	0.5121	2.1039	15.4660	62.3263
3.76	0.0091	0.2613	0.0349	8.6302	0.5111	2.1053	15.4239	62.4708
3.77	0.0090	0.2602	0.0345	8.7093	0.5101	2.1068	15.3819	62.6149
3.78	0.0089	0.2592	0.0342	8.7891	0.5091	2.1082	15.3402	62.7584
3.79	0.0087	0.2582	0.0339	8.8695	0.5081	2.1097	15.2988	62.9013
3.80	0.0086	0.2572	0.0335	8.9506	0.5072	2.1111	15.2575	63.0438

M	$\frac{p}{p_0}$	$\frac{T}{T_0}$	$\frac{\rho}{\rho_0}$	$\frac{A}{A^*}$	$\frac{a}{a_0}$	M^*	μ	ν
3.81	0.0085	0.2562	0.0332	9.0323	0.5062	2.1125	15.2165	63.1857
3.82	0.0084	0.2552	0.0329	9.1148	0.5052	2.1140	15.1757	63.3271
3.83	0.0083	0.2542	0.0326	9.1979	0.5042	2.1154	15.1352	63.4679
3.84	0.0082	0.2532	0.0323	9.2816	0.5032	2.1168	15.0948	63.6082
3.85	0.0081	0.2522	0.0320	9.3661	0.5022	2.1182	15.0547	63.7481
3.86	0.0080	0.2513	0.0316	9.4513	0.5013	2.1195	15.0147	63.8874
3.87	0.0078	0.2503	0.0313	9.5371	0.5003	2.1209	14.9750	64.0262
3.88	0.0077	0.2493	0.0310	9.6237	0.4993	2.1223	14.9355	64.1644
3.89	0.0076	0.2484	0.0307	9.7110	0.4984	2.1236	14.8963	64.3022
3.90	0.0075	0.2474	0.0304	9.7989	0.4974	2.1250	14.8572	64.4395
3.91	0.0074	0.2465	0.0302	9.8876	0.4964	2.1263	14.8183	64.5762
3.92	0.0073	0.2455	0.0299	9.9770	0.4955	2.1277	14.7796	64.7125
3.93	0.0072	0.2446	0.0296	10.0672	0.4945	2.1290	14.7412	64.8483
3.94	0.0071	0.2436	0.0293	10.1580	0.4936	2.1303	14.7029	64.9835
3.95	0.0070	0.2427	0.0290	10.2496	0.4926	2.1316	14.6649	65.1183
3.96	0.0069	0.2418	0.0287	10.3419	0.4917	2.1329	14.6270	65.2526
3.97	0.0069	0.2408	0.0285	10.4350	0.4908	2.1342	14.5893	65.3863
3.98	0.0068	0.2399	0.0282	10.5288	0.4898	2.1355	14.5519	65.5196
3.99	0.0067	0.2390	0.0279	10.6234	0.4889	2.1368	14.5146	65.6525
4.00	0.0066	0.2381	0.0277	10.7187	0.4880	2.1381	14.4775	65.7848
4.01	0.0065	0.2372	0.0274	10.8148	0.4870	2.1394	14.4406	65.9166
4.02	0.0064	0.2363	0.0271	10.9117	0.4861	2.1406	14.4039	66.0480
4.03	0.0063	0.2354	0.0269	11.0093	0.4852	2.1419	14.3674	66.1789
4.04	0.0062	0.2345	0.0266	11.1077	0.4843	2.1431	14.3311	66.3093
4.05	0.0062	0.2336	0.0264	11.2069	0.4833	2.1444	14.2950	66.4393
4.06	0.0061	0.2327	0.0261	11.3068	0.4824	2.1456	14.2590	66.5687
4.07	0.0060	0.2319	0.0259	11.4076	0.4815	2.1468	14.2232	66.6978
4.08	0.0059	0.2310	0.0256	11.5091	0.4806	2.1480	14.1876	66.8263
4.09	0.0058	0.2301	0.0254	11.6115	0.4797	2.1493	14.1522	66.9544
4.10	0.0058	0.2293	0.0252	11.7146	0.4788	2.1505	14.1170	67.0820
4.11	0.0057	0.2284	0.0249	11.8186	0.4779	2.1517	14.0819	67.2092
4.12	0.0056	0.2275	0.0247	11.9234	0.4770	2.1529	14.0470	67.3359
4.13	0.0055	0.2267	0.0245	12.0290	0.4761	2.1540	14.0123	67.4621
4.14	0.0055	0.2258	0.0242	12.1354	0.4752	2.1552	13.9778	67.5879
4.15	0.0054	0.2250	0.0240	12.2427	0.4743	2.1564	13.9434	67.7132
4.16	0.0053	0.2242	0.0238	12.3508	0.4735	2.1576	13.9092	67.8381
4.17	0.0053	0.2233	0.0236	12.4597	0.4726	2.1587	13.8752	67.9626
4.18	0.0052	0.2225	0.0234	12.5695	0.4717	2.1599	13.8414	68.0866
4.19	0.0051	0.2217	0.0231	12.6802	0.4708	2.1610	13.8077	68.2101
4.20	0.0051	0.2208	0.0229	12.7917	0.4699	2.1622	13.7741	68.3333
4.21	0.0050	0.2200	0.0227	12.9040	0.4691	2.1633	13.7408	68.4559
4.22	0.0049	0.2192	0.0225	13.0172	0.4682	2.1644	13.7076	68.5782
4.23	0.0049	0.2184	0.0223	13.1313	0.4673	2.1655	13.6745	68.7000

M	$\frac{p}{p_0}$	$\frac{T}{T_0}$	$\frac{\rho}{\rho_0}$	$\frac{A}{A^*}$	$\frac{a}{a_0}$	M*	μ	ν
4.24	0.0048	0.2176	0.0221	13.2463	0.4665	2.1667	13.6417	68.8214
4.25	0.0047	0.2168	0.0219	13.3622	0.4656	2.1678	13.6090	68.9423
4.26	0.0047	0.2160	0.0217	13.4789	0.4648	2.1689	13.5764	69.0628
4.27	0.0046	0.2152	0.0215	13.5966	0.4639	2.1700	13.5440	69.1829
4.28	0.0046	0.2144	0.0213	13.7151	0.4631	2.1711	13.5117	69.3026
4.29	0.0045	0.2136	0.0211	13.8346	0.4622	2.1721	13.4797	69.4219
4.30	0.0044	0.2129	0.0209	13.9550	0.4614	2.1732	13.4477	69.5407
4.31	0.0044	0.2121	0.0207	14.0762	0.4605	2.1743	13.4159	69.6591
4.32	0.0043	0.2113	0.0205	14.1984	0.4597	2.1754	13.3843	69.7771
4.33	0.0043	0.2105	0.0203	14.3216	0.4588	2.1764	13.3528	69.8947
4.34	0.0042	0.2098	0.0202	14.4456	0.458	2.1775	13.3215	70.0118
4.35	0.0042	0.2090	0.0200	14.5706	0.4572	2.1785	13.2903	70.1286
4.36	0.0041	0.2082	0.0198	14.6966	0.4563	2.1796	13.2592	70.2449
4.37	0.0041	0.2075	0.0196	14.8235	0.4555	2.1806	13.2284	70.3608
4.38	0.0040	0.2067	0.0194	14.9513	0.4547	2.1816	13.1976	70.4764
4.39	0.0040	0.2060	0.0193	15.0802	0.4539	2.1827	13.1670	70.5915
4.40	0.0039	0.2053	0.0191	15.2100	0.4530	2.1837	13.1365	70.7062
4.41	0.0039	0.2045	0.0189	15.3407	0.4522	2.1847	13.1062	70.8206
4.42	0.0038	0.2038	0.0187	15.4724	0.4514	2.1857	13.076	70.9345
4.43	0.0038	0.2030	0.0186	15.6052	0.4506	2.1867	13.0460	71.0480
4.44	0.0037	0.2023	0.0184	15.7389	0.4498	2.1877	13.0161	71.1612
4.45	0.0037	0.2016	0.0182	15.8736	0.4490	2.1887	12.9863	71.2739
4.46	0.0036	0.2009	0.0181	16.0093	0.4482	2.1897	12.9567	71.3863
4.47	0.0036	0.2002	0.0179	16.1460	0.4474	2.1907	12.9272	71.4982
4.48	0.0035	0.1994	0.0178	16.2838	0.4466	2.1917	12.8979	71.6098
4.49	0.0035	0.1987	0.0176	16.4225	0.4458	2.1926	12.8687	71.7210
4.50	0.0035	0.1980	0.0174	16.5623	0.4450	2.1936	12.8396	71.8318
4.51	0.0034	0.1973	0.0173	16.7031	0.4442	2.1946	12.8106	71.9423
4.52	0.0034	0.1966	0.0171	16.8450	0.4434	2.1955	12.7818	72.0523
4.53	0.0033	0.1959	0.0170	16.9879	0.4426	2.1965	12.7531	72.1620
4.54	0.0033	0.1952	0.0168	17.1318	0.4418	2.1974	12.7245	72.2713
4.55	0.0032	0.1945	0.0167	17.2768	0.4411	2.1984	12.6961	72.3802
4.56	0.0032	0.1938	0.0165	17.4229	0.4403	2.1993	12.6678	72.4888
4.57	0.0032	0.1932	0.0164	17.5701	0.4395	2.2002	12.6396	72.5969
4.58	0.0031	0.1925	0.0163	17.7183	0.4387	2.2012	12.6116	72.7048
4.59	0.0031	0.1918	0.0161	17.8676	0.4380	2.2021	12.5836	72.8122
4.60	0.0031	0.1911	0.0160	18.0180	0.4372	2.2030	12.5558	72.9193
4.61	0.0030	0.1905	0.0158	18.1694	0.4364	2.2039	12.5281	73.0260
4.62	0.0030	0.1898	0.0157	18.3220	0.4357	2.2048	12.5006	73.1323
4.63	0.0029	0.1891	0.0156	18.4757	0.4349	2.2057	12.4732	73.2383
4.64	0.0029	0.1885	0.0154	18.6305	0.4341	2.2066	12.4458	73.3439
4.65	0.0029	0.1878	0.0153	18.7864	0.4334	2.2075	12.4186	73.4492
4.66	0.0028	0.1872	0.0152	18.9435	0.4326	2.2084	12.3916	73.5541

M	$\frac{p}{p_0}$	$\frac{T}{T_0}$	$\frac{\rho}{\rho_0}$	$\frac{A}{A^*}$	$\frac{a}{a_0}$	M*	μ	ν
4.67	0.0028	0.1865	0.0150	19.1017	0.4319	2.2093	12.3646	73.6587
4.68	0.0028	0.1859	0.0149	19.2610	0.4311	2.2102	12.3378	73.7629
4.69	0.0027	0.1852	0.0148	19.4214	0.4304	2.2110	12.3111	73.8668
4.70	0.0027	0.1846	0.0146	19.583	0.4296	2.2119	12.2845	73.9703
4.71	0.0027	0.1839	0.0145	19.7458	0.4289	2.2128	12.2580	74.0734
4.72	0.0026	0.1833	0.0144	19.9098	0.4281	2.2136	12.2316	74.1762
4.73	0.0026	0.1827	0.0143	20.0749	0.4274	2.2145	12.2053	74.2787
4.74	0.0026	0.1820	0.0141	20.2411	0.4267	2.2154	12.1792	74.3808
4.75	0.0025	0.1814	0.0140	20.4086	0.4259	2.2162	12.1532	74.4826
4.76	0.0025	0.1808	0.0139	20.5773	0.4252	2.2171	12.1272	74.584
4.77	0.0025	0.1802	0.0138	20.7471	0.4245	2.2179	12.1014	74.6851
4.78	0.0025	0.1795	0.0137	20.9182	0.4237	2.2187	12.0757	74.7859
4.79	0.0024	0.1789	0.0135	21.0905	0.4230	2.2196	12.0501	74.8863
4.80	0.0024	0.1783	0.0134	21.264	0.4223	2.2204	12.0247	74.9864
4.81	0.0024	0.1777	0.0133	21.4387	0.4216	2.2212	11.9993	75.0862
4.82	0.0023	0.1771	0.0132	21.6147	0.4208	2.2220	11.9740	75.1856
4.83	0.0023	0.1765	0.0131	21.7918	0.4201	2.2228	11.9489	75.2847
4.84	0.0023	0.1759	0.0130	21.9703	0.4194	2.2236	11.9238	75.3835
4.85	0.0023	0.1753	0.0129	22.1500	0.4187	2.2245	11.8989	75.4819
4.86	0.0022	0.1747	0.0128	22.3309	0.4180	2.2253	11.8740	75.5801
4.87	0.0022	0.1741	0.0126	22.5131	0.4173	2.2261	11.8493	75.6779
4.88	0.0022	0.1735	0.0125	22.6966	0.4166	2.2268	11.8247	75.7753
4.89	0.0022	0.1729	0.0124	22.8814	0.4159	2.2276	11.8001	75.8725
4.90	0.0021	0.1724	0.0123	23.0675	0.4152	2.2284	11.7757	75.9693
4.91	0.0021	0.1718	0.0122	23.2548	0.4145	2.2292	11.7514	76.0658
4.92	0.0021	0.1712	0.0121	23.4435	0.4138	2.2300	11.7271	76.1620
4.93	0.0021	0.1706	0.0120	23.6334	0.4131	2.2308	11.7030	76.2579
4.94	0.0020	0.1700	0.0119	23.8247	0.4124	2.2315	11.6790	76.3535
4.95	0.0020	0.1695	0.0118	24.0173	0.4117	2.2323	11.6551	76.4487
4.96	0.0020	0.1689	0.0117	24.2112	0.4110	2.2331	11.6312	76.5437
4.97	0.0020	0.1683	0.0116	24.4065	0.4103	2.2338	11.6075	76.6383
4.98	0.0019	0.1678	0.0115	24.6031	0.4096	2.2346	11.5839	76.7327
4.99	0.0019	0.1672	0.0114	24.8011	0.4089	2.2353	11.5604	76.8267
5.00	0.0019	0.1667	0.0113	25.0004	0.4082	2.2361	11.5369	76.9204

C Normal Shock Table ($\gamma = 1.4$)

M_1	M_2	$\frac{p_2}{p_1}$	$\frac{\rho_2}{\rho_1}$	$\frac{T_2}{T_1}$	$\frac{a_2}{a_1}$	$\frac{p_{02}}{p_{01}}$
1.01	0.9901	1.0234	1.0167	1.0066	1.0033	1.0000
1.02	0.9805	1.0471	1.0334	1.0132	1.0066	1.0000
1.03	0.9712	1.0710	1.0502	1.0198	1.0099	1.0000
1.04	0.9620	1.0952	1.0671	1.0263	1.0131	0.9999
1.05	0.9531	1.1196	1.0840	1.0328	1.0163	0.9999
1.06	0.9444	1.1442	1.1009	1.0393	1.0195	0.9998
1.07	0.9360	1.1690	1.1179	1.0458	1.0226	0.9996
1.08	0.9277	1.1941	1.1349	1.0522	1.0258	0.9994
1.09	0.9196	1.2194	1.1520	1.0586	1.0289	0.9992
1.10	0.9118	1.2450	1.1691	1.0649	1.0320	0.9989
1.11	0.9041	1.2708	1.1862	1.0713	1.0350	0.9986
1.12	0.8966	1.2968	1.2034	1.0776	1.0381	0.9982
1.13	0.8892	1.3230	1.2206	1.0840	1.0411	0.9978
1.14	0.8820	1.3495	1.2378	1.0903	1.0442	0.9973
1.15	0.8750	1.3762	1.2550	1.0966	1.0472	0.9967
1.16	0.8682	1.4032	1.2723	1.1029	1.0502	0.9961
1.17	0.8615	1.4304	1.2896	1.1092	1.0532	0.9953
1.18	0.8549	1.4578	1.3069	1.1154	1.0561	0.9946
1.19	0.8485	1.4854	1.3243	1.1217	1.0591	0.9937
1.20	0.8422	1.5133	1.3416	1.1280	1.0621	0.9928
1.21	0.8360	1.5414	1.3590	1.1343	1.0650	0.9918
1.22	0.8300	1.5698	1.3764	1.1405	1.0680	0.9907
1.23	0.8241	1.5984	1.3938	1.1468	1.0709	0.9896
1.24	0.8183	1.6272	1.4112	1.1531	1.0738	0.9884
1.25	0.8126	1.6562	1.4286	1.1594	1.0767	0.9871
1.26	0.8071	1.6855	1.4460	1.1657	1.0797	0.9857
1.27	0.8016	1.7150	1.4634	1.1720	1.0826	0.9842
1.28	0.7963	1.7448	1.4808	1.1783	1.0855	0.9827
1.29	0.7911	1.7748	1.4983	1.1846	1.0884	0.9811
1.30	0.7860	1.8050	1.5157	1.1909	1.0913	0.9794
1.31	0.7809	1.8354	1.5331	1.1972	1.0942	0.9776
1.32	0.7760	1.8661	1.5505	1.2035	1.0971	0.9758
1.33	0.7712	1.8970	1.5680	1.2099	1.0999	0.9738
1.34	0.7664	1.9282	1.5854	1.2162	1.1028	0.9718

DOI: 10.1201/9781003139447-C

M_1	M_2	$\frac{p_2}{p_1}$	$\frac{\rho_2}{\rho_1}$	$\frac{T_2}{T_1}$	$\frac{a_2}{a_1}$	$\frac{p_{02}}{p_{01}}$
1.35	0.7618	1.9596	1.6028	1.2226	1.1057	0.9697
1.36	0.7572	1.9912	1.6202	1.2290	1.1086	0.9676
1.37	0.7527	2.0230	1.6376	1.2354	1.1115	0.9653
1.38	0.7483	2.0551	1.6549	1.2418	1.1144	0.9630
1.39	0.7440	2.0874	1.6723	1.2482	1.1172	0.9607
1.40	0.7397	2.1200	1.6897	1.2547	1.1201	0.9582
1.41	0.7355	2.1528	1.7070	1.2612	1.1230	0.9557
1.42	0.7314	2.1858	1.7243	1.2676	1.1259	0.9531
1.43	0.7274	2.2190	1.7416	1.2741	1.1288	0.9504
1.44	0.7235	2.2525	1.7589	1.2807	1.1317	0.9476
1.45	0.7196	2.2862	1.7761	1.2872	1.1346	0.9448
1.46	0.7157	2.3202	1.7934	1.2938	1.1374	0.9420
1.47	0.7120	2.3544	1.8106	1.3003	1.1403	0.9390
1.48	0.7083	2.3888	1.8278	1.3069	1.1432	0.9360
1.49	0.7047	2.4234	1.8449	1.3136	1.1461	0.9329
1.50	0.7011	2.4583	1.8621	1.3202	1.1490	0.9298
1.51	0.6976	2.4934	1.8792	1.3269	1.1519	0.9266
1.52	0.6941	2.5288	1.8963	1.3336	1.1548	0.9233
1.53	0.6907	2.5644	1.9133	1.3403	1.1577	0.9200
1.54	0.6874	2.6002	1.9303	1.3470	1.1606	0.9166
1.55	0.6841	2.6362	1.9473	1.3538	1.1635	0.9132
1.56	0.6809	2.6725	1.9643	1.3606	1.1664	0.9097
1.57	0.6777	2.7090	1.9812	1.3674	1.1694	0.9062
1.58	0.6746	2.7458	1.9981	1.3742	1.1723	0.9026
1.59	0.6715	2.7828	2.0149	1.3811	1.1752	0.8989
1.60	0.6684	2.8200	2.0317	1.3880	1.1781	0.8952
1.61	0.6655	2.8574	2.0485	1.3949	1.1811	0.8915
1.62	0.6625	2.8951	2.0653	1.4018	1.1840	0.8877
1.63	0.6596	2.9330	2.0820	1.4088	1.1869	0.8838
1.64	0.6568	2.9712	2.0986	1.4158	1.1899	0.8799
1.65	0.6540	3.0096	2.1152	1.4228	1.1928	0.8760
1.66	0.6512	3.0482	2.1318	1.4299	1.1958	0.8720
1.67	0.6485	3.0870	2.1484	1.4369	1.1987	0.8680
1.68	0.6458	3.1261	2.1649	1.4440	1.2017	0.8639
1.69	0.6431	3.1654	2.1813	1.4512	1.2046	0.8599
1.70	0.6405	3.2050	2.1977	1.4583	1.2076	0.8557
1.71	0.6380	3.2448	2.2141	1.4655	1.2106	0.8516
1.72	0.6355	3.2848	2.2304	1.4727	1.2136	0.8474
1.73	0.6330	3.3250	2.2467	1.4800	1.2165	0.8431
1.74	0.6305	3.3655	2.2629	1.4873	1.2195	0.8389
1.75	0.6281	3.4062	2.2791	1.4946	1.2225	0.8346
1.76	0.6257	3.4472	2.2952	1.5019	1.2255	0.8302
1.77	0.6234	3.4884	2.3113	1.5093	1.2285	0.8259

M_1	M_2	$\frac{p_2}{p_1}$	$\frac{\rho_2}{\rho_1}$	$\frac{T_2}{T_1}$	$\frac{a_2}{a_1}$	$\frac{p_{02}}{p_{01}}$
1.78	0.6210	3.5298	2.3273	1.5167	1.2315	0.8215
1.79	0.6188	3.5714	2.3433	1.5241	1.2346	0.8171
1.80	0.6165	3.6133	2.3592	1.5316	1.2376	0.8127
1.81	0.6143	3.6554	2.3751	1.5391	1.2406	0.8082
1.82	0.6121	3.6978	2.3909	1.5466	1.2436	0.8038
1.83	0.6099	3.7404	2.4067	1.5541	1.2467	0.7993
1.84	0.6078	3.7832	2.4228	1.5617	1.2497	0.7948
1.85	0.6057	3.8262	2.4381	1.5693	1.2527	0.7902
1.86	0.6036	3.8695	2.4537	1.5770	1.2558	0.7857
1.87	0.6016	3.9130	2.4693	1.5847	1.2588	0.7811
1.88	0.5996	3.9568	2.4848	1.5924	1.2619	0.7765
1.89	0.5976	4.0008	2.5003	1.6001	1.2650	0.7720
1.90	0.5956	4.0450	2.5157	1.6079	1.2680	0.7674
1.91	0.5937	4.0894	2.5310	1.6157	1.2711	0.7627
1.92	0.5918	4.1341	2.5463	1.6236	1.2742	0.7581
1.93	0.5899	4.1790	2.5616	1.6314	1.2773	0.7535
1.94	0.5880	4.2242	2.5767	1.6394	1.2804	0.7488
1.95	0.5862	4.2696	2.5919	1.6473	1.2835	0.7442
1.96	0.5844	4.3152	2.6069	1.6553	1.2866	0.7395
1.97	0.5826	4.3610	2.6220	1.6633	1.2897	0.7349
1.98	0.5808	4.4071	2.6369	1.6713	1.2928	0.7302
1.99	0.5791	4.4534	2.6518	1.6794	1.2959	0.7255
2.00	0.5774	4.5000	2.6667	1.6875	1.2990	0.7209
2.01	0.5757	4.5468	2.6815	1.6956	1.3022	0.7162
2.02	0.5740	4.5938	2.6962	1.7038	1.3053	0.7115
2.03	0.5723	4.6410	2.7108	1.7120	1.3084	0.7069
2.04	0.5707	4.6885	2.7255	1.7203	1.3116	0.7022
2.05	0.5691	4.7362	2.7400	1.7285	1.3147	0.6975
2.06	0.5675	4.7842	2.7545	1.7369	1.3179	0.6928
2.07	0.5659	4.8324	2.7689	1.7452	1.3211	0.6882
2.08	0.5643	4.8808	2.7833	1.7536	1.3242	0.6835
2.09	0.5628	4.9294	2.7976	1.7620	1.3274	0.6789
2.10	0.5613	4.9783	2.8119	1.7704	1.3306	0.6742
2.11	0.5598	5.0274	2.8261	1.7789	1.3338	0.6696
2.12	0.5583	5.0768	2.8402	1.7875	1.3370	0.6649
2.13	0.5568	5.1264	2.8543	1.7960	1.3402	0.6603
2.14	0.5554	5.1762	2.8683	1.8046	1.3434	0.6557
2.15	0.5540	5.2262	2.8823	1.8132	1.3466	0.6511
2.16	0.5525	5.2765	2.8962	1.8219	1.3498	0.6464
2.17	0.5511	5.3270	2.9100	1.8306	1.3530	0.6419
2.18	0.5498	5.3778	2.9238	1.8393	1.3562	0.6373
2.19	0.5484	5.4288	2.9376	1.8481	1.3594	0.6327
2.20	0.5471	5.4800	2.9512	1.8569	1.3627	0.6281

M_1	M_2	$\frac{p_2}{p_1}$	$\frac{\rho_2}{\rho_1}$	$\frac{T_2}{T_1}$	$\frac{a_2}{a_1}$	$\frac{p_{02}}{p_{01}}$
2.22	0.5444	5.5831	2.9784	1.8746	1.3691	0.6191
2.23	0.5431	5.6350	2.9918	1.8835	1.3724	0.6145
2.24	0.5418	5.6872	3.0053	1.8924	1.3756	0.6100
2.25	0.5406	5.7396	3.0186	1.9014	1.3789	0.6055
2.26	0.5393	5.7922	3.0319	1.9104	1.3822	0.6011
2.27	0.5381	5.8450	3.0452	1.9194	1.3854	0.5966
2.28	0.5368	5.8981	3.0584	1.9285	1.3887	0.5921
2.29	0.5356	5.9514	3.0715	1.9376	1.3920	0.5877
2.30	0.5344	6.0050	3.0845	1.9468	1.3953	0.5833
2.31	0.5332	6.0588	3.0975	1.9560	1.3986	0.5789
2.32	0.5321	6.1128	3.1105	1.9652	1.4019	0.5745
2.33	0.5309	6.1670	3.1234	1.9745	1.4052	0.5702
2.34	0.5297	6.2215	3.1362	1.9838	1.4085	0.5658
2.35	0.5286	6.2762	3.1490	1.9931	1.4118	0.5615
2.36	0.5275	6.3312	3.1617	2.0025	1.4151	0.5572
2.37	0.5264	6.3864	3.1743	2.0119	1.4184	0.5529
2.38	0.5253	6.4418	3.1869	2.0213	1.4217	0.5486
2.39	0.5242	6.4974	3.1994	2.0308	1.4251	0.5444
2.40	0.5231	6.5533	3.2119	2.0403	1.4284	0.5401
2.41	0.5221	6.6094	3.2243	2.0499	1.4317	0.5359
2.42	0.5210	6.6658	3.2367	2.0595	1.4351	0.5317
2.43	0.5200	6.7224	3.2489	2.0691	1.4384	0.5276
2.44	0.5189	6.7792	3.2612	2.0788	1.4418	0.5234
2.45	0.5179	6.8362	3.2733	2.0885	1.4451	0.5193
2.46	0.5169	6.8935	3.2855	2.0982	1.4485	0.5152
2.47	0.5159	6.9510	3.2975	2.1080	1.4519	0.5111
2.48	0.5149	7.0088	3.3095	2.1178	1.4553	0.5071
2.49	0.5140	7.0668	3.3215	2.1276	1.4586	0.5030
2.50	0.5130	7.1250	3.3333	2.1375	1.4620	0.4990
2.51	0.5120	7.1834	3.3452	2.1474	1.4654	0.4950
2.52	0.5111	7.2421	3.3569	2.1574	1.4688	0.4911
2.53	0.5102	7.3010	3.3686	2.1674	1.4722	0.4871
2.54	0.5092	7.3602	3.3803	2.1774	1.4756	0.4832
2.55	0.5083	7.4196	3.3919	2.1875	1.4790	0.4793
2.56	0.5074	7.4792	3.4034	2.1976	1.4824	0.4754
2.57	0.5065	7.5390	3.4149	2.2077	1.4858	0.4715
2.58	0.5056	7.5991	3.4263	2.2179	1.4893	0.4677
2.59	0.5047	7.6594	3.4377	2.2281	1.4927	0.4639
2.60	0.5039	7.7200	3.4490	2.2383	1.4961	0.4601
2.61	0.5030	7.7808	3.4602	2.2486	1.4995	0.4564
2.62	0.5022	7.8418	3.4714	2.2590	1.5030	0.4526
2.63	0.5013	7.9030	3.4826	2.2693	1.5064	0.4489
2.64	0.5005	7.9645	3.4936	2.2797	1.5099	0.4452

M_1	M_2	$\frac{p_2}{p_1}$	$\frac{\rho_2}{\rho_1}$	$\frac{T_2}{T_1}$	$\frac{a_2}{a_1}$	$\frac{p_{02}}{p_{01}}$
2.65	0.4996	8.0262	3.5047	2.2902	1.5133	0.4416
2.66	0.4988	8.0882	3.5156	2.3006	1.5168	0.4379
2.67	0.4980	8.1504	3.5266	2.3111	1.5202	0.4343
2.68	0.4972	8.2128	3.5374	2.3217	1.5237	0.4307
2.69	0.4964	8.2754	3.5482	2.3323	1.5272	0.4271
2.70	0.4956	8.3383	3.5590	2.3429	1.5307	0.4236
2.71	0.4949	8.4014	3.5697	2.3536	1.5341	0.4201
2.72	0.4941	8.4648	3.5803	2.3642	1.5376	0.4166
2.73	0.4933	8.5284	3.5909	2.3750	1.5411	0.4131
2.74	0.4926	8.5922	3.6015	2.3858	1.5446	0.4097
2.75	0.4918	8.6562	3.6119	2.3966	1.5481	0.4062
2.76	0.4911	8.7205	3.6224	2.4074	1.5516	0.4028
2.77	0.4903	8.7850	3.6327	2.4183	1.5551	0.3994
2.78	0.4896	8.8498	3.6431	2.4292	1.5586	0.3961
2.79	0.4889	8.9148	3.6533	2.4402	1.5621	0.3928
2.80	0.4882	8.9800	3.6635	2.4512	1.5656	0.3895
2.81	0.4875	9.0454	3.6737	2.4622	1.5691	0.3862
2.82	0.4868	9.1111	3.6838	2.4733	1.5727	0.3829
2.83	0.4861	9.1770	3.6939	2.4844	1.5762	0.3797
2.84	0.4854	9.2432	3.7039	2.4955	1.5797	0.3765
2.85	0.4847	9.3096	3.7138	2.5067	1.5833	0.3733
2.86	0.4840	9.3762	3.7238	2.5179	1.5868	0.3701
2.87	0.4833	9.4430	3.7336	2.5292	1.5903	0.3670
2.88	0.4827	9.5101	3.7434	2.5405	1.5939	0.3639
2.89	0.4820	9.5774	3.7532	2.5518	1.5974	0.3608
2.90	0.4814	9.6450	3.7629	2.5632	1.6010	0.3577
2.91	0.4807	9.7128	3.7725	2.5746	1.6046	0.3547
2.92	0.4801	9.7808	3.7821	2.5861	1.6081	0.3517
2.93	0.4795	9.8490	3.7917	2.5975	1.6117	0.3487
2.94	0.4788	9.9175	3.8012	2.6091	1.6153	0.3457
2.95	0.4782	9.9862	3.8106	2.6206	1.6188	0.3428
2.96	0.4776	10.0552	3.8200	2.6322	1.6224	0.3398
2.97	0.4770	10.1244	3.8294	2.6439	1.6260	0.3369
2.98	0.4764	10.1938	3.8387	2.6555	1.6296	0.3340
2.99	0.4758	10.2634	3.8479	2.6673	1.6332	0.3312
3.00	0.4752	10.3333	3.8571	2.6790	1.6368	0.3283
3.01	0.4746	10.4034	3.8663	2.6908	1.6404	0.3255
3.02	0.4740	10.4738	3.8754	2.7026	1.6440	0.3227
3.03	0.4734	10.5444	3.8845	2.7145	1.6476	0.3200
3.04	0.4729	10.6152	3.8935	2.7264	1.6512	0.3172
3.05	0.4723	10.6862	3.9025	2.7383	1.6548	0.3145
3.06	0.4717	10.7575	3.9114	2.7503	1.6584	0.3118
3.07	0.4712	10.8290	3.9203	2.7623	1.6620	0.3091

M_1	M_2	$\frac{p_2}{p_1}$	$\frac{\rho_2}{\rho_1}$	$\frac{T_2}{T_1}$	$\frac{a_2}{a_1}$	$\frac{p_{02}}{p_{01}}$
3.08	0.4706	10.9008	3.9291	2.7744	1.6656	0.3065
3.09	0.4701	10.9728	3.9379	2.7865	1.6693	0.3038
3.10	0.4695	11.0450	3.9466	2.7986	1.6729	0.3012
3.11	0.4690	11.1174	3.9553	2.8108	1.6765	0.2986
3.12	0.4685	11.1901	3.9639	2.8230	1.6802	0.2960
3.13	0.4679	11.2630	3.9725	2.8352	1.6838	0.2935
3.14	0.4674	11.3362	3.9811	2.8475	1.6875	0.2910
3.15	0.4669	11.4096	3.9896	2.8598	1.6911	0.2885
3.16	0.4664	11.4832	3.9981	2.8722	1.6947	0.2860
3.17	0.4659	11.5570	4.0065	2.8846	1.6984	0.2835
3.18	0.4654	11.6311	4.0149	2.8970	1.7021	0.2811
3.19	0.4648	11.7054	4.0232	2.9095	1.7057	0.2786
3.20	0.4643	11.7800	4.0315	2.9220	1.7094	0.2762
3.21	0.4639	11.8548	4.0397	2.9345	1.7130	0.2738
3.22	0.4634	11.9298	4.0479	2.9471	1.7167	0.2715
3.23	0.4629	12.0050	4.0561	2.9597	1.7204	0.2691
3.24	0.4624	12.0805	4.0642	2.9724	1.7241	0.2668
3.25	0.4619	12.1562	4.0723	2.9851	1.7277	0.2645
3.26	0.4614	12.2322	4.0803	2.9979	1.7314	0.2622
3.27	0.4610	12.3084	4.0883	3.0106	1.7351	0.2600
3.28	0.4605	12.3848	4.0963	3.0234	1.7388	0.2577
3.29	0.4600	12.4614	4.1042	3.0363	1.7425	0.2555
3.30	0.4596	12.5383	4.1120	3.0492	1.7462	0.2533
3.31	0.4591	12.6154	4.1198	3.0621	1.7499	0.2511
3.32	0.4587	12.6928	4.1276	3.0751	1.7536	0.2489
3.33	0.4582	12.7704	4.1354	3.0881	1.7513	0.2468
3.34	0.4578	12.8482	4.1431	3.1011	1.7610	0.2446
3.35	0.4573	12.9262	4.1507	3.1142	1.7647	0.2425
3.36	0.4569	13.0045	4.1583	3.1273	1.7684	0.2404
3.37	0.4565	13.0830	4.1659	3.1405	1.7721	0.2383
3.38	0.4560	13.1618	4.1734	3.1537	1.7759	0.2363
3.39	0.4556	13.2408	4.1809	3.1669	1.7796	0.2342
3.40	0.4552	13.3200	4.1884	3.1802	1.7833	0.2322
3.41	0.4548	13.3994	4.1958	3.1935	1.7870	0.2302
3.42	0.4544	13.4791	4.2032	3.2069	1.7908	0.2282
3.43	0.4540	13.5590	4.2105	3.2203	1.7945	0.2263
3.44	0.4535	13.6392	4.2178	3.2337	1.7982	0.2243
3.45	0.4531	13.7196	4.2251	3.2471	1.8020	0.2224
3.46	0.4527	13.8002	4.2323	3.2607	1.8057	0.2205
3.47	0.4523	13.8810	4.2395	3.2742	1.8095	0.2186
3.48	0.4519	13.9621	4.2467	3.2878	1.8132	0.2167
3.49	0.4515	14.0434	4.2538	3.3014	1.8170	0.2148
3.50	0.4512	14.1250	4.2609	3.3150	1.8207	0.2129

M_1	M_2	$\frac{p_2}{p_1}$	$\frac{\rho_2}{\rho_1}$	$\frac{T_2}{T_1}$	$\frac{a_2}{a_1}$	$\frac{p_{02}}{p_{01}}$
3.51	0.4508	14.2068	4.2679	3.3287	1.8245	0.2111
3.52	0.4504	14.2888	4.2749	3.3425	1.8282	0.2093
3.53	0.4500	14.3710	4.2819	3.3562	1.8320	0.2075
3.54	0.4496	14.4535	4.2888	3.3701	1.8358	0.2057
3.55	0.4492	14.5362	4.2957	3.3839	1.8395	0.2039
3.56	0.4489	14.6192	4.3026	3.3978	1.8433	0.2022
3.57	0.4485	14.7024	4.3094	3.4117	1.8471	0.2004
3.58	0.4481	14.7858	4.3162	3.4257	1.8509	0.1987
3.59	0.4478	14.8694	4.3229	3.4397	1.8546	0.1970
3.60	0.4474	14.9533	4.3296	3.4537	1.8584	0.1953
3.61	0.4471	15.0374	4.3363	3.4678	1.8622	0.1936
3.62	0.4467	15.1218	4.3429	3.4819	1.8660	0.1920
3.63	0.4463	15.2064	4.3496	3.4961	1.8698	0.1903
3.64	0.4460	15.2912	4.3561	3.5103	1.8736	0.1887
3.65	0.4456	15.3762	4.3627	3.5245	1.8774	0.1871
3.66	0.4453	15.4615	4.3692	3.5388	1.8812	0.1855
3.67	0.4450	15.5470	4.3756	3.5531	1.8850	0.1839
3.68	0.4446	15.6328	4.3821	3.5674	1.8888	0.1823
3.69	0.4443	15.7188	4.3885	3.5818	1.8926	0.1807
3.70	0.4439	15.8050	4.3949	3.5962	1.8964	0.1792
3.71	0.4436	15.8914	4.4012	3.6107	1.9002	0.1777
3.72	0.4433	15.9781	4.4075	3.6252	1.9040	0.1761
3.73	0.4430	16.0650	4.4138	3.6397	1.9078	0.1746
3.74	0.4426	16.1522	4.4200	3.6549	1.9116	0.1731
3.75	0.4423	16.2396	4.4262	3.6689	1.9154	0.1717
3.76	0.4420	16.3272	4.4324	3.6836	1.9193	0.1702
3.77	0.4417	16.4150	4.4385	3.6983	1.9231	0.1687
3.78	0.4414	16.5031	4.4447	3.7130	1.9269	0.1673
3.79	0.4410	16.5914	4.4507	3.7278	1.9307	0.1659
3.80	0.4407	16.6800	4.4568	3.7426	1.9346	0.1645
3.81	0.4404	16.7688	4.4628	3.7574	1.9384	0.1631
3.82	0.4401	16.8578	4.4688	3.7723	1.9422	0.1617
3.83	0.4398	16.9470	4.4747	3.7873	1.9461	0.1603
3.84	0.4395	17.0365	4.4807	3.8022	1.9499	0.1589
3.85	0.4392	17.1262	4.4866	3.8172	1.9538	0.1576
3.86	0.4389	17.2162	4.4924	3.8323	1.9576	0.1563
3.87	0.4386	17.3063	4.4983	3.8473	1.9615	0.1549
3.88	0.4383	17.3968	4.5041	3.8625	1.9653	0.1536
3.89	0.4380	17.4874	4.5098	3.8776	1.9692	0.1523
3.90	0.4377	17.5783	4.5156	3.8928	1.9730	0.1510
3.91	0.4375	17.6694	4.5213	3.9080	1.9769	0.1497
3.92	0.4372	17.7608	4.5270	3.9233	1.9807	0.1485
3.93	0.4369	17.8524	4.5326	3.9386	1.9846	0.1472

M_1	M_2	$\frac{p_2}{p_1}$	$\frac{\rho_2}{\rho_1}$	$\frac{T_2}{T_1}$	$\frac{a_2}{a_1}$	$\frac{p_{02}}{p_{01}}$
3.94	0.4366	17.9442	4.5383	3.9540	1.9885	0.1460
3.95	0.4363	18.0362	4.5439	3.9694	1.9923	0.1448
3.96	0.4360	18.1285	4.5494	3.9848	1.9962	0.1435
3.97	0.4358	18.2210	4.5550	4.0002	2.0001	0.1423
3.98	0.4355	18.3138	4.5605	4.0157	2.0039	0.1411
3.99	0.4352	18.4068	4.5660	4.0313	2.0078	0.1399
4.00	0.4350	18.5000	4.5714	4.0469	2.0117	0.1388
4.01	0.4347	18.5934	4.5769	4.0625	2.0156	0.1376
4.02	0.4344	18.6871	4.5823	4.0781	2.0194	0.1364
4.03	0.4342	18.7810	4.5876	4.0938	2.0233	0.1353
4.04	0.4339	18.8752	4.5930	4.1096	2.0272	0.1342
4.05	0.4336	18.9696	4.5983	4.1253	2.0311	0.1330
4.06	0.4334	19.0642	4.6036	4.1412	2.0350	0.1319
4.07	0.4331	19.1590	4.6089	4.1570	2.0389	0.1308
4.08	0.4329	19.2541	4.6141	4.1729	2.0428	0.1297
4.09	0.4326	19.3494	4.6193	4.1888	2.0467	0.1286
4.10	0.4324	19.4450	4.6245	4.2048	2.0506	0.1276
4.11	0.4321	19.5408	4.6296	4.2208	2.0545	0.1265
4.12	0.4319	19.6368	4.6348	4.2368	2.0584	0.1254
4.13	0.4316	19.7331	4.6399	4.2529	2.0623	0.1244
4.14	0.4314	19.8295	4.6450	4.2690	2.0662	0.1234
4.15	0.4311	19.9262	4.6500	4.2852	2.0701	0.1223
4.16	0.4309	20.0232	4.6550	4.3014	2.0740	0.1213
4.17	0.4306	20.1204	4.6601	4.3176	2.0779	0.1203
4.18	0.4304	20.2178	4.6650	4.3339	2.0818	0.1193
4.19	0.4302	20.3155	4.6700	4.3502	2.0857	0.1183
4.20	0.4299	20.4133	4.6749	4.3666	2.0896	0.1173
4.21	0.4297	20.5115	4.6798	4.383	2.0936	0.1164
4.22	0.4295	20.6098	4.6847	4.3994	2.0975	0.1154
4.23	0.4292	20.7084	4.6896	4.4159	2.1014	0.1144
4.24	0.4290	20.8072	4.6944	4.4324	2.1053	0.1135
4.25	0.4288	20.9063	4.6992	4.4489	2.1092	0.1126
4.26	0.4286	21.0056	4.704	4.4655	2.1132	0.1116
4.27	0.4283	21.1051	4.7087	4.4821	2.1171	0.1107
4.28	0.4281	21.2048	4.7135	4.4988	2.1210	0.1098
4.29	0.4279	21.3048	4.7182	4.5155	2.1250	0.1089
4.30	0.4277	21.4050	4.7229	4.5322	2.1289	0.1080
4.31	0.4275	21.5055	4.7275	4.5490	2.1328	0.1071
4.32	0.4272	21.6062	4.7322	4.5658	2.1368	0.1062
4.33	0.4270	21.7071	4.7368	4.5827	2.1407	0.1054
4.34	0.4268	21.8082	4.7414	4.5996	2.1447	0.1045
4.35	0.4266	21.9096	4.746	4.6165	2.1486	0.1036
4.36	0.4264	22.0113	4.7505	4.6335	2.1525	0.1028

M_1	M_2	$\frac{p_2}{p_1}$	$\frac{\rho_2}{\rho_1}$	$\frac{T_2}{T_1}$	$\frac{a_2}{a_1}$	$\frac{p_{02}}{p_{01}}$
4.37	0.4262	22.1131	4.755	4.6505	2.1565	0.1020
4.38	0.426	22.2152	4.7595	4.6675	2.1604	0.1011
4.39	0.4258	22.3175	4.764	4.6846	2.1644	0.1003
4.40	0.4255	22.4201	4.7685	4.7017	2.1683	0.0995
4.41	0.4253	22.5229	4.7729	4.7189	2.1723	0.0987
4.42	0.4251	22.6259	4.7773	4.7361	2.1763	0.0979
4.43	0.4249	22.7291	4.7817	4.7533	2.1802	0.0971
4.44	0.4247	22.8326	4.7861	4.7706	2.1842	0.0963
4.45	0.4245	22.9363	4.7904	4.7879	2.1881	0.0955
4.46	0.4243	23.0403	4.7948	4.8053	2.1921	0.0947
4.47	0.4241	23.1445	4.7991	4.8227	2.1961	0.094
4.48	0.4239	23.2489	4.8034	4.8401	2.2000	0.0932
4.49	0.4237	23.3535	4.8076	4.8576	2.2040	0.0924
4.50	0.4236	23.4584	4.8119	4.8751	2.2080	0.0917
4.51	0.4234	23.5635	4.8161	4.8927	2.2119	0.0910
4.52	0.4232	23.6689	4.8203	4.9102	2.2159	0.0902
4.53	0.423	23.7745	4.8245	4.9279	2.2199	0.0895
4.54	0.4228	23.8803	4.8287	4.9455	2.2239	0.0888
4.55	0.4226	23.9864	4.8328	4.9632	2.2278	0.0881
4.56	0.4224	24.0926	4.8369	4.981	2.2318	0.0874
4.57	0.4222	24.1992	4.841	4.9988	2.2358	0.0867
4.58	0.422	24.3059	4.8451	5.0166	2.2398	0.0860
4.59	0.4219	24.4129	4.8492	5.0344	2.2438	0.0853
4.60	0.4217	24.5201	4.8532	5.0523	2.2477	0.0846
4.61	0.4215	24.6276	4.8572	5.0703	2.2517	0.0839
4.62	0.4213	24.7353	4.8612	5.0883	2.2557	0.0832
4.63	0.4211	24.8432	4.8652	5.1063	2.2597	0.0826
4.64	0.421	24.9513	4.8692	5.1243	2.2637	0.0819
4.65	0.4208	25.0597	4.8731	5.1424	2.2677	0.0813
4.66	0.4206	25.1683	4.8771	5.1605	2.2717	0.0806
4.67	0.4204	25.2772	4.881	5.1787	2.2757	0.0800
4.68	0.4203	25.3863	4.8849	5.1969	2.2797	0.0793
4.69	0.4201	25.4956	4.8887	5.2152	2.2837	0.0787
4.70	0.4199	25.6051	4.8926	5.2335	2.2877	0.0781
4.71	0.4197	25.7149	4.8964	5.2518	2.2917	0.0775
4.72	0.4196	25.8249	4.9002	5.2701	2.2957	0.0769
4.73	0.4194	25.9352	4.9040	5.2885	2.2997	0.0762
4.74	0.4192	26.0457	4.9078	5.3070	2.3037	0.0756
4.75	0.4191	26.1564	4.9116	5.3255	2.3077	0.075
4.76	0.4189	26.2674	4.9153	5.3440	2.3117	0.0745
4.77	0.4187	26.3785	4.9190	5.3625	2.3157	0.0739
4.78	0.4186	26.4900	4.9227	5.3811	2.3197	0.0733
4.79	0.4184	26.6016	4.9264	5.3998	2.3237	0.0727

M_1	M_2	$\frac{p_2}{p_1}$	$\frac{\rho_2}{\rho_1}$	$\frac{T_2}{T_1}$	$\frac{a_2}{a_1}$	$\frac{p_{02}}{p_{01}}$
4.80	0.4183	26.7135	4.9301	5.4184	2.3278	0.0721
4.81	0.4181	26.8256	4.9338	5.4372	2.3318	0.0716
4.82	0.4179	26.9380	4.9374	5.4559	2.3358	0.0710
4.83	0.4178	27.0506	4.9410	5.4747	2.3398	0.0705
4.84	0.4176	27.1634	4.9446	5.4935	2.3438	0.0699
4.85	0.4175	27.2764	4.9482	5.5124	2.3478	0.0694
4.86	0.4173	27.3897	4.9518	5.5313	2.3519	0.0688
4.87	0.4172	27.5032	4.9553	5.5502	2.3559	0.0683
4.88	0.4170	27.6170	4.9589	5.5692	2.3599	0.0677
4.89	0.4169	27.7310	4.9624	5.5882	2.3639	0.0672
4.9	0.4167	27.8452	4.9659	5.6073	2.3680	0.0667
4.91	0.4165	27.9597	4.9694	5.6264	2.3720	0.0662
4.92	0.4164	28.0743	4.9728	5.6455	2.3760	0.0657
4.93	0.4162	28.1893	4.9763	5.6647	2.3801	0.0652
4.94	0.4161	28.3044	4.9797	5.6839	2.3841	0.0647
4.95	0.4160	28.4198	4.9831	5.7032	2.3881	0.0642
4.96	0.4158	28.5354	4.9865	5.7225	2.3922	0.0637
4.97	0.4157	28.6513	4.9899	5.7418	2.3962	0.0632
4.98	0.4155	28.7674	4.9933	5.7612	2.4002	0.0627
4.99	0.4154	28.8837	4.9967	5.7806	2.4043	0.0622
5.00	0.4152	29.0002	5.0000	5.8000	2.4083	0.0617

D Oblique Shock Chart ($\gamma = 1.4$)

The $\theta - \beta - M$ diagram. (Source: From NACA Report 1135, Ames Research Staff, Equations, Tables, and Carts for Compressible Flow, 1953)

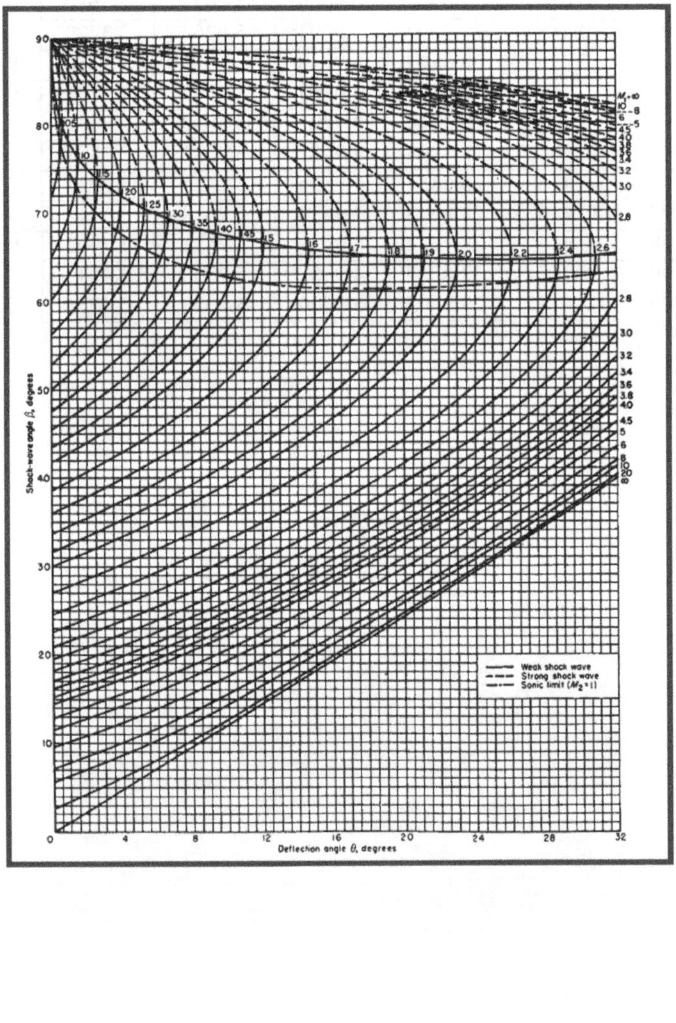

DOI: 10.1201/9781003139447-D

E One-Dimensional Flow with Friction ($\gamma = 1.4$)

M	$\frac{T}{T^*}$	$\frac{p}{p^*}$	$\frac{p_0}{p_0^*}$	$\frac{\rho}{\rho^*}$	$\frac{F}{F^*}$	$\frac{4fL_{max}}{D}$
0.02000	1.19990	54.77006	28.94213	0.02191	22.83364	1778.44988
0.04000	1.19962	27.38175	14.48149	0.04381	11.43462	440.35221
0.06000	1.19914	18.25085	9.66591	0.06570	7.64285	193.03108
0.08000	1.19847	13.68431	7.26161	0.08758	5.75288	106.71822
0.10000	1.19760	10.94351	5.82183	0.10944	4.62363	66.92156
0.12000	1.19655	9.11559	4.86432	0.13126	3.87473	45.40796
0.14000	1.19531	7.80932	4.18240	0.15306	3.34317	32.51131
0.16000	1.19389	6.82907	3.67274	0.17482	2.94743	24.19783
0.18000	1.19227	6.06618	3.27793	0.19654	2.64223	18.54265
0.20000	1.19048	5.45545	2.96352	0.21822	2.40040	14.53327
0.22000	1.18850	4.95537	2.70760	0.23984	2.20464	11.59605
0.24000	1.18633	4.53829	2.49556	0.26141	2.04344	9.38648
0.26000	1.18399	4.18505	2.31729	0.28291	1.90880	7.68757
0.28000	1.18147	3.88199	2.16555	0.30435	1.79503	6.35721
0.30000	1.17878	3.61906	2.03507	0.32572	1.69794	5.29925
0.32000	1.17592	3.38874	1.92185	0.34701	1.61440	4.44674
0.34000	1.17288	3.18529	1.82288	0.36822	1.54200	3.75195
0.36000	1.16968	3.00422	1.73578	0.38935	1.47888	3.18012
0.38000	1.16632	2.84200	1.65870	0.41039	1.42356	2.70545
0.40000	1.16279	2.69582	1.59014	0.43133	1.37487	2.30849
0.42000	1.15911	2.56338	1.52890	0.45218	1.33185	1.97437
0.44000	1.15527	2.44280	1.47401	0.47293	1.29371	1.69152
0.46000	1.15128	2.33256	1.42463	0.49357	1.25981	1.45091
0.48000	1.14714	2.23135	1.38010	0.51410	1.22962	1.24534
0.50000	1.14286	2.13809	1.33984	0.53452	1.20268	1.06906
0.52000	1.13843	2.05187	1.30339	0.55483	1.17860	0.91742
0.54000	1.13387	1.97192	1.27032	0.57501	1.15705	0.78663
0.56000	1.12918	1.89755	1.24029	0.59507	1.13777	0.67357
0.58000	1.12435	1.82820	1.21301	0.61501	1.12050	0.57568
0.60000	1.11940	1.76336	1.18820	0.63481	1.10504	0.49082
0.62000	1.11433	1.70261	1.16565	0.65448	1.09120	0.41720
0.64000	1.10914	1.64556	1.14515	0.67402	1.07883	0.35330
0.66000	1.10383	1.59187	1.12654	0.69342	1.06777	0.29785
0.68000	1.09842	1.54126	1.10965	0.71268	1.05792	0.24978
0.70000	1.09290	1.49345	1.09437	0.73179	1.04915	0.20814

DOI: 10.1201/9781003139447-E

M	$\frac{T}{T^*}$	$\frac{p}{p^*}$	$\frac{p_0}{p_0^*}$	$\frac{\rho}{\rho^*}$	$\frac{F}{F^*}$	$\frac{4fL_{max}}{D}$
0.72000	1.08727	1.44823	1.08057	0.75076	1.04137	0.17215
0.74000	1.08155	1.40537	1.06814	0.76958	1.03449	0.14112
0.76000	1.07573	1.36470	1.05700	0.78825	1.02844	0.11447
0.78000	1.06982	1.32605	1.04705	0.80677	1.02314	0.09167
0.80000	1.06383	1.28928	1.03823	0.82514	1.01853	0.07229
0.82000	1.05775	1.25423	1.03046	0.84335	1.01455	0.05593
0.84000	1.05160	1.22080	1.02370	0.86140	1.01115	0.04226
0.86000	1.04537	1.18888	1.01787	0.87929	1.00829	0.03097
0.88000	1.03907	1.15835	1.01294	0.89703	1.00591	0.02179
0.90000	1.03270	1.12913	1.00886	0.91460	1.00399	0.01451
0.92000	1.02627	1.10114	1.00560	0.93201	1.00248	0.00891
0.94000	1.01978	1.07430	1.00311	0.94925	1.00136	0.00482
0.96000	1.01324	1.04854	1.00136	0.96633	1.00059	0.00206
0.98000	1.00664	1.02379	1.00034	0.98325	1.00014	0.00049
1.00000	1.00000	1.00000	1.00000	1.00000	1.00000	0.00000
1.02000	0.99331	0.97711	1.00033	1.01658	1.00014	0.00046
1.04000	0.98658	0.95507	1.00131	1.03300	1.00053	0.00177
1.06000	0.97982	0.93383	1.00291	1.04925	1.00116	0.00384
1.08000	0.97302	0.91335	1.00512	1.06533	1.00200	0.00658
1.10000	0.96618	0.89359	1.00793	1.08124	1.00305	0.00994
1.12000	0.95932	0.87451	1.01131	1.09699	1.00429	0.01382
1.14000	0.95244	0.85608	1.01527	1.11256	1.00569	0.01819
1.16000	0.94554	0.83826	1.01978	1.12797	1.00726	0.02298
1.18000	0.93861	0.82103	1.02484	1.14321	1.00897	0.02814
1.20000	0.93168	0.80436	1.03044	1.15828	1.01081	0.03364
1.22000	0.92473	0.78822	1.03657	1.17319	1.01278	0.03943
1.24000	0.91777	0.77258	1.04323	1.18792	1.01486	0.04547
1.26000	0.91080	0.75743	1.05041	1.20249	1.01705	0.05174
1.28000	0.90383	0.74274	1.05810	1.21690	1.01933	0.05820
1.30000	0.89686	0.72848	1.06630	1.23114	1.02170	0.06483
1.32000	0.88989	0.71465	1.07502	1.24521	1.02414	0.07161
1.34000	0.88292	0.70122	1.08424	1.25912	1.02666	0.07850
1.36000	0.87596	0.68818	1.09396	1.27286	1.02925	0.08550
1.38000	0.86901	0.67551	1.10419	1.28645	1.03189	0.09259
1.40000	0.86207	0.66320	1.11493	1.29987	1.03459	0.09974
1.42000	0.85514	0.65122	1.12616	1.31313	1.03733	0.10694
1.44000	0.84822	0.63958	1.13790	1.32623	1.04012	0.11419
1.46000	0.84133	0.62825	1.15015	1.33917	1.04295	0.12146
1.48000	0.83445	0.61722	1.16290	1.35195	1.04581	0.12875
1.50000	0.82759	0.60648	1.17617	1.36458	1.04870	0.13605
1.52000	0.82075	0.59602	1.18994	1.37705	1.05162	0.14335
1.54000	0.81393	0.58583	1.20423	1.38936	1.05456	0.15063
1.56000	0.80715	0.57591	1.21904	1.40152	1.05752	0.15790

M	$\frac{T}{T^*}$	$\frac{p}{p^*}$	$\frac{p_0}{p_0^*}$	$\frac{\rho}{\rho^*}$	$\frac{F}{F^*}$	$\frac{4\bar{f}L_{max}}{D}$
1.58000	0.80038	0.56623	1.23438	1.41353	1.06049	0.16514
1.60000	0.79365	0.55679	1.25024	1.42539	1.06348	0.17236
1.62000	0.78695	0.54759	1.26663	1.43710	1.06647	0.17954
1.64000	0.78027	0.53862	1.28355	1.44866	1.06948	0.18667
1.66000	0.77363	0.52986	1.30102	1.46008	1.07249	0.19377
1.68000	0.76703	0.52131	1.31904	1.47135	1.07550	0.20081
1.70000	0.76046	0.51297	1.33761	1.48247	1.07851	0.20780
1.72000	0.75392	0.50482	1.35674	1.49345	1.08152	0.21474
1.74000	0.74742	0.49686	1.37643	1.50429	1.08453	0.22162
1.76000	0.74096	0.48909	1.39670	1.51499	1.08753	0.22844
1.78000	0.73454	0.48149	1.41755	1.52555	1.09053	0.23519
1.80000	0.72816	0.47407	1.43898	1.53598	1.09351	0.24189
1.82000	0.72181	0.46681	1.46101	1.54626	1.09649	0.24851
1.84000	0.71551	0.45972	1.48365	1.55642	1.09946	0.25507
1.86000	0.70925	0.45278	1.50689	1.56644	1.10242	0.26156
1.88000	0.70304	0.44600	1.53076	1.57633	1.10536	0.26798
1.90000	0.69686	0.43936	1.55526	1.58609	1.10829	0.27433
1.92000	0.69073	0.43287	1.58039	1.59572	1.11120	0.28061
1.94000	0.68465	0.42651	1.60617	1.60523	1.11410	0.28681
1.96000	0.67861	0.42029	1.63261	1.61460	1.11698	0.29295
1.98000	0.67262	0.41421	1.65972	1.62386	1.11984	0.29901
2.00000	0.66667	0.40825	1.68750	1.63299	1.12268	0.30500
2.02000	0.66076	0.40241	1.71597	1.64201	1.12551	0.31091
2.04000	0.65491	0.39670	1.74514	1.65090	1.12831	0.31676
2.06000	0.64910	0.39110	1.77502	1.65967	1.13110	0.32253
2.08000	0.64334	0.38562	1.80561	1.66833	1.13387	0.32822
2.10000	0.63762	0.38024	1.83694	1.67687	1.13661	0.33385
2.12000	0.63195	0.37498	1.86902	1.68530	1.13933	0.33940
2.14000	0.62633	0.36982	1.90184	1.69362	1.14204	0.34489
2.16000	0.62076	0.36476	1.93544	1.70183	1.14471	0.35030
2.18000	0.61523	0.35980	1.96981	1.70992	1.14737	0.35564
2.20000	0.60976	0.35494	2.00497	1.71791	1.15001	0.36091
2.22000	0.60433	0.35017	2.04094	1.72579	1.15262	0.36611
2.24000	0.59895	0.34550	2.07773	1.73357	1.15521	0.37124
2.26000	0.59361	0.34091	2.11535	1.74125	1.15777	0.37631
2.28000	0.58833	0.33641	2.15381	1.74882	1.16032	0.38130
2.30000	0.58309	0.33200	2.19313	1.75629	1.16284	0.38623
2.32000	0.57790	0.32767	2.23332	1.76366	1.16533	0.39109
2.34000	0.57276	0.32342	2.27440	1.77093	1.16780	0.39589
2.36000	0.56767	0.31925	2.31638	1.77811	1.17025	0.40062
2.38000	0.56262	0.31516	2.35928	1.78519	1.17268	0.40529
2.40000	0.55762	0.31114	2.40310	1.79218	1.17508	0.40989
2.42000	0.55267	0.30720	2.44787	1.79907	1.17746	0.41443

M	$\frac{T}{T^*}$	$\frac{p}{p^*}$	$\frac{p_0}{p_0^*}$	$\frac{\rho}{\rho^*}$	$\frac{F}{F^*}$	$\frac{4fL_{max}}{D}$
2.44000	0.54777	0.30332	2.49360	1.80587	1.17981	0.41891
2.46000	0.54291	0.29952	2.54031	1.81258	1.18214	0.42332
2.48000	0.53810	0.29579	2.58801	1.81921	1.18445	0.42768
2.50000	0.53333	0.29212	2.63672	1.82574	1.18673	0.43198
2.52000	0.52862	0.28852	2.68645	1.83219	1.18899	0.43621
2.54000	0.52394	0.28498	2.73723	1.83855	1.19123	0.44039
2.56000	0.51932	0.28150	2.78906	1.84483	1.19344	0.44451
2.58000	0.51474	0.27808	2.84197	1.85103	1.19563	0.44858
2.60000	0.51020	0.27473	2.89598	1.85714	1.19780	0.45259
2.62000	0.50571	0.27143	2.95109	1.86318	1.19995	0.45654
2.64000	0.50127	0.26818	3.00733	1.86913	1.20207	0.46044
2.66000	0.49687	0.26500	3.06472	1.87501	1.20417	0.46429
2.68000	0.49251	0.26186	3.12327	1.88081	1.20625	0.46808
2.70000	0.48820	0.25878	3.18301	1.88653	1.20830	0.47182
2.72000	0.48393	0.25575	3.24395	1.89218	1.21033	0.47551
2.74000	0.47971	0.25278	3.30611	1.89775	1.21235	0.47915
2.76000	0.47553	0.24985	3.36952	1.90325	1.21433	0.48273
2.78000	0.47139	0.24697	3.43418	1.90868	1.21630	0.48627
2.80000	0.46729	0.24414	3.50012	1.91404	1.21825	0.48976
2.82000	0.46323	0.24135	3.56737	1.91933	1.22017	0.49321
2.84000	0.45922	0.23861	3.63593	1.92455	1.22208	0.49660
2.86000	0.45525	0.23592	3.70584	1.92970	1.22396	0.49995
2.88000	0.45132	0.23326	3.77711	1.93479	1.22582	0.50326
2.90000	0.44743	0.23066	3.84977	1.93981	1.22766	0.50652
2.92000	0.44358	0.22809	3.92383	1.94477	1.22948	0.50973
2.94000	0.43977	0.22556	3.99932	1.94966	1.23128	0.51290
2.96000	0.43600	0.22307	4.07625	1.95449	1.23307	0.51603
2.98000	0.43226	0.22063	4.15466	1.95925	1.23483	0.51912
3.00000	0.42857	0.21822	4.23457	1.96396	1.23657	0.52216
3.02000	0.42492	0.21585	4.31599	1.96861	1.23829	0.52516
3.04000	0.42130	0.21351	4.39895	1.97319	1.23999	0.52813
3.06000	0.41772	0.21121	4.48347	1.97772	1.24168	0.53105
3.08000	0.41418	0.20895	4.56959	1.98219	1.24334	0.53393
3.10000	0.41068	0.20672	4.65731	1.98661	1.24499	0.53678
3.12000	0.40721	0.20453	4.74667	1.99097	1.24662	0.53958
3.14000	0.40378	0.20237	4.83769	1.99527	1.24823	0.54235
3.16000	0.40038	0.20024	4.93039	1.99952	1.24982	0.54509
3.18000	0.39702	0.19814	5.02481	2.00372	1.25139	0.54778
3.20000	0.39370	0.19608	5.12096	2.00786	1.25295	0.55044
3.22000	0.39041	0.19405	5.21887	2.01195	1.25449	0.55307
3.24000	0.38716	0.19204	5.31857	2.01599	1.25601	0.55566
3.26000	0.38394	0.19007	5.42008	2.01998	1.25752	0.55822
3.28000	0.38075	0.18812	5.52343	2.02392	1.25901	0.56074

M	$\frac{T}{T^*}$	$\frac{p}{p^*}$	$\frac{p_0}{p_0^*}$	$\frac{\rho}{\rho^*}$	$\frac{F}{F^*}$	$\frac{4fL_{max}}{D}$
3.30000	0.37760	0.18621	5.62865	2.02781	1.26048	0.56323
3.32000	0.37448	0.18432	5.73576	2.03165	1.26193	0.56569
3.34000	0.37139	0.18246	5.84479	2.03545	1.26337	0.56812
3.36000	0.36833	0.18063	5.95577	2.03920	1.26479	0.57051
3.38000	0.36531	0.17882	6.06873	2.04290	1.26620	0.57287
3.40000	0.36232	0.17704	6.18370	2.04656	1.26759	0.57521
3.42000	0.35936	0.17528	6.30070	2.05017	1.26897	0.57751
3.44000	0.35643	0.17355	6.41976	2.05374	1.27033	0.57978
3.46000	0.35353	0.17185	6.54092	2.05727	1.27167	0.58203
3.48000	0.35066	0.17016	6.66419	2.06075	1.27300	0.58424
3.50000	0.34783	0.16851	6.78962	2.06419	1.27432	0.58643
3.52000	0.34502	0.16687	6.91723	2.06759	1.27562	0.58859
3.54000	0.34224	0.16526	7.04705	2.07094	1.27691	0.59072
3.56000	0.33949	0.16367	7.17912	2.07426	1.27818	0.59282
3.58000	0.33677	0.16210	7.31346	2.07754	1.27944	0.59490
3.60000	0.33408	0.16055	7.45011	2.08077	1.28068	0.59695
3.62000	0.33141	0.15903	7.58910	2.08397	1.28191	0.59898
3.64000	0.32877	0.15752	7.73045	2.08713	1.28313	0.60098
3.66000	0.32616	0.15604	7.87421	2.09026	1.28433	0.60296
3.68000	0.32358	0.15458	8.02040	2.09334	1.28552	0.60491
3.70000	0.32103	0.15313	8.16907	2.09639	1.28670	0.60684
3.72000	0.31850	0.15171	8.32023	2.09941	1.28787	0.60874
3.74000	0.31600	0.15030	8.47393	2.10238	1.28902	0.61062
3.76000	0.31352	0.14892	8.63020	2.10533	1.29016	0.61247
3.78000	0.31107	0.14755	8.78907	2.10824	1.29128	0.61431
3.80000	0.30864	0.14620	8.95059	2.11111	1.29240	0.61612
3.82000	0.30624	0.14487	9.11477	2.11395	1.29350	0.61791
3.84000	0.30387	0.14355	9.28167	2.11676	1.29459	0.61968
3.86000	0.30151	0.14225	9.45131	2.11954	1.29567	0.62142
3.88000	0.29919	0.14097	9.62373	2.12228	1.29674	0.62315
3.90000	0.29688	0.13971	9.79897	2.12499	1.29779	0.62485
3.92000	0.29460	0.13846	9.97707	2.12767	1.29883	0.62653
3.94000	0.29235	0.13723	10.15806	2.13032	1.29987	0.62819
3.96000	0.29011	0.13602	10.34197	2.13294	1.30089	0.62984
3.98000	0.28790	0.13482	10.52886	2.13553	1.30190	0.63146
4.00000	0.28571	0.13363	10.71875	2.13809	1.30290	0.63306
4.02000	0.28355	0.13246	10.91168	2.14062	1.30389	0.63465
4.04000	0.28140	0.13131	11.10770	2.14312	1.30487	0.63622
4.06000	0.27928	0.13017	11.30684	2.14560	1.30583	0.63776
4.08000	0.27718	0.12904	11.50915	2.14804	1.30679	0.63929
4.10000	0.27510	0.12793	11.71465	2.15046	1.30774	0.64080
4.12000	0.27304	0.12683	11.92340	2.15285	1.30868	0.64230
4.14000	0.27101	0.12574	12.13543	2.15522	1.30960	0.64377

M	$\frac{T}{T^*}$	$\frac{p}{p^*}$	$\frac{p_0}{p_0^*}$	$\frac{\rho}{\rho^*}$	$\frac{F}{F^*}$	$\frac{4fL_{max}}{D}$
4.16000	0.26899	0.12467	12.35079	2.15756	1.31052	0.64523
4.18000	0.26699	0.12362	12.56951	2.15987	1.31143	0.64668
4.20000	0.26502	0.12257	12.79164	2.16215	1.31233	0.64810
4.22000	0.26306	0.12154	13.01722	2.16442	1.31322	0.64951
4.24000	0.26112	0.12052	13.24629	2.16665	1.31410	0.65090
4.26000	0.25921	0.11951	13.47890	2.16886	1.31497	0.65228
4.28000	0.25731	0.11852	13.71509	2.17105	1.31583	0.65364
4.30000	0.25543	0.11753	13.95490	2.17321	1.31668	0.65499
4.32000	0.25357	0.11656	14.19838	2.17535	1.31752	0.65632
4.34000	0.25172	0.11560	14.44557	2.17747	1.31836	0.65763
4.36000	0.24990	0.11466	14.69652	2.17956	1.31919	0.65893
4.38000	0.24809	0.11372	14.95127	2.18163	1.32000	0.66022
4.40000	0.24631	0.11279	15.20987	2.18368	1.32081	0.66149
4.42000	0.24453	0.11188	15.47236	2.18571	1.32161	0.66275
4.44000	0.24278	0.11097	15.73879	2.18771	1.32241	0.66399
4.46000	0.24105	0.11008	16.00921	2.18970	1.32319	0.66522
4.48000	0.23933	0.10920	16.28366	2.19166	1.32397	0.66643
4.50000	0.23762	0.10833	16.56219	2.19360	1.32474	0.66763
4.52000	0.23594	0.10746	16.84486	2.19552	1.32550	0.66882
4.54000	0.23427	0.10661	17.13170	2.19742	1.32625	0.67000
4.56000	0.23262	0.10577	17.42277	2.19930	1.32700	0.67116
4.58000	0.23098	0.10494	17.71812	2.20116	1.32773	0.67231
4.60000	0.22936	0.10411	18.01779	2.20300	1.32846	0.67345
4.62000	0.22775	0.10330	18.32185	2.20482	1.32919	0.67457
4.64000	0.22616	0.10249	18.63032	2.20662	1.32990	0.67569
4.66000	0.22459	0.10170	18.94328	2.20841	1.33061	0.67679
4.68000	0.22303	0.10091	19.26076	2.21017	1.33131	0.67788
4.70000	0.22148	0.10013	19.58283	2.21192	1.33201	0.67895
4.72000	0.21995	0.09936	19.90953	2.21365	1.33269	0.68002
4.74000	0.21844	0.09860	20.24091	2.21536	1.33338	0.68107
4.76000	0.21694	0.09785	20.57703	2.21705	1.33405	0.68211
4.78000	0.21545	0.09711	20.91795	2.21872	1.33472	0.68315
4.80000	0.21398	0.09637	21.26371	2.22038	1.33538	0.68417
4.82000	0.21252	0.09564	21.61437	2.22202	1.33603	0.68518
4.84000	0.21108	0.09492	21.96999	2.22365	1.33668	0.68618
4.86000	0.20965	0.09421	22.33061	2.22526	1.33732	0.68717
4.88000	0.20823	0.09351	22.69631	2.22685	1.33796	0.68814
4.90000	0.20683	0.09281	23.06712	2.22842	1.33859	0.68911
4.92000	0.20543	0.09212	23.44311	2.22998	1.33921	0.69007
4.94000	0.20406	0.09144	23.82434	2.23153	1.33983	0.69102
4.96000	0.20269	0.09077	24.21086	2.23306	1.34044	0.69196
4.98000	0.20134	0.09010	24.60272	2.23457	1.34104	0.69288
5.00000	0.20000	0.08944	25.00000	2.23607	1.34164	0.69380

F One-Dimensional Flow with Heat Transfer ($\gamma = 1.4$)

M	$\frac{T_0}{T_0^*}$	$\frac{T}{T^*}$	$\frac{p}{p^*}$	$\frac{p_0}{p_0^*}$	$\frac{\rho}{\rho^*}$
0.00000	0.00000	0.00000	2.40000	1.26788	0.00000
0.02000	0.00192	0.00230	2.39866	1.26752	0.00096
0.04000	0.00765	0.00917	2.39464	1.26646	0.00383
0.06000	0.01712	0.02053	2.38796	1.26470	0.00860
0.08000	0.03022	0.03621	2.37869	1.26226	0.01522
0.10000	0.04678	0.05602	2.36686	1.25915	0.02367
0.12000	0.06661	0.07970	2.35257	1.25539	0.03388
0.14000	0.08947	0.10695	2.33590	1.25103	0.04578
0.16000	0.11511	0.13743	2.31696	1.24608	0.05931
0.18000	0.14324	0.17078	2.29586	1.24059	0.07439
0.20000	0.17355	0.20661	2.27273	1.23460	0.09091
0.22000	0.20574	0.24452	2.24770	1.22814	0.10879
0.24000	0.23948	0.28411	2.22091	1.22126	0.12792
0.26000	0.27446	0.32496	2.19250	1.21400	0.14821
0.28000	0.31035	0.36667	2.16263	1.20642	0.16955
0.30000	0.34686	0.40887	2.13144	1.19855	0.19183
0.32000	0.38369	0.45119	2.09908	1.19045	0.21495
0.34000	0.42056	0.49327	2.06569	1.18215	0.23879
0.36000	0.45723	0.53482	2.03142	1.17371	0.26327
0.38000	0.49346	0.57553	1.99641	1.16517	0.28828
0.40000	0.52903	0.61515	1.96078	1.15658	0.31373
0.42000	0.56376	0.65346	1.92468	1.14796	0.33951
0.44000	0.59748	0.69025	1.88822	1.13936	0.36556
0.46000	0.63007	0.72538	1.85151	1.13082	0.39178
0.48000	0.66139	0.75871	1.81466	1.12238	0.41810
0.50000	0.69136	0.79012	1.77778	1.11405	0.44444
0.52000	0.71990	0.81955	1.74095	1.10588	0.47075
0.54000	0.74695	0.84695	1.70425	1.09789	0.49696
0.56000	0.77249	0.87227	1.66778	1.09011	0.52302

DOI: 10.1201/9781003139447-F

M	$\frac{T_0}{T_0^*}$	$\frac{T}{T^*}$	$\frac{p}{p^*}$	$\frac{p_0}{p_0^*}$	$\frac{\rho}{\rho^*}$
0.58000	0.79648	0.89552	1.63159	1.08256	0.54887
0.60000	0.81892	0.91670	1.59574	1.07525	0.57447
0.62000	0.83983	0.93584	1.56031	1.06822	0.59978
0.64000	0.85920	0.95298	1.52532	1.06147	0.62477
0.66000	0.87708	0.96816	1.49083	1.05503	0.64941
0.68000	0.89350	0.98144	1.45688	1.04890	0.67366
0.70000	0.90850	0.99290	1.42349	1.04310	0.69751
0.72000	0.92212	1.00260	1.39069	1.03764	0.72093
0.74000	0.93442	1.01062	1.35851	1.03253	0.74392
0.76000	0.94546	1.01706	1.32696	1.02777	0.76645
0.78000	0.95528	1.02198	1.29606	1.02337	0.78853
0.80000	0.96395	1.02548	1.26582	1.01934	0.81013
0.82000	0.97152	1.02763	1.23625	1.01569	0.83125
0.84000	0.97807	1.02853	1.20734	1.01241	0.85190
0.86000	0.98363	1.02826	1.17911	1.00951	0.87207
0.88000	0.98828	1.02689	1.15154	1.00699	0.89175
0.90000	0.99207	1.02452	1.12465	1.00486	0.91097
0.92000	0.99506	1.02120	1.09842	1.00311	0.92970
0.94000	0.99729	1.01702	1.07285	1.00175	0.94797
0.96000	0.99883	1.01205	1.04793	1.00078	0.96577
0.98000	0.99971	1.00636	1.02365	1.00019	0.98311
1.00000	1.00000	1.00000	1.00000	1.00000	1.00000
1.02000	0.99973	0.99304	0.97698	1.00019	1.01645
1.04000	0.99895	0.98554	0.95456	1.00078	1.03246
1.06000	0.99769	0.97755	0.93275	1.00175	1.04804
1.08000	0.99601	0.96913	0.91152	1.00311	1.06320
1.10000	0.99392	0.96031	0.89087	1.00486	1.07795
1.12000	0.99148	0.95115	0.87078	1.00699	1.09230
1.14000	0.98871	0.94169	0.85123	1.00952	1.10626
1.16000	0.98564	0.93196	0.83222	1.01243	1.11984
1.18000	0.98230	0.92200	0.81374	1.01573	1.13305
1.20000	0.97872	0.91185	0.79576	1.01942	1.14589
1.22000	0.97492	0.90153	0.77827	1.02349	1.15838
1.24000	0.97092	0.89108	0.76127	1.02795	1.17052
1.26000	0.96675	0.88052	0.74473	1.03280	1.18233
1.28000	0.96243	0.86988	0.72865	1.03803	1.19382
1.30000	0.95798	0.85917	0.71301	1.04366	1.20499
1.32000	0.95341	0.84843	0.69780	1.04968	1.21585
1.34000	0.94873	0.83766	0.68301	1.05608	1.22642

M	$\frac{T_0}{T_0^*}$	$\frac{T}{T^*}$	$\frac{p}{p^*}$	$\frac{p_0}{p_0^*}$	$\frac{\rho}{\rho^*}$
1.36000	0.94398	0.82689	0.66863	1.06288	1.23669
1.38000	0.93914	0.81613	0.65464	1.07007	1.24669
1.40000	0.93425	0.80539	0.64103	1.07765	1.25641
1.42000	0.92931	0.79469	0.62779	1.08563	1.26587
1.44000	0.92434	0.78405	0.61491	1.09401	1.27507
1.46000	0.91933	0.77346	0.60237	1.10278	1.28402
1.48000	0.91431	0.76294	0.59018	1.11196	1.29273
1.50000	0.90928	0.75250	0.57831	1.12155	1.30120
1.52000	0.90424	0.74215	0.56676	1.13153	1.30945
1.54000	0.89920	0.73189	0.55552	1.14193	1.31748
1.56000	0.89418	0.72173	0.54458	1.15274	1.32530
1.58000	0.88917	0.71168	0.53393	1.16397	1.33291
1.60000	0.88419	0.70174	0.52356	1.17561	1.34031
1.62000	0.87922	0.69190	0.51346	1.18768	1.34753
1.64000	0.87429	0.68219	0.50363	1.20017	1.35455
1.66000	0.86939	0.67259	0.49405	1.21309	1.36140
1.68000	0.86453	0.66312	0.48472	1.22644	1.36806
1.70000	0.85971	0.65377	0.47562	1.24024	1.37455
1.72000	0.85493	0.64455	0.46677	1.25447	1.38088
1.74000	0.85019	0.63545	0.45813	1.26915	1.38705
1.76000	0.84551	0.62649	0.44972	1.28428	1.39306
1.78000	0.84087	0.61765	0.44152	1.29987	1.39891
1.80000	0.83628	0.60894	0.43353	1.31592	1.40462
1.82000	0.83174	0.60036	0.42573	1.33244	1.41019
1.84000	0.82726	0.59191	0.41813	1.34943	1.41562
1.86000	0.82283	0.58359	0.41072	1.36690	1.42092
1.88000	0.81845	0.57540	0.40349	1.38486	1.42608
1.90000	0.81414	0.56734	0.39643	1.40330	1.43112
1.92000	0.80987	0.55941	0.38955	1.42224	1.43604
1.94000	0.80567	0.55160	0.38283	1.44168	1.44083
1.96000	0.80152	0.54392	0.37628	1.46164	1.44551
1.98000	0.79742	0.53636	0.36988	1.48210	1.45008
2.00000	0.79339	0.52893	0.36364	1.50310	1.45455
2.02000	0.78941	0.52161	0.35754	1.52462	1.45890
2.04000	0.78549	0.51442	0.35158	1.54668	1.46315
2.06000	0.78162	0.50735	0.34577	1.56928	1.46731
2.08000	0.77782	0.50040	0.34009	1.59244	1.47136
2.10000	0.77406	0.49356	0.33454	1.61616	1.47533
2.12000	0.77037	0.48684	0.32912	1.64045	1.47920

M	$\frac{T_0}{T_0^*}$	$\frac{T}{T^*}$	$\frac{p}{p^*}$	$\frac{p_0}{p_0^*}$	$\frac{\rho}{\rho^*}$
2.14000	0.76673	0.48023	0.32382	1.66531	1.48298
2.16000	0.76314	0.47373	0.31865	1.69076	1.48668
2.18000	0.75961	0.46734	0.31359	1.71680	1.49029
2.20000	0.75613	0.46106	0.30864	1.74345	1.49383
2.22000	0.75271	0.45488	0.30381	1.77070	1.49728
2.24000	0.74934	0.44882	0.29908	1.79858	1.50066
2.26000	0.74602	0.44285	0.29446	1.82708	1.50396
2.28000	0.74276	0.43698	0.28993	1.85623	1.50719
2.30000	0.73954	0.43122	0.28551	1.88602	1.51035
2.32000	0.73638	0.42555	0.28118	1.91647	1.51344
2.34000	0.73326	0.41998	0.27695	1.94759	1.51646
2.36000	0.73020	0.41451	0.27281	1.97939	1.51942
2.38000	0.72718	0.40913	0.26875	2.01187	1.52232
2.40000	0.72421	0.40384	0.26478	2.04505	1.52515
2.42000	0.72129	0.39864	0.26090	2.07895	1.52793
2.44000	0.71842	0.39352	0.25710	2.11356	1.53065
2.46000	0.71558	0.38850	0.25337	2.14891	1.53331
2.48000	0.71280	0.38356	0.24973	2.18499	1.53591
2.50000	0.71006	0.37870	0.24615	2.22183	1.53846
2.52000	0.70736	0.37392	0.24266	2.25944	1.54096
2.54000	0.70471	0.36923	0.23923	2.29782	1.54341
2.56000	0.70210	0.36461	0.23587	2.33699	1.54581
2.58000	0.69952	0.36007	0.23258	2.37696	1.54816
2.60000	0.69700	0.35561	0.22936	2.41774	1.55046
2.62000	0.69451	0.35122	0.22620	2.45935	1.55272
2.64000	0.69206	0.34691	0.22310	2.50179	1.55493
2.66000	0.68964	0.34266	0.22007	2.54509	1.55710
2.68000	0.68727	0.33849	0.21709	2.58925	1.55922
2.70000	0.68494	0.33439	0.21417	2.63429	1.56131
2.72000	0.68264	0.33035	0.21131	2.68021	1.56335
2.74000	0.68037	0.32638	0.20850	2.72704	1.56536
2.76000	0.67815	0.32248	0.20575	2.77478	1.56732
2.78000	0.67595	0.31864	0.20305	2.82346	1.56925
2.80000	0.67380	0.31486	0.20040	2.87308	1.57114
2.82000	0.67167	0.31114	0.19780	2.92366	1.57300
2.84000	0.66958	0.30749	0.19525	2.97521	1.57482
2.86000	0.66752	0.30389	0.19275	3.02775	1.57661
2.88000	0.66550	0.30035	0.19029	3.08129	1.57836
2.90000	0.66350	0.29687	0.18788	3.13585	1.58008

M	$\frac{T_0}{T_0^*}$	$\frac{T}{T^*}$	$\frac{p}{p^*}$	$\frac{p_0}{p_0^*}$	$\frac{\rho}{\rho^*}$
2.92000	0.66154	0.29344	0.18551	3.19145	1.58178
2.94000	0.65960	0.29007	0.18319	3.24809	1.58343
2.96000	0.65770	0.28675	0.18091	3.30579	1.58506
2.98000	0.65583	0.28349	0.17867	3.36457	1.58666
3.00000	0.65398	0.28028	0.17647	3.42445	1.58824
3.02000	0.65216	0.27711	0.17431	3.48544	1.58978
3.04000	0.65037	0.27400	0.17219	3.54756	1.59129
3.06000	0.64861	0.27094	0.17010	3.61082	1.59278
3.08000	0.64687	0.26792	0.16806	3.67524	1.59425
3.10000	0.64516	0.26495	0.16604	3.74084	1.59568
3.12000	0.64348	0.26203	0.16407	3.80764	1.59709
3.14000	0.64182	0.25915	0.16212	3.87565	1.59848
3.16000	0.64018	0.25632	0.16022	3.94488	1.59985
3.18000	0.63857	0.25353	0.15834	4.01537	1.60119
3.20000	0.63699	0.25078	0.15649	4.08712	1.60250
3.22000	0.63543	0.24808	0.15468	4.16015	1.60380
3.24000	0.63389	0.24541	0.15290	4.23449	1.60507
3.26000	0.63237	0.24279	0.15115	4.31014	1.60632
3.28000	0.63088	0.24021	0.14942	4.38714	1.60755
3.30000	0.62940	0.23766	0.14773	4.46549	1.60877
3.32000	0.62795	0.23515	0.14606	4.54522	1.60996
3.34000	0.62652	0.23268	0.14442	4.62635	1.61113
3.36000	0.62512	0.23025	0.14281	4.70889	1.61228
3.38000	0.62373	0.22785	0.14122	4.79287	1.61341
3.40000	0.62236	0.22549	0.13966	4.87830	1.61453
3.42000	0.62101	0.22317	0.13813	4.96521	1.61562
3.44000	0.61968	0.22087	0.13662	5.05362	1.61670
3.46000	0.61837	0.21861	0.13513	5.14355	1.61776
3.48000	0.61708	0.21639	0.13367	5.23501	1.61881
3.50000	0.61580	0.21419	0.13223	5.32804	1.61983
3.52000	0.61455	0.21203	0.13081	5.42264	1.62085
3.54000	0.61331	0.20990	0.12942	5.51885	1.62184
3.56000	0.61209	0.20780	0.12805	5.61668	1.62282
3.58000	0.61089	0.20573	0.12670	5.71615	1.62379
3.60000	0.60970	0.20369	0.12537	5.81730	1.62474
3.62000	0.60853	0.20167	0.12406	5.92013	1.62567
3.64000	0.60738	0.19969	0.12277	6.02468	1.62660
3.66000	0.60624	0.19773	0.12150	6.13097	1.62750
3.68000	0.60512	0.19581	0.12024	6.23902	1.62840

M	$\frac{T_0}{T_0^*}$	$\frac{T}{T^*}$	$\frac{p}{p^*}$	$\frac{p_0}{p_0^*}$	$\frac{\rho}{\rho^*}$
3.70000	0.60401	0.19390	0.11901	6.34884	1.62928
3.72000	0.60292	0.19203	0.11780	6.46048	1.63014
3.74000	0.60184	0.19018	0.11660	6.57394	1.63100
3.76000	0.60078	0.18836	0.11543	6.68926	1.63184
3.78000	0.59973	0.18656	0.11427	6.80646	1.63267
3.80000	0.59870	0.18478	0.11312	6.92557	1.63348
3.82000	0.59768	0.18303	0.11200	7.04660	1.63429
3.84000	0.59667	0.18131	0.11089	7.16958	1.63508
3.86000	0.59568	0.17961	0.10979	7.29454	1.63586
3.88000	0.59470	0.17793	0.10871	7.42151	1.63663
3.90000	0.59373	0.17627	0.10765	7.55050	1.63739
3.92000	0.59278	0.17463	0.10661	7.68156	1.63814
3.94000	0.59184	0.17302	0.10557	7.81469	1.63888
3.96000	0.59091	0.17143	0.10456	7.94993	1.63960
3.98000	0.58999	0.16986	0.10355	8.08731	1.64032
4.00000	0.58909	0.16831	0.10256	8.22685	1.64103
4.02000	0.58819	0.16678	0.10159	8.36858	1.64172
4.04000	0.58731	0.16527	0.10063	8.51252	1.64241
4.06000	0.58644	0.16378	0.09968	8.65872	1.64309
4.08000	0.58558	0.16231	0.09875	8.80718	1.64375
4.10000	0.58473	0.16086	0.09782	8.95794	1.64441
4.12000	0.58390	0.15943	0.09691	9.11104	1.64506
4.14000	0.58307	0.15802	0.09602	9.26649	1.64570
4.16000	0.58225	0.15662	0.09513	9.42433	1.64633
4.18000	0.58145	0.15524	0.09426	9.58459	1.64696
4.20000	0.58065	0.15388	0.09340	9.74729	1.64757
4.22000	0.57987	0.15254	0.09255	9.91247	1.64818
4.24000	0.57909	0.15121	0.09171	10.08015	1.64878
4.26000	0.57832	0.14990	0.09089	10.25037	1.64937
4.28000	0.57757	0.14861	0.09007	10.42316	1.64995
4.30000	0.57682	0.14734	0.08927	10.59854	1.65052
4.32000	0.57608	0.14607	0.08847	10.77656	1.65109
4.34000	0.57535	0.14483	0.08769	10.95723	1.65165
4.36000	0.57463	0.14360	0.08691	11.14060	1.65220
4.38000	0.57392	0.14239	0.08615	11.32669	1.65275
4.40000	0.57322	0.14119	0.08540	11.51554	1.65329
4.42000	0.57252	0.14000	0.08465	11.70717	1.65382
4.44000	0.57183	0.13883	0.08392	11.90163	1.65434
4.46000	0.57116	0.13767	0.08319	12.09894	1.65486

M	$\frac{T_0}{T_0^*}$	$\frac{T}{T^*}$	$\frac{p}{p^*}$	$\frac{p_0}{p_0^*}$	$\frac{\rho}{\rho^*}$
4.48000	0.57049	0.13653	0.08248	12.29914	1.65537
4.50000	0.56982	0.13540	0.08177	12.50226	1.65588
4.52000	0.56917	0.13429	0.08107	12.70834	1.65638
4.54000	0.56852	0.13319	0.08039	12.91740	1.65687
4.56000	0.56789	0.13210	0.07970	13.12949	1.65735
4.58000	0.56726	0.13102	0.07903	13.34464	1.65783
4.60000	0.56663	0.12996	0.07837	13.56288	1.65831
4.62000	0.56602	0.12891	0.07771	13.78425	1.65878
4.64000	0.56541	0.12787	0.07707	14.00879	1.65924
4.66000	0.56480	0.12685	0.07643	14.23653	1.65969
4.68000	0.56421	0.12583	0.07580	14.46750	1.66014
4.70000	0.56362	0.12483	0.07517	14.70174	1.66059
4.72000	0.56304	0.12384	0.07456	14.93930	1.66103
4.74000	0.56246	0.12286	0.07395	15.18020	1.66146
4.76000	0.56190	0.12190	0.07335	15.42449	1.66189
4.78000	0.56133	0.12094	0.07275	15.67220	1.66232
4.80000	0.56078	0.12000	0.07217	15.92337	1.66274
4.82000	0.56023	0.11906	0.07159	16.17803	1.66315
4.84000	0.55969	0.11814	0.07101	16.43624	1.66356
4.86000	0.55915	0.11722	0.07045	16.69801	1.66397
4.88000	0.55862	0.11632	0.06989	16.96341	1.66436
4.90000	0.55809	0.11543	0.06934	17.23245	1.66476
4.92000	0.55758	0.11455	0.06879	17.50519	1.66515
4.94000	0.55706	0.11367	0.06825	17.78167	1.66554
4.96000	0.55655	0.11281	0.06772	18.06192	1.66592
4.98000	0.55605	0.11196	0.06719	18.34598	1.66629
5.00000	0.55556	0.11111	0.06667	18.63390	1.66667

G Letter of Admittance

A letter of admittance to the author's book Essentials of Aircraft Armaments, published by Springer Nature, from the Honorable Defense Minister of India, Late Mr. Manohar Parrikar (2016).

D.O No. 01/OD/RM/2016

सत्यमेव जयते

रक्षा मंत्री
भारत
MINISTER OF DEFENCE
INDIA

3rd November, 2016

Dear Shri. Mrinal Kaushik,

I acknowledge the receipt of your book title 'Essentials of Aircraft Armaments'. Thank you very much for sending me the same.

With Regards,

Yours Sincerely,

Parrikar.

(Manohar Parrikar)

Shri. Mrinal Kaushik,
Department of Aerospace Engineering,
Indian Institute of Technology,
Kharagpur.

104, South Block, New Delhi-110011. Ph. : 23012286, 23019030 Fax : 23015403

Bibliography

1. W. Heiser, D. Pratt, D. Daley, and U. Mehta (1994), Hypersonic Air Breathing Propulsion, AIAA Education Series, Washington D.C., USA.
2. Hansen, C.F. (1959), Approximations for the Thermodynamic and Transport Properties of High-Temperature Air, NACA TR R-50.
3. Kline, S., and Mcclintock, F. (1953), Describing Uncertainties in Single-Sample Experiments, Mechanical Engineering, 75: 3–8.
4. Bell, J.H., and Mehta, R.D. (1989), Boundary-layer predictions for small low-speed contractions, AIAA Journal, 27(3): 372–374.
5. Collar, A.R. (1939), The effect of a gauge on the velocity distribution in a uniform duct, ARC R&M, 1867.
6. W.T. Eckert, K.W. Mort, and J. Jope (1976), Aerodynamic design guidelines and computer program for estimation of subsonic wind tunnel performance.
7. Wattendorf, F. (1969), AGARD History 1952–1968, NATO.
8. A. Pope and K.L. Goin (1965), High-Speed Wind Tunnel Testing, John Wiley & Sons, New York, USA.
9. M. Kaushik (2022), Fundamentals of Gas Dynamics, Springer Nature, Singapore.
10. M. Kaushik (2019), Theoretical and Experimental Aerodynamics, Springer Nature, Singapore.
11. M. Kaushik (2016), Essentials of Aircraft Armaments, Springer Nature, Singapore.
12. M. Kaushik (2012), Innovative Passive Control Techniques for Supersonic Jet Mixing, Lambert Academic Publishing, Saarbrücken, Germany
13. Tam, C.K.W., and Chen, P. (1994), Turbulent Mixing Noise from Supersonic Jets, AIAA Journal, 32(9): 1774–1780.
14. Mahapatra, D., and Jagadeesh, G. (2008), "Shock tunnel studies on cowl/ramp shock interactions in a generic scramjet inlet," Proc. Inst. Mech. Eng., Part G 222, 1183–1191.
15. Oswatitsch, K. (1947), "Pressure Recovery for Missiles with Reaction Propulsion at High Supersonic Speeds (The Efficiency of Shock Diffusers)," NACA Technical Memorandum 1140.
16. Smart, M. (1999), "Optimization of Two-Dimensional Scramjet Inlets," Journal of Aircraft, 36:430–433.
17. Idris, A.C., Saad, M.R., Behtash, H.Z., and Kontis, K. (2014), "Luminescent Measurement Systems for the Investigation of a Scramjet Inlet-Isolator," Sensors, 14:6606–6632.
18. Kantrowitz, A., and Donaldson, C.Dup. (1945), "Preliminary Investigation of Supersonic Diffusers," Technical Report L5D20, National Advisory Committee for Aeronautics.
19. Ran, H. and Mavris, D., (2005) "Preliminary design of a 2D supersonic inlet to maximize total pressure recovery," AIAA 5th Aviation, Technology, Integration, and Operations Conference, Arlington, Virginia, USA.

20. Kaushik, M. (2019), "Experimental Studies on Micro-Vortex Generator Controlled Shock/Boundary-Layer Interactions in Mach 2.2 Intake," International Journal of Aeronautical and Space Sciences, 20: 584–595.
21. Jana, T., and Kaushik, M. (2022), "Survey of control techniques to alleviate repercussions of shock-wave and boundary-layer interactions," Adv. Aerodyn. 4, 27.
22. Kaushik, M., Thakur, P.S., and Rathakrishnan, E. (2006), "Studies on the Effect of Notches on Circular Sonic Jet Mixing," Journal of Propulsion and Power, 22(1): 211–214.
23. Kaushik, M. and Rathakrishnan, E. (2010), "Mixing Characteristics of a Supersonic Jet in the Presence of Two Neighboring Jets," International Review of Aerospace Engineering, 3(4):180–185.

Index

For Product Safety Concerns and Information please contact our EU representative GPSR@taylorandfrancis.com Taylor & Francis Verlag GmbH, Kaufingerstraße 24, 80331 München, Germany